BIOLOGICAL SMALL ANGLE SCATTERING

INTERNATIONAL UNION OF CRYSTALLOGRAPHY
BOOK SERIES

IUCr Monographs on Crystallography

1 *Accurate molecular structures*
A. Domenicano, I. Hargittai, editors
2 *P.P. Ewald and his dynamical theory of X-ray diffraction*
D.W.J. Cruickshank, H.J. Juretschke, N. Kato, editors
3 *Electron diffraction techniques, Vol. 1*
J.M. Cowley, editor
4 *Electron diffraction techniques, Vol. 2*
J.M. Cowley, editor
5 *The Rietveld method*
R.A. Young, editor
6 *Introduction to crystallographic statistics*
U. Shmueli, G.H. Weiss
7 *Crystallographic instrumentation*
L.A. Aslanov, G.V. Fetisov, J.A.K. Howard
8 *Direct phasing in crystallography*
C. Giacovazzo
9 *The weak hydrogen bond*
G.R. Desiraju, T. Steiner
10 *Defect and microstructure analysis by diffraction*
R.L. Snyder, J. Fiala, H.J. Bunge
11 *Dynamical theory of X-ray diffraction*
A. Authier
12 *The chemical bond in inorganic chemistry*
I.D. Brown

13 *Structure determination from powder diffraction data*
W.I.F. David, K. Shankland, L.B. McCusker, Ch. Baerlocher, editors
14 *Polymorphism in molecular crystals*
J. Bernstein
15 *Crystallography of modular materials*
G. Ferraris, E. Makovicky, S. Merlino
16 *Diffuse X-ray scattering and models of disorder*
T.R. Welberry
17 *Crystallography of the polymethylene chain: an inquiry into the structure of waxes*
D.L. Dorset
18 *Crystalline molecular complexes and compounds: structure and principles*
F.H. Herbstein
19 *Molecular aggregation: structure analysis and molecular simulation of crystals and liquids*
A. Gavezzotti
20 *Aperiodic crystals: from modulated phases to quasicrystals*
T. Janssen, G. Chapuis, M. de Boissieu
21 *Incommensurate crystallography*
S. van Smaalen
22 *Structural crystallography of inorganic oxysalts*
S.V. Krivovichev
23 *The nature of the hydrogen bond: outline of a comprehensive hydrogen bond theory*
G. Gilli, P. Gilli
24 *Macromolecular crystallization and crystal perfection*
N.E. Chayen, J.R. Helliwell, E.H.Snell
25 *Neutron protein crystallography: hydrogen, protons, and hydration in bio-macromolecules*
N. Niimura, A. Podjarny
26 *Intermetallics: structures, properties, and statistics*
W. Steurer, J. Dshemuchadse
27 *The chemical bond in inorganic chemistry: the bond valence model*, second edition
I.D. Brown
28 *Aperiodic crystals: from modulated phases to quasicrystals: structure and properties*, second edition
T. Janssen, G. Chapuis, M. de Boissieu
29 *Biological small angle scattering: theory and practice*
E.E. Lattman, T.D. Grant, E.H. Snell

IUCr Texts on Crystallography

1 *The solid state*
A. Guinier, R. Julien

4 *X-ray charge densities and chemical bonding*
P. Coppens

8 *Crystal structure refinement: a crystallographer's guide to SHELXL*
P. Müller, editor

9 *Theories and techniques of crystal structure determination*
U. Shmueli

10 *Advanced structural inorganic chemistry*
Wai-Kee Li, Gong-Du Zhou, Thomas Mak

11 *Diffuse scattering and defect structure simulations: a cook book using the program DISCUS*
R.B. Neder, T. Proffen

13 *Crystal structure analysis: principles and practice, second edition*
W. Clegg, editor

14 *Crystal structure analysis: a primer, third edition*
J.P. Glusker, K.N. Trueblood

15 *Fundamentals of crystallography, third edition*
C. Giacovazzo, editor

16 *Electron crystallography: electron microscopy and electron diffraction*
X. Zou, S. Hovmöller, P. Oleynikov

17 *Symmetry in crystallography: understanding the International Tables*
P.G. Radaelli

18 *Symmetry relationships between crystal structures: applications of crystallographic group theory in crystal chemistry*
U. Müller

19 *Small angle X-ray and neutron scattering from solutions of biological macromolecules*
D.I. Svergun, M.H.J. Koch, P.A. Timmins, R.P. May

20 *Phasing in crystallography: a modern perspective*
C. Giacovazzo

21 *The basics of crystallography and diffraction, fourth edition*
C. Hammond

Biological Small Angle Scattering

Theory and Practice

Eaton E. Lattman
Thomas D. Grant
Edward H. Snell

OXFORD
UNIVERSITY PRESS

OXFORD
UNIVERSITY PRESS

Great Clarendon Street, Oxford, OX2 6DP,
United Kingdom

Oxford University Press is a department of the University of Oxford.
It furthers the University's objective of excellence in research, scholarship,
and education by publishing worldwide. Oxford is a registered trade mark of
Oxford University Press in the UK and in certain other countries

First Edition published in 2018

Published in the United States of America by Oxford University Press
198 Madison Avenue, New York, NY 10016, United States of America

British Library Cataloguing in Publication Data
Data available

Library of Congress Control Number: 2018933085

ISBN 978–0–19–967087–1
DOI: 10.1093/oso/9780199670871.001.0001

Preface

Small angle solution scattering is seeing increased applications to biological problems. It is a complementary technique that, when used in appropriate circumstances with carefully structured questions, can provide unique information not available from other methods. While small angle solution scattering has been around for many decades, a confluence of recent developments has dramatically enhanced its power. Intense third-generation X-ray sources, low noise detectors, new algorithms, and the computational power to take advantage of these have all appeared in the past decade. Beamtime at synchrotron sources is in demand, workshops on the subject are popular, and the group of researchers adopting the technique as part of their repertoire is growing. With these in mind, we have written this book to guide structural biologists who may wish to adopt the technique and understand its strengths and weaknesses or who just have a general interest in its potential. We focus on three areas: the theory; practical aspects and applications; and the potential, with a glimpse of developing areas. We intend it to be an introduction and guide to the field but not a comprehensive treatment.

Eaton E. Lattman, Thomas D. Grant, and Edward H. Snell

Contents

Part 3 Practical Aspects of Small Angle Scattering

Acknowledgments

We would like to dedicate this book to the memory of Dr. Hirotsugu Tsuruta (Hiro), a collaborator, mentor, and friend. Without him, this book would not have happened. He gave his time and knowledge, guiding our first steps in the field, and always pointing us forward. He was helped during that process by his colleagues at Stanford Synchrotron Radiation Lightsource (SSRL), Drs. Anne Martel, Tsutomu Matsui, and Thomas Weiss. We miss Hiro, his impact on us, and his impact on the field.

Closer to home, we would like to acknowledge the help and support of the staff at the Hauptman–Woodward Medical Research Institute in pursuing this research, in particular, Joseph Luft, who accompanied us on many SAXS trips, and the initial financial support of Dr. George DeTitta, with this enabling our first efforts. We thank Dr. Gaetano Montelione for providing many of our initial samples and Drs. Eric Phizicky and Elizabeth Grayhack as collaborators on structural studies incorporating small angle scattering. Outside of our institute, we have benefited from many useful discussions, including those with Drs. Richard Gililan, Angela Criswell, Robert Rambo, Michal Hammel, Greg Hura, and Arwen Pearson. We thank them and, in the case of Dr. Pearson, thank her students for so many comments and questions that we hope we have answered here. We thank Drs. Thomas Weiss, Lois Pollack, and Arwen Pearson for constructive comments along the way. We have been very appreciative of the Small Angle Scattering community in general for sharing much of their knowledge, advice, and encouragement. Our work has been enabled by funding from the Oishei Foundation and support from the National Institutes of Health and National Science Foundation.

Part 1

Introduction

1

Introduction

Biological macromolecules are the machinery of life; however, despite the term "macro," they are small. For example, a red blood cell is about ~6–8 μm in size, but the hemoglobin molecule inside the cell (built up from two heterodimers) is ~0.007 μm—approximately two orders of magnitude smaller than the wavelength of visible light. A larger example, the ribosome complex contains almost 100 macromolecules and has a largest dimension of ~0.27 μm, approaching the wavelength of visible light. Optical microscopy cannot resolve these key components of life, and X-ray crystallography has been the major technique used to provide structural knowledge to date, with other techniques, including nuclear magnetic resonance (NMR) and electron microscopy, also providing structural data. The weakness of crystallography is revealed in its name: the growth of the crystals needed for diffraction studies is a difficult and chancy process. Huge resources have been focused on the problem. For example, there were many efforts as part of the Protein Structure Initiative (a multi-institute high-throughput approach to structural biology). In this case, detailed tracking was carried out during studies on thousands of different proteins. Crystallization failure as well as success was monitored, but despite this multimillion-dollar, state-of-the art effort, only ~20% of purified, soluble proteins led to structure determination. There remain a large number of samples, the majority of those studied, where no structural knowledge is available. X-ray crystallography provides the most accurate information available about the atomic structure of molecules, and, increasingly, such information can be provided with cryo-electron microscopy . However, information is also available from the solution phase. As far back as the early 1930s, therefore, investigators began to explore the use of solutions rather than crystals as specimens for X-ray diffraction, or X-ray scattering, as it is more commonly called in this case.

X-ray solution scattering is applicable whether structural information is available from crystallography or not. When structural information is available, the technique can study oligomerization or structural transitions in the solution state. It can be applied to piece together separate components into one complete structure or to describe the structural relationship between known and unknown components. Over the last couple of decades, the complete structures of many large cellular molecular assemblies have been built up by a combination of techniques spanning different resolution ranges. X-ray scattering can play a major role in the study of

Biological Small Angle Scattering: Theory and Practice. Eaton E. Lattman, Thomas D. Grant, and Edward H. Snell.

DOI: 10.1093/oso/9780199670871.001.0001

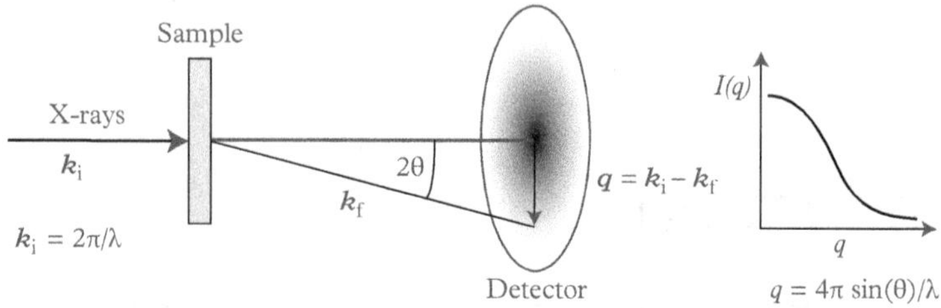

Figure 1.1 Geometry of SAXS experimental setup. Incident X-rays corresponding to wave vector $\boldsymbol{k}_i$ with magnitude $2\pi/\lambda$ pass through the sample chamber and scatter with an angle of 2θ. The isotropic detector image is integrated and circularly averaged to produce a one-dimensional scattering profile of intensity (I) versus momentum transfer (q).

complexes of various types, and, in particular, in generating hypotheses about large-scale motions or rearrangements that occur within them. The related technique of neutron scattering plays an important, complementary role to X-ray scattering in studies of large complexes. The method of selective deuteration of particular subunits allows contrast matching using deuterated water, which can make some subunits effectively invisible in the scattering pattern. Historically, this method was used by Engelman and Moore (1972) to measure the distances between various pairs of ribosomal protein subunits, for example, and the set of many inter-subunit distances was used in a kind of triangulation protocol to develop a model of protein subunit structure within the ribosome. Although crude, it proved compatible with crystallographic studies about two decades later.

A solution scattering experiment is simple in outline (see Figure 1.1). One illuminates a solution with a parallel, monochromatic X-ray beam and then collects the scattered radiation on a detector placed far back from the specimen. Because the solution is homogeneous and isotropic, the observed scattering pattern is circularly symmetric. The full pattern, a three-dimensional function in diffraction space, is spherically symmetric. After suitable corrections and calibrations, the data is usually displayed as a one-dimensional curve in which intensity I is plotted against q (where $q = 4\pi \sin(\theta)/\lambda$, 2θ is the scattering angle, and λ the incident X-ray wavelength); $I(q)$ represents a radial trace through the full, three-dimensional scattering pattern. Because of the large range of intensities measured, log I is often plotted instead. The $I(q)$ curve is often collected multiple times, using several concentrations to detect radiation effects or the presence of aggregation. The term "solution scattering" is applied to the general phenomenon, with the term "small angle scattering" reserved for the most common application in which observations are confined to radiation scattered within a small angular cone around the main beam.

Solution scattering can be related to crystallographic techniques by consideration of the molecular transform, the diffraction pattern from a single molecule. Figure 1.2 shows the molecular transform for a biological macromolecule on the

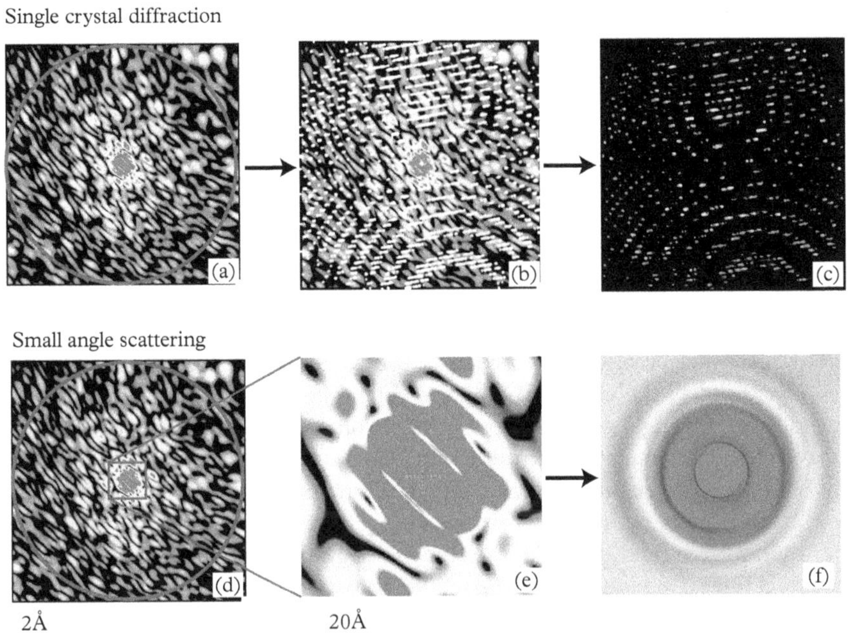

Figure 1.2 An illustration of how the molecular transform is sampled by single crystal diffraction and small angle solution scattering. The information contained in small angle scattering is commonly masked behind the beamstop in single crystal studies.

left side (a). On the top row, the effect that the study of a crystal has on this transform is shown. Constructive interference from the crystal lattice means that only certain points in the transform remain observable (Bragg sampling). Sampling of the transform is marked by white spots (b), and the resulting diffraction image is shown in (c). For illustrative purposes, no attempt has been made to scale intensity in each case. Below the single crystal diffraction case, the small angle scattering case is illustrated (d), starting from the same molecular transform. While the single crystal case samples the molecular transform to, for example, at 2 Å resolution, the small angle scattering case captures information at much lower resolution. In (d), the small square with corners at 20 Å indicates a region that is mostly obscured behind the beamstop in single crystal studies. The sampling is from randomly orientated tumbling particles, so the transform can be averaged to give (f), an isotropic scattering pattern.

Solution scattering provides more information than intuition may suggest, as we describe in Chapter 3. Obtaining useful information from the smooth and rather featureless $I(q)$ curve is a non-trivial task. This curve can be thought of as representing the scattering from a single molecule as it passes through all possible orientations within the beam (we will discuss exceptions to this later). Intuitively, the rotational average of a particle seems to have lost most of the structural detail that we would like to obtain: however, the variety and depth of the information that can be obtained from solution scattering is often surprising to those encountering it for the first time.

It is profitable to preview some of the insights that can be obtained from these experiments. In doing so, we will use the term "particles" for the chemical units of interest that comprise the solute of the solution, even when these units are (say) expanded polymer coils that do not seem particle-like. We will also show that at larger values of θ, the scattering intensity from typical particles falls off as q^{-4} and is thus very weak: hence, the name "small angle scattering." With or without other structural information, small angle solution scattering provides unique results.

- The particle molecular mass can be obtained from $I(0)$. In the forward direction, all the electrons (for X-rays) in the particle scatter in phase, so that at zero angle, the intensity is proportional to the square of the number of electrons in the particle. In practice, $I(0)$ is masked by the direct beamstop (used to protect the detector and minimize noise in the scattering), but extrapolation of $I(\theta \to 0)$ works well to estimate the needed value. The molecular weight can also be estimated from the volume estimated by the total integrated intensity of the scattering pattern (Porod's law), providing a check for self-consistency.
- The radius of gyration, R_g (the root-mean-square value of distances of all the scattering units from the center of mass), can be derived from a so-called Guinier plot. For very small q, the $I(q)$ curve has a Gaussian form whose variance is proportional to $1/R_g^2$. The R_g is usually smaller than a typical dimension of the particle; for example, for a ball of radius r, $R_g^2 = (3/5)r^2$.
- The pair distance distribution (closely related to the Patterson function in crystallography) can be obtained from the Fourier transform of $I(q)$. The pair distribution function sums the number of scattering centers in the particle where the centers are separated by a distance r. The value of r at which the pair distribution function falls to zero thus corresponds to the maximum dimension of the particle. The pair distribution function provides the first glimpse of shape. For example, a bimodal pair distance distribution likely corresponds to a particle comprising two well-separated domains.
- For a mid-range value of q, $I(q)$ is usually proportional to $q^{-\alpha}$, where the value of α provides information about the surface of the particle. Globular particles with well-defined surfaces have $\alpha = 4$, while unfolded proteins or polymers in good solvent give $\alpha = 1$. This analysis is embodied in the so-called Kratky plot.
- In favorable cases, one can construct a model of the molecular envelope of the particle. The rotational averaging of the particle diffraction and the deceptively simple appearance of the $I(q)$ curve make it intuitively

implausible that one could build an asymmetric model. However, such reconstructions are now commonplace. They are aided by the application of constraints such as positivity of the electron density (for X-rays) and compactness of the particle. In addition, the large dynamic range of intensities represents a powerful constraint on modeling. Other experimental constraints, such as knowledge of distances or contacts between elements in the particle, both stabilize the reconstruction process and serve to validate the resulting model.

Solution scattering can provide information about systems having more than one component or state. A glass sphere seems invisible when surrounded by a fluid that has the same refractive index as the glass (a technique used with effect in smuggling diamonds). Similarly, the signal in solution scattering arises from the difference in scattering density between the solvent and the particle. A protein molecule dissolved in a solvent of identical scattering density has a vanishingly small angle solution scattering signal. This can be used to effect in biological studies; imagine, for example, a particle composed of two phases, each of different scattering densities (e.g., lipid, and the protein within a lipoprotein). By exploiting solution scattering experiments at various solvent densities, information can be obtained that reveals the whole particle, the lipid only, or the protein only. Such an experiment can provide much more information than the solution scattering from the whole particle alone.

In the same vein, dynamic states can be revealed; consider, for example, an experiment examining protein unfolding by solution scattering. At zero denaturant concentration, the scattering curve from the whole protein is seen. As denaturant is added, the curve evolves, until at high denaturant concentration, it represents the fully unfolded state. For a two-state process, the experiment is a titration of folding. Cases that are more complex exist (e.g., a protein displaying a folding intermediate) so that the solution comprises three protein phases: folded, unfolded, and intermediate. This intermediate is never present in high concentration. Its presence can be determined with small angle scattering data using singular value decomposition (SVD). Similarly, there may be several conformations present in solution. Ensemble optimization methods (EOMs) can be used with the small angle scattering data to determine the number of these conformers and reconstruct an envelope for each.

We have just discussed what X-ray solution scattering can do as an independent technique, and how it can extend the results of crystallography—for example, in the case of large complexes. There is an important additional synergy between the two methods: the establishment of consistency of structures as determined by crystallography and small angle scattering. One always has to question the conformation in the crystal. Is it the same as that in solution? Have we crystallized the particular state of some molecule or system that we think we have? Is a dimeric structure seen in the crystal of physiological relevance? Many of these questions can be probed effectively by solution scattering experiments. The radius of gyration and/or molecular mass determined by solution scattering can distinguish a monomer from a dimer. The maximum molecular dimension measured by solution scattering can often distinguish different structural states of an ensemble. Thus, solution scattering plays a powerful role in developing a sense of certainty about the physiological or functional meaning of structures observed in crystals and can provide information complementary to other structural techniques.

We have touched on some of the information that can be extracted using solution scattering but have to caution the reader. One has to be aware of the limitations of the technique in the level of detail that it provides. Most solution scattering experiments are low resolution. The information produced can often determine certain sample characteristics with high precision (e.g., the radius of gyration, R_g), but when molecular envelopes are modeled, the resolution is typically on the scale of 15–20 Å scale. Care has to be taken to avoid overinterpretation of this information, and the technique itself is ideally suited to test or generate hypotheses for further investigation or to serve as a complementary partner to other experimental information. There are exciting developments on the horizon. The development of direct electron density determination described later is one of these. It extends the technique beyond simple envelope reconstruction, provides a quantitative indication of resolution, and can access dynamic information in the data directly.

Solution scattering methods were introduced starting in the late 1930s. The laboratories of pioneer investigators such as André Giunier and Otto Kratky began a long process of developing both the theoretical framework for analyzing solution scattering data, and the apparatus needed for the delicate and difficult job of collecting such data. It is here that we add a second caution to the reader: the experimental obstacles are formidable. Typical solution specimens contain 1% by weight of the material of interest, resulting in very weak scattering. The X-ray beam must be collimated to an angular spread smaller than the angles one wishes to observe. In addition, optical elements such as slits and collimators become luminous sources that scatter in the forward direction, with the capacity to overwhelm the scattering from the particles. Scattering from the air and the solvent also contribute to the background.

Today we see developments in synchrotron sources, in sensitive and low-noise detectors, in computational capabilities unimagined when solution scattering was first formalized, and in high-flux neutron sources and the development of the X-ray free electron laser (XFEL). Each of these on its own opens new possibilities in solution scattering. Furthermore, the focus on the ease of use of new and maturing beamline facilities, and of the experimental control and data analysis software, extends the application to those who are knowledgeable of the potential information that small angle scattering data can provide but who are not necessarily experts themselves in the technique. These developments combined, the future of the field is exciting. A highly recommended review of the applications of solution small angle scattering is given elsewhere (Svergun et al., 2013). We build on this work in a complementary manner, with a focus on practical aspects. We aim for a book that guides new users in the methods application and provides an understanding of the quality of, and information present in, the resulting data. The book is directed at those interested in understanding why and how they might effectively utilize the technique, those on the frontline engaged in experiments, and those who may have an interest in critically understanding the information that results. We predominately discuss small angle X-ray scattering (SAXS), the most basic application, and devote a chapter to the specific case of neutrons and their use in biological studies. The field of application and development is wide, and we hope that the text serves as a gateway to growth in the field and introduces the possibilities available with more advanced applications.

Part 2

Theory of Small Angle Scattering

2

Theoretical Background

2.1 Introduction

A detailed theoretical background, although very useful, is not required to carry out a successful small angle scattering experiment. Practical details needed for a successful experiment are readily available in Part III. However, a good theoretical knowledge will enhance the understanding of the analysis and interpretation of results and is important where those results are not straightforward. We seek to provide intuitive and heuristic approaches, rather than rigorously formal ones (which are covered in detail elsewhere), whenever possible.

In an X-ray scattering experiment, whether small angle scattering or crystallography, a monochromatic, parallel beam of X-rays is shone on a specimen, and the pattern of scattered or diffracted X-rays is collected on a detector. The method of small angle scattering, which is the subject of this book, deals with the case in which the specimens for the X-ray diffraction experiments are individual particles or molecules. Unlike the situation with a crystal, these are rotationally averaged during the course of the experiment. In this chapter, we work through the basic theory of solution scattering. At a number of points, we make use of results from crystallography, which we introduce carefully for those not familiar with that area.

2.2 Scattering Basics

Figure 2.1 illustrates the basic geometry and notation of scattering. In (a) the vectors $\boldsymbol{k}_\mathrm{i}$ and $\boldsymbol{k}_\mathrm{f}$ represent the directions of the incident beam and the scattered beam, respectively. In the conventions of small angle scattering, they have lengths $|\boldsymbol{k}_\mathrm{i}| = |\boldsymbol{k}_\mathrm{f}| = 2\pi/\lambda$, where λ is the wavelength of the radiation. The vector $\boldsymbol{q} = \boldsymbol{k}_\mathrm{i} - \boldsymbol{k}_\mathrm{f}$ is called the scattering vector. It represents the momentum change undergone by a scattered photon and is key to describing the phase difference (the phase angle by which one wave leads or lags another). In (b) two particles at points $\boldsymbol{r}_1$ and $\boldsymbol{r}_2$ are shown, separated by the distance $\boldsymbol{r}$. Here, the path length difference between waves scattered by points $\boldsymbol{r}_1$ and $\boldsymbol{r}_2$ is $(\lambda/2\pi)\boldsymbol{r} \cdot \boldsymbol{q}$, and the phase difference is given by $\Delta\varphi = \boldsymbol{r} \cdot \mathbf{q}$. These are explained further in the figure caption.

Biological Small Angle Scattering: Theory and Practice. Eaton E. Lattman, Thomas D. Grant, and Edward H. Snell.

Published in 2018 by Oxford University Press.
DOI 10.1093/oso/9780199670871.001.0001

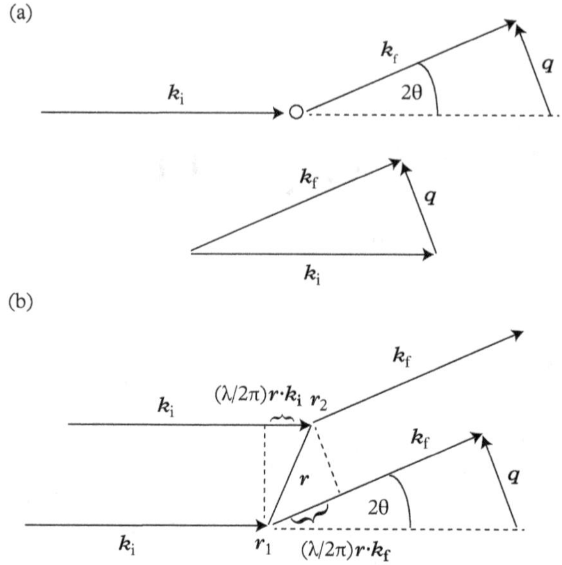

Figure 2.1 The basic diffraction geometry of scattering, where (a) an incident wave $\boldsymbol{k}_i$ is scattered by an object producing a forward scattered wave $\boldsymbol{k}_f$. The difference between these vectors is termed the scattering vector and notated by $\boldsymbol{q}$. In (b) two points are shown, $\boldsymbol{r}_1$ and $\boldsymbol{r}_2$, which are separated by $\boldsymbol{r}$. The scattered wave shown is similar to that shown in (a).

A distant observer (i.e., one at the detector) sees the scattering from these two points as the sum of the waves they emit. The complex amplitude of this sum is given by

$$\boldsymbol{F}(\boldsymbol{q}) = f_1 \exp(i\boldsymbol{q}\cdot\boldsymbol{r}_1) + f_2 \exp(i\boldsymbol{q}\cdot\boldsymbol{r}_2), \tag{2.1}$$

which can also be expressed as $\boldsymbol{F}(\boldsymbol{q}) = Fe^{i\alpha}$, where f_1 and f_2 represent the weight of the scattering from each point, respectively, and the Euler notation e^{ix} (where $e^{ix} = \cos x + i \sin x$) is used to encode phase as a polar angle. The scattered waves can be represented on a vector diagram in which f represents the amplitude of the scattered wave, and the exponential factor marks phase. These vectors can be added to give the total scattering represented by $\boldsymbol{F}(\boldsymbol{q})$ (see Figure 2.2). This argument can be generalized in several ways. First, equation 2.1 is valid for any number of scattering centers, not just the two that we have in our diagram. The theory is linear.

Second, we can provide more detail about the nature of the scattering centers. If these are atoms, then the factor f becomes a function of the scattering vector $\boldsymbol{q}$. This factor represents the scattering that would take place from a single atom of the chosen type placed at the origin of the coordinate system. Such factors are called atomic scattering factors and are tabulated. This specialization to atomic scatters

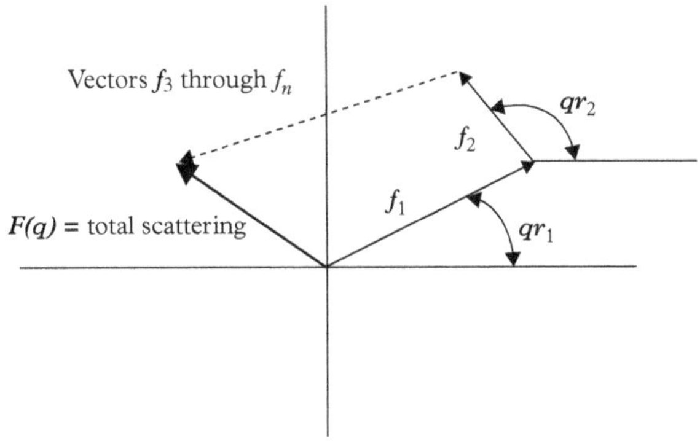

Figure 2.2 Addition of complex amplitudes of waves scattered by centers at points $\boldsymbol{r}_1$ and $\boldsymbol{r}_2$, to yield the amplitude and the phase of the total scattered wave.

gives rise to the so-called structure factor equation, a summation of the scattering from the individual atoms:

$$\boldsymbol{F}(\boldsymbol{q}) = \sum_{j=1}^{N} f_j(\boldsymbol{q})\exp(i\boldsymbol{q}\cdot\boldsymbol{r}_j). \tag{2.2}$$

This form of the equation is widely used in crystallography, where $\boldsymbol{r}_j$ represents atomic coordinates. If we regard the object as a continuous distribution of scattering density, $\rho(\boldsymbol{r})$, we can pass to a continuous limit yielding

$$\boldsymbol{F}(\boldsymbol{q}) = \int \rho(\boldsymbol{r})\exp(i\boldsymbol{q}\cdot\boldsymbol{r})dV. \tag{2.3}$$

This equation is readily recognizable as a Fourier transform. Thus, the mathematical process of Fourier transformation represents the scattering of electromagnetic radiation. This puts the mathematical toolkit of Fourier transforms at our disposal. We will make use of a number of these tools.

The Fourier inversion theorem states that it is possible to recover a function from its Fourier transform. Applying this to equation 2.3 allows us to calculate the scattering density (and therefore the structure) from the complex diffraction pattern:

$$\rho(\boldsymbol{r}) = \int \boldsymbol{F}(\boldsymbol{q})\exp(-i\boldsymbol{q}\cdot\boldsymbol{r})d\boldsymbol{r}. \tag{2.4}$$

Thus, if we can find $\boldsymbol{F}$, we can create an image of our specimen. In the realm of X-ray scattering, we have no mechanism for preserving the phase information in the diffraction pattern. The waves just pile up at the detector surface, depositing energy. Instead, we observe the intensity I, the square of the magnitude of $\boldsymbol{F}$:

$$I(\boldsymbol{q}) = |\boldsymbol{F}(\boldsymbol{q})|^2 = \boldsymbol{F}(\boldsymbol{q})\boldsymbol{F}(\boldsymbol{q})^{\star}. \tag{2.5}$$

Here, the superscript ⋆ represents complex conjugation (equal real part but imaginary part opposite in sign): if $\boldsymbol{F} = F\exp(i\alpha)$, then $\boldsymbol{F}^{\star} = F\exp(-i\alpha)$.

The convolution theorem states that, under certain conditions, the Fourier transform of a convolution is the product of the Fourier transform of its individual terms. We can see that the intensity is the product of the diffraction pattern $\boldsymbol{F}$ and its complex conjugate $\boldsymbol{F}^{\star}$. The Fourier transform of this product is termed the Patterson function $\gamma(\boldsymbol{u})$ and is given by

$$\gamma(\boldsymbol{u}) = \int \rho(\boldsymbol{r}+\boldsymbol{u})dV = \rho(\boldsymbol{r}) \otimes \rho(-\boldsymbol{r}). \tag{2.6}$$

The operator $\otimes$ represents the convolution product shown in the integral.

The Patterson function is a complete representation of the experimental data, which can always be recovered from it by inverse Fourier transformation. It was originally created for use in crystallography. It is sometimes called a vector map, because peaks at a position $\boldsymbol{r}$ mean that the specimen has one or more pairs of scatterers separated by the vector $\boldsymbol{r}$. However, the concept of decomposing the Patterson function into vectors between scatterers loses meaning for large specimens, since there are N^2 Patterson peaks if there are N scattering centers. Such a large number of Patterson peaks are heavily overlapped, leading to a relatively smooth and featureless function. Clearly, much information has been lost by losing the phase of $\boldsymbol{F}$, since there is no straightforward way to recover ρ from γ.

In solution scattering from randomly oriented particles, there is an additional level of information loss. What we observe is not the intensity function $I(\boldsymbol{q})$, but its rotational average that arises from the superposition of scattering from molecules in all possible orientations. We symbolize this by using the scalar symbol q instead of $\boldsymbol{q}$:

$$I(q) = \langle I(\boldsymbol{q})\rangle_{\Omega}, \tag{2.7}$$

where Ω represents the set of all orientations. The Fourier transform of $I(q)$ also yields a rotational average, that of the Patterson function, $\langle\gamma(\boldsymbol{r})\rangle_{\Omega} = \gamma(r)$. This is called the characteristic function $\gamma(r)$, which is a spherically symmetric three-dimensional function. The number of vectors having lengths between r and $r + dr$ is proportional to $r^2\gamma(r)dr$.

This relation suggests another convenient function, the pair distance distribution function (sometimes termed the distance distribution function) $P(r) = r^2\,\gamma(r)$; $P(r)$ is directly proportional to the number of vectors of length r within the particle. In fact, $P(r)$ plays a major role in the analysis of SAXS data. It contains all the information available from $I(q)$ and is often much easier to interpret, as will be seen in section 2.4.

We will elaborate on these functions in the following sections.

Table 2.1 and Figure 2.3 summarize the various functions and operations described in this section. It also includes related topics from crystallography. These are not essential for understanding small angle scattering but are included for the insight they may provide to those who come to this book by way of

Table 2.1 *Connections between specimens, diffraction patterns, and Patterson functions*

	Crystal	Single Particle	Solution
Shape	Unit cell content $\rho(\boldsymbol{r})$ convolved with a lattice to make periodic structure	Single molecule $\rho(\boldsymbol{r})$	Many molecules in random orientations
Output	X-ray scattering = Fourier transform	X-ray scattering = Fourier transform	X-ray scattering
Result	Complex discrete diffraction pattern $\boldsymbol{F}(\boldsymbol{h})$	Complex continuous diffraction pattern $\boldsymbol{F}(\boldsymbol{q})$	Rotational average of $I(\boldsymbol{q})$
Limitation	Loss of phase in detector $\lvert \boldsymbol{F} \rvert^2 = I(\boldsymbol{h})$	Loss of phase in detector $\lvert \boldsymbol{F} \rvert^2 = I(\boldsymbol{q})$	Loss of three-dimensional phase and amplitude
Treatment	Fourier transform	Fourier transform	Fourier transform
Direct space information recoverable	Patterson function = map of inter-scatterer vectors $\gamma(\boldsymbol{u}) = \int \rho(\boldsymbol{r})\, \rho(\boldsymbol{r}+\boldsymbol{u}) dV$ Single molecule Patterson not recoverable	Patterson function = map of inter-scatterer vectors $\gamma(\boldsymbol{u}) = \int \rho(\boldsymbol{r})\, \rho(\boldsymbol{r}+\boldsymbol{u}) dV$	Characteristic function = rotation average $r^2\gamma(\boldsymbol{u})$ = pair distance distribution function

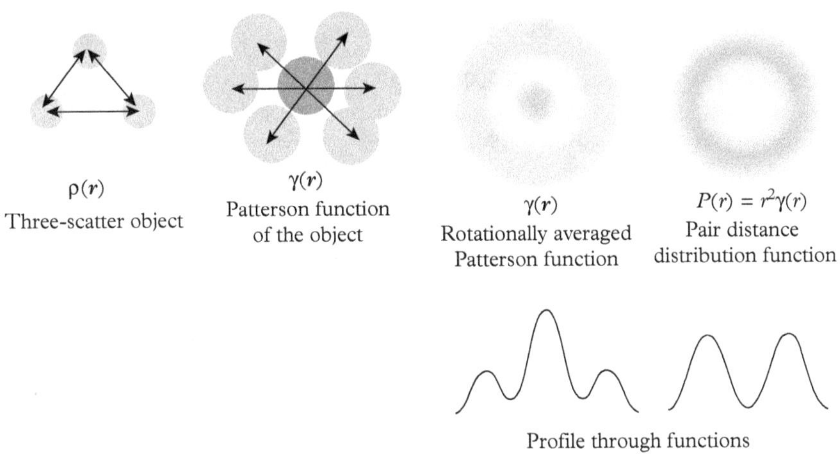

Figure 2.3 The top row shows examples of a three-scatter object, the Patterson function of the object, the rotationally averaged Patterson function, and the pair distance distribution function. The bottom row shows cross-sections through the Patterson function and the pair distance distribution function; in practice, these are radially integrated about the center.

crystallography, a common pathway, given the complementary nature of the techniques.

The function $I(q)$ can be calculated from an atomic model of a molecular structure. In the same way that the scattering by an atom represents a fundamental way of decomposing the scattering from a molecule, the scattering from a pair of centers, separated by a vector distance $\boldsymbol{r}$, is an insightful way to decompose the scattering from an object in solution. This is then generalized to an object containing N scattering elements or N^2 pairs.

We use the structure factor equation 2.2 to write the scattering $\boldsymbol{F}$ and corresponding intensity for a two-atom unit with atoms located at points $\boldsymbol{r}_1$ and $\boldsymbol{r}_2$ (see Figure 2.1); $\boldsymbol{F}^*$ is the complex conjugate of $\boldsymbol{F}$:

$$\begin{aligned} \boldsymbol{F}(q) &= f_1\exp(i\boldsymbol{q}\boldsymbol{r}_1) + f_2\exp(i\boldsymbol{q}\boldsymbol{r}_2), \\ \boldsymbol{F}^*(q) &= f_1\exp(-i\boldsymbol{q}\boldsymbol{r}_1) + f_2\exp(-i\boldsymbol{q}\boldsymbol{r}_2), \\ \boldsymbol{F}\boldsymbol{F}^*(q) &= (f_1\exp(i\boldsymbol{q}\boldsymbol{r}_1) + f_2\exp(i\boldsymbol{q}\boldsymbol{r}_2)) \cdot (f_1\exp(-i\boldsymbol{q}\boldsymbol{r}_1) + f_2\exp(-i\boldsymbol{q}\boldsymbol{r}_2)), \\ \boldsymbol{F}\boldsymbol{F}^* &= I = f_1^2 + f_2^2 + 2f_1f_2\cos(\boldsymbol{q}\boldsymbol{r}), \end{aligned} \tag{2.8}$$

where $\boldsymbol{r} = \boldsymbol{r}_1 - \boldsymbol{r}_2$. Here, f is the scattering factor for the unit in question, atoms in the most common case. This formula assumes a particular direction for $\boldsymbol{r}$; because small angle scattering is a solution technique, with particles freely tumbling during the exposure time, we need to average over all orientations between the scattering vector $\boldsymbol{q}$ (sometimes called the diffraction vector) and the interatomic vector $\boldsymbol{r}$ to obtain the intensity for the scattering:

$$\begin{aligned} \cos(\boldsymbol{q}\boldsymbol{r}) &= \cos(qr\cos\theta), \\ \langle\cos(qr\cos\theta)\rangle &= \frac{\int_0^\pi \cos(qr\cos\theta)dA}{\int_0^\pi dA} \\ &= \frac{1}{qr}\sin(qr), \end{aligned} \tag{2.9}$$

where $dA = 2\pi \sin\theta$ arises because for a given θ, the tip of the vector $\boldsymbol{q}$ lies on a circle of latitude of a sphere, as shown in Figure 2.4. Using equation 2.9 in equation 2.8, we find that the rotationally averaged scattering from the pair of centers is given by

$$\langle \boldsymbol{F}\boldsymbol{F}^* \rangle_{\text{rot}} = I(q) = f_1^2 + f_2^2 + 2f_1f_2\frac{\sin(qr)}{qr}. \tag{2.10}$$

This produces a bull's-eye pattern with a large peak at the origin, and a series of decreasing ripples radiating out. These ripples oscillate more frequently as the

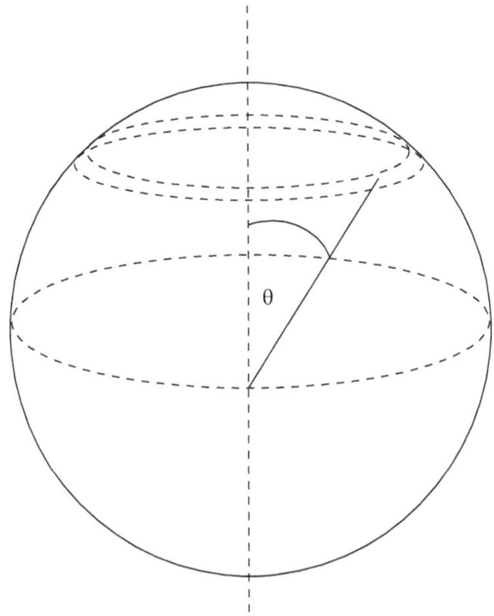

Figure 2.4 A schematic illustration of the sin(θ)/2 factor arising in the derivation of equation 2.9. The number of orientations of the scatterer between θ and θ+ *d*θ is equal to the area of the stripe on the unit sphere between the dotted latitude lines. For a unit sphere, this is 2π sin θ.

distance between the centers increases. The $\sin(x)/x$ function is commonly called $\mathrm{sinc}(x)$. An illustration of this point is given by Figure 2.5.

2.3 Multi-Scatterer Systems: The Debye Equation

It is easy to generalize equation 2.10 to the case where there are *N* scatterers. In that case, equation 2.1 would have one f_2 -type term for each value $i = 1, N$, and one $\mathrm{sinc}(qr_{ij})$ term for each pair of distinct scatterers, $r_{ij} = |\boldsymbol{r}_i - \boldsymbol{r}_j|$. The solution scattering profile is determined by the set of inter-scatterer distances. We can extend equation 2.10 to account for *N* centers as follows:

$$I(q) = \sum_{i=1}^{N}\sum_{j=1}^{N} f_i f_j \frac{\sin(qr_{ij})}{qr_{ij}}. \tag{2.11}$$

This is called the Debye equation. Note that when $i = j$, $\mathrm{sinc}(qr_{ij}) = 1$, giving rise to the f^2-type terms.

The overall intensity curve is therefore determined by the set of inter-scatter distances r_{ij} and is represented by a sum of sinc functions, one for each pair *i*, *j*. This equation is expressed using atomic scattering factors. We shall see in

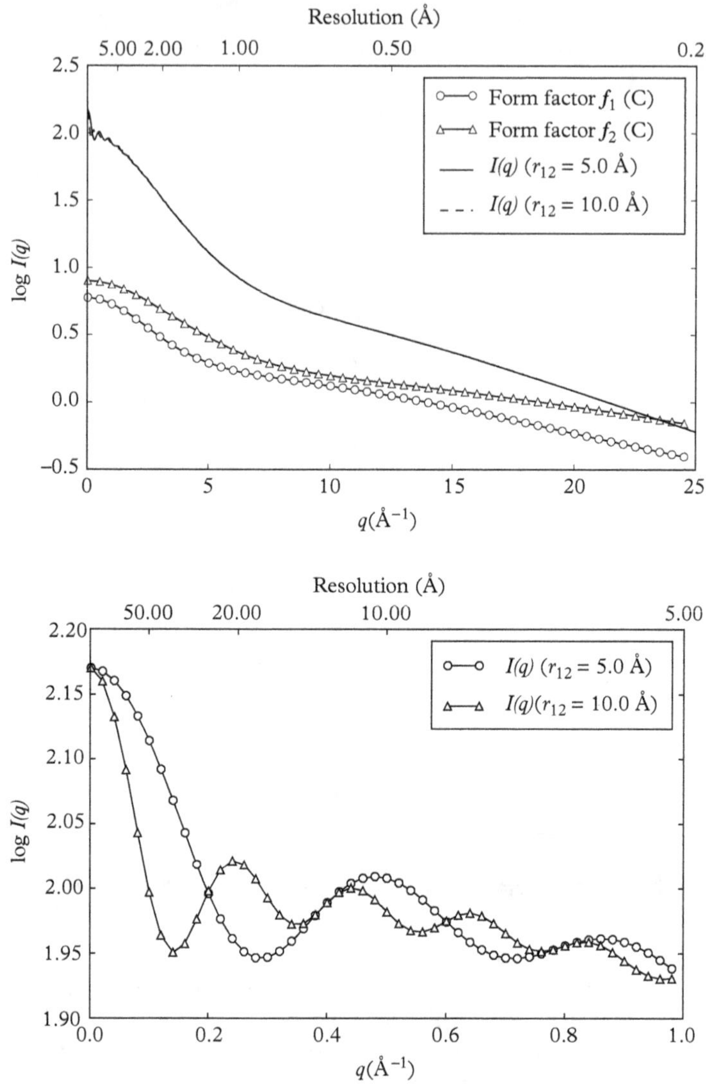

Figure 2.5 Scattering profiles from a pair of atoms at two distances. The top panel shows the scattering profile (i.e., the form factor) for carbon (C) and oxygen (O) atoms. Profiles are also shown for two C–O pairs, corresponding to 5 Å and 10 Å distance between the C and O atoms. Note that at high resolution ($q > \sim 1.0$ Å^{-1}), the scattering profiles of the two systems are nearly identical, reflecting the highly similar density functions at high resolution. The bottom panel shows the same two scattering profiles at low resolution, where differences are more prominent. Note that the larger distance (10 Å) between the atoms results in a steeper initial decline in intensity and a higher overall frequency of oscillations for the sinc function.

Chapter 4 that structures are often modeled at low resolution by packing spheres to fill a molecular envelope. The formalism is equally valid in that case, with f representing the form factor for the sphere size being used.

This discussion allows another view of the rotational averaging process inherent in solution scattering. In direct space, if we could observe the full diffraction pattern of an isolated molecule, we could readily calculate the Patterson function of this object. As the diffraction pattern is rotationally averaged, so is the direct space structure to which it corresponds. Thus, the direct space function determined by the scattering profile $I(q)$ is the rotationally averaged Patterson function of the particle. In the same way that we built up the Debye equation from scattering of pairs of atoms, we can build up the rotational average of the Patterson function by averaging the patterns from single inter-scatterer vectors. Placing one end of the Patterson vector at the origin rotating the other end through all possible angles gives rise to a spherical shell thickness determined by the size of the two atoms. The entire rotationally averaged Patterson function is the sum of such shells from all Patterson vectors.

2.4 The Pair Distance Distribution Function, *P*(*r*)

In sections 2.2 and 2.3, we introduced the pair distance distribution function, $P(r)$. This function can be derived from the Debye equation, providing additional insight into the meaning of this function.

Imagine that the terms in equation 2.11 are reordered from the smallest to the largest distance r_{ij} and placed into bins of narrow size. For $i = j$ there will be N terms corresponding to zero distance. These will be followed by a series of bins corresponding to larger lengths, with the last bin labeled by a distance representing the maximum dimension (D_{max}) of the particle under consideration. We can write

$$I(q) = \sum_{k=1}^{K} b_k f_k^2 \frac{\sin(qr_k)}{qr_k}, \tag{2.12}$$

where K is the number of bins, b_k is the number of distances r_{ij} in bin k, and r_k is the mean distance within the bin; f_k^2 is the mean-square value of $f_i f_j$ in the k-th interval.

This reordering by distance is a manipulation that allows distance to become the independent variable, leading to the pair distance distribution function.

As previously mentioned, in the Patterson function $\gamma(\mathbf{r})$ the number of weighted vectors within the particle in the interval $\mathbf{r}$, $\mathbf{r} + \Delta\mathbf{r}$ is proportional to $\gamma(\mathbf{r})dV$. If we perform a spherical average on $\gamma(\mathbf{r})$ to obtain $\gamma(r)$ (based on the assumption of an isotropic system), then $\gamma(r)dV$ gives the number of vectors of length r, $r + \Delta r$, independent of direction. We can then equate $b_k f_k^2$ (equation 2.12) with $\gamma(r_k)dV$. Expressing dV as $4\pi r^2 dr$ in spherical coordinates, we extrapolate to continuous variables to arrive at a scattering intensity given by

$$I(q) = \int_0^{D_{\max}} 4\pi r^2 \gamma(r) \frac{\sin(qr)}{qr} dr. \tag{2.13}$$

In the discussion of equation 2.7, we showed that

$$P(r) = r^2 \gamma(r). \tag{2.14}$$

Substituting $P(r)$ into equation 2.13, we have

$$I(q) = 4\pi \int_0^{D_{\max}} P(r) \frac{\sin(qr)}{qr} dr = \frac{4\pi}{q} \int_0^{D_{\max}} \frac{P(r)}{r} \sin(qr) dr. \tag{2.15}$$

Equation 2.15 is, in fact, a Fourier transform, which can be readily inverted:

$$P(r) = \frac{r}{2\pi^2} \int_0^{\infty} qI(q) \sin(qr). \tag{2.16}$$

Equations 2.15 and 2.16 allow one to pass back and forth readily between P and I, the pair distance distribution and the recorded intensity, respectively.

The pair distance distribution function is conceptually easy to picture and allows a rapid interpretation of scattering results. Practically, the pair distance distribution function (describing the distribution of the interatomic scattering distances) should equal zero both at $r = 0$ and at any r greater than the maximum particle dimension ($D_{\max}$). These limits can be used to assess the quality of the data in the determination of $D_{\max}$. In addition, $P(r)$ can be illustrated in two dimensions with reference to Figure 2.6, where the distances between spheres representing atoms are shown. In the case of Figure 2.6(a), the $P(r)$ function is a unit value at distance r, representing a single interatomic scattering vector. When three atoms are considered, as in Figure 2.6(b), each adjacent pair has an interatomic scattering vector of length r, and the outer pair contributes a single vector of length $2r$. For the purposes of illustrating the concept, we assume point scattering. In reality, there is a continuum, but the concept is useful in understanding the difference between elongated particles, as shown in Figure 2.6(c), and ones that are more globular, as shown in Figure 2.6(d).

If we extend our view in Figure 2.6 from two to three dimensions, as shown in Figure 2.7, we can begin to see how shape influences $P(r)$. Elongated shapes will progressively have more and more inter-scatterer distances that become longer and longer, as shown in Figure 2.7(a), (c), and (e). For systems that are more compact, the inter-scatter distances do not become longer but the number increases, as shown in Figure 2.7(b), (d), and (f).

Equation 2.16 is given by the spherical average over all r, which leads to ambiguity in analysis of the SAXS profile. The result is that SAXS data can be fit equally well by multiple models, leading to a problem of uniqueness (Volkov & Svergun, 2003).

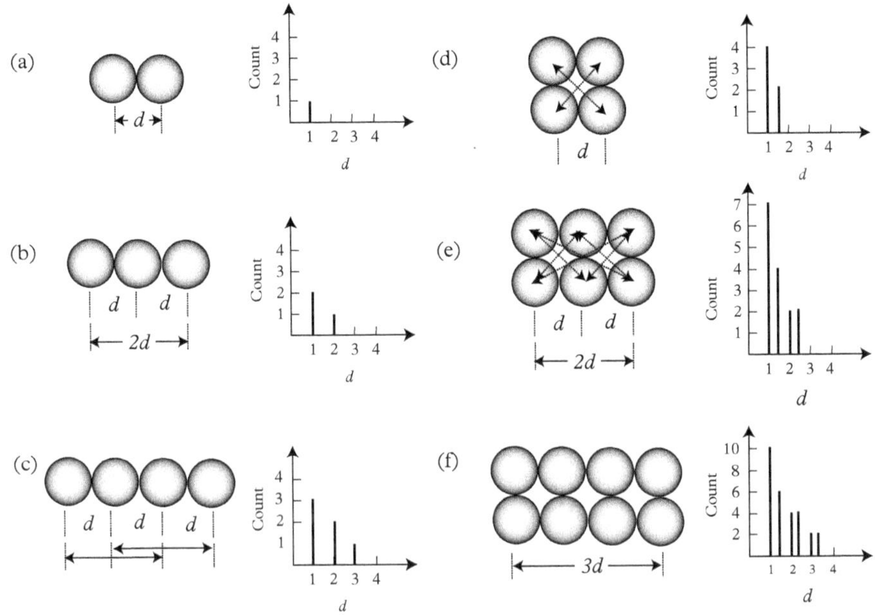

Figure 2.6 Two-dimensional example of how an object with N centers can be divided into N^2 pairs of inter-scatterer distances, with the histogram of scattering distances illustrating a simplified view of the pair distance distribution function $P(r)$. These histograms assume that all the scattering occurs from the center of the scatterer.

The ambiguity can be illustrated using a similar construction to Figures 2.5 and 2.6 in Figure 2.8, where we have two different particles but the same pair distance distribution function.

While experimentally the low-resolution portion of a SAXS curve appears to define the particle shape uniquely (Sokolova et al., 2003), theoretically small angle scattering data alone cannot distinguish between multiple possible particle shapes. Unless there is a specific hypothesis to be tested, envelope reconstructions from small angle scattering data should not be relied upon without complementary information from other techniques. We will address the pair distance distribution function in more detail in Chapter 3.

Svergun and Koch (2003) clearly illustrate the scattering profile and resulting pair distance distribution function, $P(r)$, for the simple geometric shapes of a long rod, a flat disk, a dumbbell, a hollow sphere, and a solid sphere (see Figure 2.9). The solid sphere shows a sinc function, with the hollow sphere producing an offset equivalent. The recorded scattering provides the same information as $P(r)$ but in practice, $P(r)$ provides more intuitive information on the shape. Globular particles (the sphere, in this example) display bell-shaped $P(r)$ functions with a maximum around $D_{max}/2$. A hollow sphere shifts the maximum to distances larger than $D_{max}/2$. Elongated particles (e.g., the rod) display skewed distributions with a

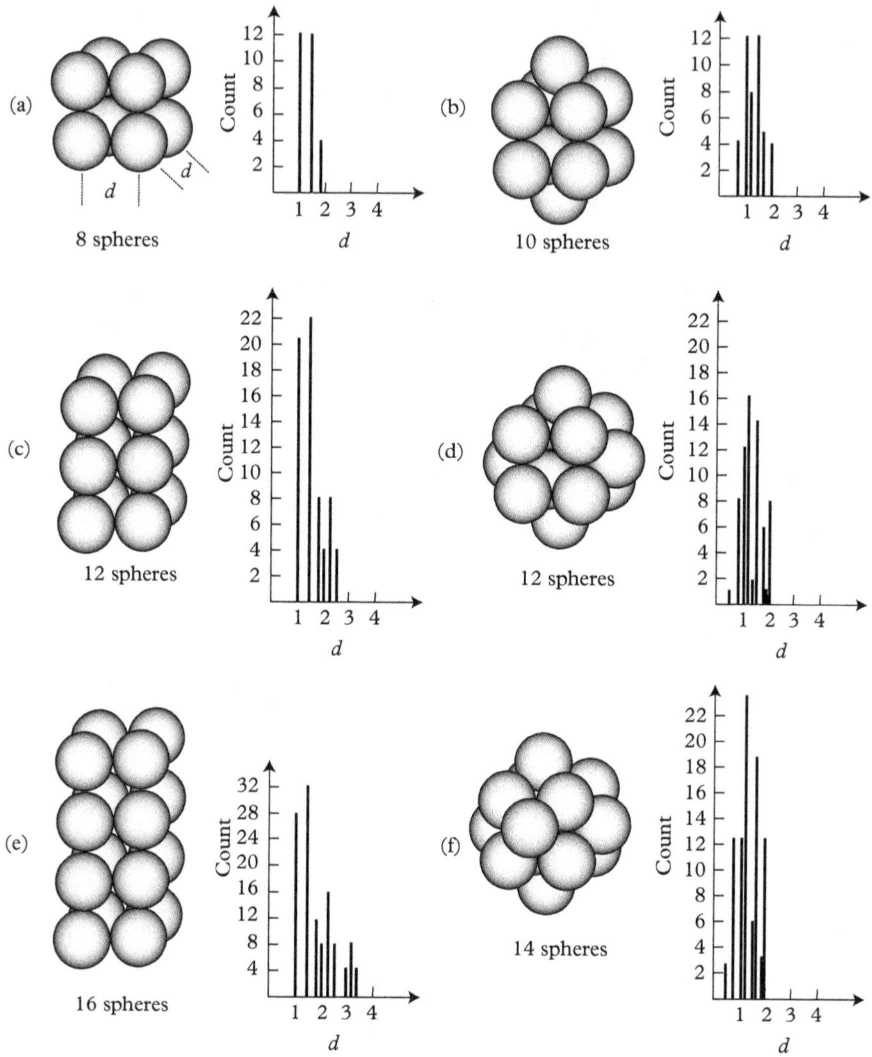

Figure 2.7 Schematic of three-dimensional particles and their distance distributions. As in Figure 2.6, these histograms assume that all the scattering occurs from the center of the scatterer.

maximum corresponding to the radius of the cross section. Flattened particles (such as a flat disk) have a broad maximum curve shifted to distances smaller than $D_{max}/2$. Particles consisting of well-separated units (e.g., the dumbbell) display multiple maxima, the smallest corresponding to the inter-subunit distances, and the others indicating the separation between groups of subunits (Svergun & Koch, 2003).

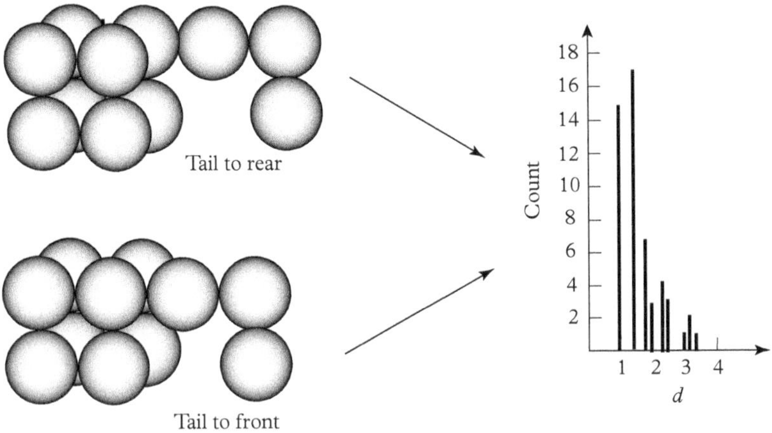

Figure 2.8 Two different particles that display the same pair distance distribution function.

2.5 The Concept of Contrast in Solution Scattering

The Debye equation (introduced in section 2.3) for $I(q)$ ignores the bulk solvent that surrounds the particle. As we will see in later chapters, the operational $I(q)$ curve represents a more realistic system in which the bulk solvent density $\bar{\rho}_s$ has been subtracted from the entire experimental volume, so that bulk solvent has zero scattering density, and the particle has a density that is reduced by the mean scattering density of the solvent $\bar{\rho}_s$. The signal from the dissolved particles is determined by their contrast above this background density value, $\Delta\bar{\rho}_p = \bar{\rho}_p - \bar{\rho}_s$. These densities are illustrated in Figure 2.10.

Bolstered by this simple approach, we can address more formally the effects of solvent, and the related issue of what structural features contribute to the scattering curve at various values of $\boldsymbol{q}$. We first need to define the meaning of "small angle." Equation 2.4 shows that the electron density $\rho(\boldsymbol{r})$ of a particle is the sum of Fourier components of the form $\exp(i\boldsymbol{qr})$. If $\boldsymbol{q}$ is small, then the spatial wavelength of $\exp(i\boldsymbol{qr})$ is correspondingly large. In the small angle approximation, we ask that half this wavelength be greater than or equal to the diameter of the particle. That is, for small angles, $|\boldsymbol{q}| \leq \frac{\pi}{D_{max}}$.

In the small angle case, there can be no appreciable fluctuation of density within the particle, and we can assume its density is essentially constant.

We have seen above that the scattering signal is proportional to the difference in electron density between the solvent and the particle. It is thus convenient to re-express the equations above based on $\Delta\rho_p$ (see Figure 2.10).

As derived above,

$$\gamma(\boldsymbol{r}) = \int \rho(\boldsymbol{r}_1)\rho(\boldsymbol{r} + \boldsymbol{r}_1)dV_{r_1}. \tag{2.17}$$

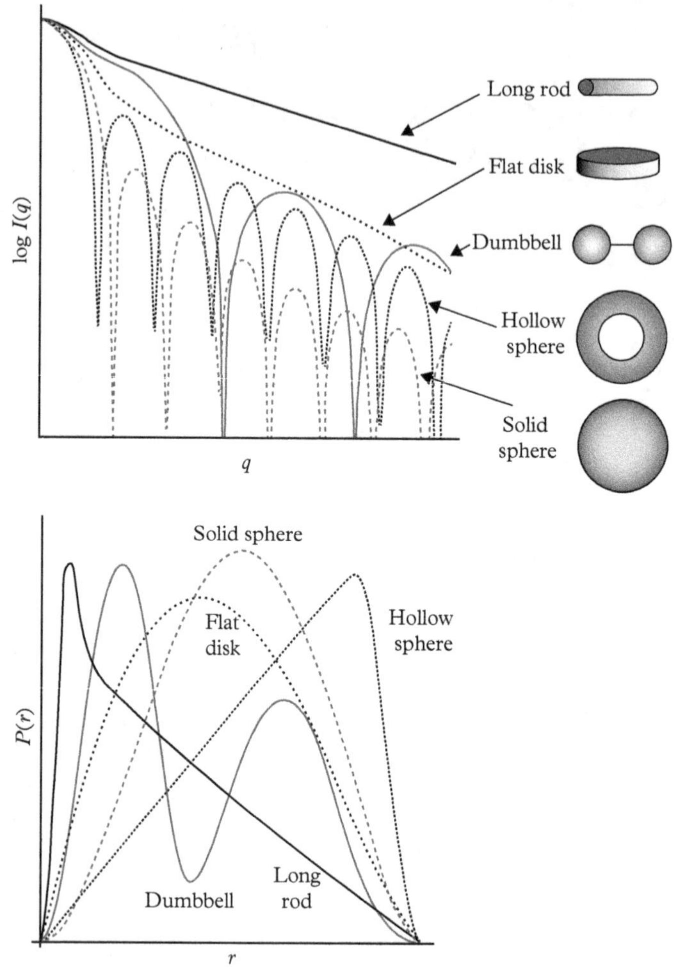

Figure 2.9 Illustration of scattering curves from three-dimensional shapes, and the corresponding pair distance distribution function.

[1] DOI 10.1088/0034-4885/66/10/R05.

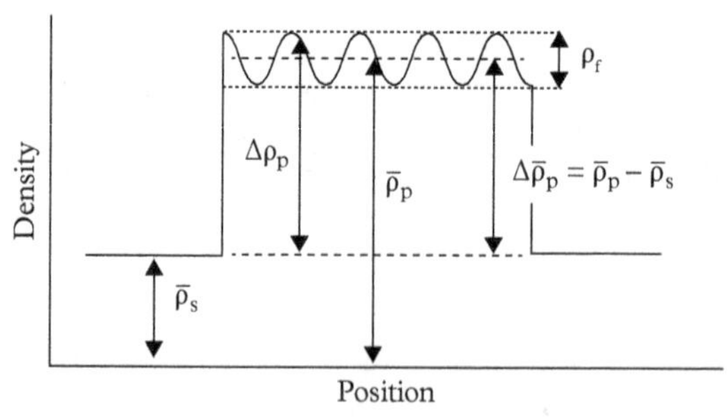

Figure 2.10 One-dimensional representation of a particle of width a and mean density $\bar{\rho}_p$ and immersed in a solvent of mean density $\bar{\rho}_s$. The particle is schematized as a rectangle with a density fluctuation of amplitude ρ_f riding on top. Difference densities are also labeled.

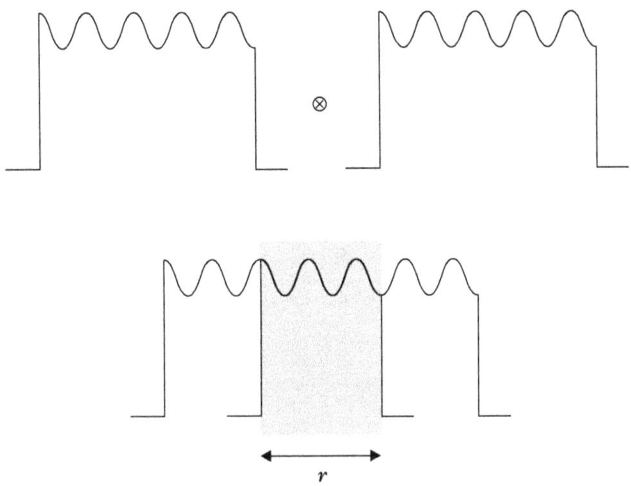

Figure 2.11 Illustration of the convolution integral in equation 2.17. The upper row shows two copies of ρ. The lower row shows $\rho(\boldsymbol{r})$ and $\rho(\boldsymbol{r}+\boldsymbol{r}_1)$ superimposed. The gray area marks the region where both are nonzero. The integral in equation 2.17 goes over their product in this region.

This is illustrated in Figure 2.11. In the small angle case, where we can ignore the internal density fluctuations in the particle,

$$\rho(\boldsymbol{r}) = \bar{\rho}_p \rho_c(\boldsymbol{r})$$

and

$$\gamma(\boldsymbol{r}) = \bar{\rho}_p^2 V(\boldsymbol{r}), \qquad (2.18)$$

where ρ_c is a shape function with value 1 inside the particle and 0 outside it, and $V(\boldsymbol{r})$ is the volume of overlap between images of the particle centered at $\boldsymbol{r}$ and at $\boldsymbol{r}+\boldsymbol{r}_1$. Setting $\boldsymbol{r} = 0$, we find $\gamma(0)$ such that

$$\begin{aligned} \gamma(0) &= \bar{\rho}_p^2 V_{\text{particle}}, \\ V_{\text{particle}} &= \frac{\gamma(0)}{\bar{\rho}_p^2}. \end{aligned} \qquad (2.19)$$

To introduce contrast, we can write, using the notation in Figure 2.9, the electron density of the particle as a constant plus the fluctuation $\rho(\boldsymbol{r}) = \Delta\bar{\rho}_p \rho_c(\boldsymbol{r}) + \rho_f(\boldsymbol{r})$. The fluctuation of density around the mean particle density is given by $\rho_f(\boldsymbol{r})$. Using this notation, we can rewrite the Patterson function as follows:

$$\gamma(\boldsymbol{r}) = \int [\Delta\bar{\rho}_p \rho_c(\boldsymbol{r}_1) + \rho_f(\boldsymbol{r}_1)][\Delta\bar{\rho}_p \rho_c(\boldsymbol{r}+\boldsymbol{r}_1) + \rho_f(\boldsymbol{r}+\boldsymbol{r}_1)]dV, \qquad (2.20a)$$

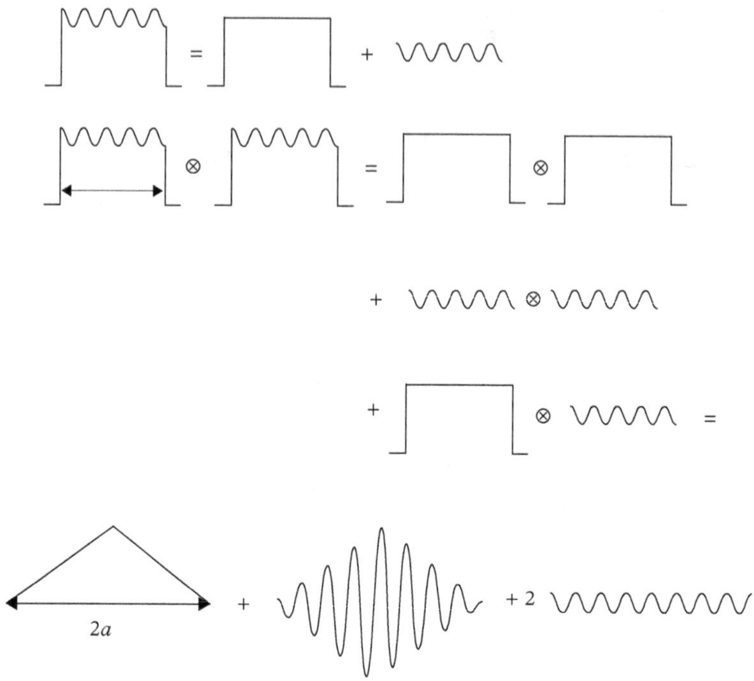

Figure 2.12 Illustration of the three terms in equation 2.20b. Convolution of the step-function part gives the triangle step. Convolution of the step with the fluctuation gives the wedge-shaped fluctuation, and convolution of the fluctuations gives a similar fluctuation.

$$\gamma(\boldsymbol{r}) = \Delta\bar{\rho}_{\mathrm{p}}{}^{2}\int \rho_{\mathrm{c}}(\boldsymbol{r})\ \rho_{\mathrm{c}}(\boldsymbol{r}+\boldsymbol{r}_1)dV + 2\Delta\bar{\rho}_{\mathrm{p}}\int \rho_{\mathrm{c}}(\boldsymbol{r})\ \rho_{\mathrm{f}}(\boldsymbol{r}+\boldsymbol{r}_1)dV + \int \rho_{\mathrm{f}}(\boldsymbol{r}+\boldsymbol{r}_1)\ \rho_{\mathrm{f}}(\boldsymbol{r}+\boldsymbol{r}_1)dV. \tag{2.20b}$$

The three terms in equation 2.20b have distinct meanings, which are pictured in Figure 2.12.

The first term can be simplified if we recognize that the integral within it is merely the function $V(\boldsymbol{r})$ we defined above. Thus,

$$\Delta\bar{\rho}_{\mathrm{p}}{}^{2}\int \rho_{\mathrm{c}}(\boldsymbol{r}_1)\ \rho_{\mathrm{c}}(\boldsymbol{r}+\boldsymbol{r}_1)dV = \Delta\bar{\rho}_{\mathrm{p}}{}^{2}\ V(\boldsymbol{r}). \tag{2.21}$$

This is closely similar to the low-resolution approximation given for $\gamma(\boldsymbol{r})$ above. This term contributes to scattering only at low resolution.

The last term is quite small, and contains information about density fluctuations that contribute to scattering only at higher angles. However, it does not depend on the contrast between solvent and mean particle density. The middle term contributes modestly over broader resolution ranges, but it is smaller than the first term. It does depend on the contrast.

This representation also clarifies the effects of contrast matching, in which electron density of the solvent is adjusted to match the mean electron density of one component of the structure—say, the protein in a protein–DNA complex. With $\Delta\bar{\rho}_p = 0$, only the third term above contributes to scattering; the low-angle part vanishes. Thus, any scattering observed at small angles arises from the components that are not contrast matched. Using X-rays, contrast matching is difficult, since most elements in biological solutions have similar electron densities and therefore similar scattering cross sections. Contrast matching is more effective in neutron scattering (Jacques & Trewhella, 2010; also see Chapter 12) and is a useful tool in particular for macromolecular complexes containing mixtures of proteins, nucleic acids, and lipids. The neutron scattering cross section is not dependent on the number of electrons in the atom but on the properties of the nucleus; for example, the magnitude and sign of the cross section of the deuterium isotope of hydrogen are significantly different from that of hydrogen. Using varying mixtures of H_2O, D_2O, and perdeuterated solutes, the scattering from single or multiple components of a complex can be isolated and studied, while the scattering of the remaining components matches the solvent and is effectively zero.

2.6 Resolution and Information Content

The ability to define details in an experiment is often termed the resolution. The nominal resolution from a scattering experiment is equal to $2\pi/q_{max}$, where q_{max} corresponds to the maximum scattering angle collected in an experiment. In equation 2.4, this corresponds to the spatial wavelength of the term $\exp(i\mathbf{q}_{max}\mathbf{r})$. In typical experimental small angle scattering setups for biological samples such as proteins, this leads to resolution estimates ranging from 10 Å to 50 Å. However, the precise definition of resolution in scattering experiments is somewhat ambiguous, owing to the loss of information due to spherical averaging, which reduces the three-dimensional anisotropic scattering function to a one-dimensional profile. Thus, a direct comparison of nominal resolution in SAXS to crystallography is imprecise.

Another way of assessing a SAXS experiment is through information content (i.e., the number of unique observations present in the data). For crystallographic data, the number of unique pieces of information is given by the number of unique reflections measured in an experiment; this can often exceed tens or hundreds of thousands of reflections. To create a viable atomic model, atoms are placed in electron density maps; the calculated diffraction of the atomic model is then compared to the experimental diffraction pattern, and the model is refined until the best agreement is found. Each atom in the model generally has four possible parameters for refinement, namely, coordinates in three dimensions, and an associated temperature factor describing the average displacement of the atom with respect to its equilibrium position. For an object consisting of N atoms, a model thus requires at least $4N$ pieces of information (i.e., measurements or reflections) to satisfy the necessary ratio of observations to parameters. However, in practice, two to five times this number of measurements is usually required to overcome issues of noise and overfitting. For atomic models, additional information is often afforded by incorporating biochemical restraints such as bond lengths and angles precisely known from other experimental methods such as small molecule crystallography. These external pieces of information act as additional observations to favorably weight the observations-to-parameters ratio. In essence, these additional restraints reduce the number of dimensions required to define an atom from three Cartesian coordinates, x, y, and z, to two dimensions defined by φ and ψ torsion angles for backbone atoms, along with favored rotamer angles for various side chains. Taken together, this collective body of information from experimental data enables the determination of atomic coordinates to precisions of tenths of angstroms for high-resolution crystal structures, better than the corresponding nominal resolution of typical data to ~2 Å.

SAXS profiles are fundamentally different from crystallographic patterns, primarily owing to two factors: the use of continuous versus discrete patterns (see Chapter 1, Figure 1.2), and the use of one-dimensional versus three-dimensional data. The latter factor results from the spherical averaging of molecules tumbling in solution over the course of an experiment, leading to a substantial loss of information in the resulting profile. Spherical averaging causes multiple problems when analyzing and modeling SAXS data. First, resulting models are inherently ambiguous, as multiple models may exist that fit the data equally well (see the example in Figure 2.8). This is directly due to the loss of three-dimensional information. Second, attempting to model one-dimensional data with three-dimensional models provides substantial opportunity for overfitting, since the information content in a one-dimensional SAXS profile is much less than the equivalent information content in a three-dimensional pattern of identical resolution. The real-space information directly obtainable from a SAXS profile is the pair distance distribution function, $P(r)$ (i.e., the set of all pair distances between electrons in a particle). The corresponding function from crystallographic data is the Patterson function, which contains the set of all pair *vectors* in three dimensions between electrons in a particle. As the directional information from the list of three-dimensional vectors is lost in a SAXS experiment, the number of unique pieces of information present in the data is also greatly reduced. As such, nominal resolution of 10 Å for crystallographic data (assuming known phases) would result in an electron density map that would be much more detailed, compared to a model derived from SAXS data at an identical resolution.

The other major factor distinguishing SAXS data from crystallographic data is that scattering profiles are continuous distributions of intensity, whereas crystallographic patterns consist of measurements at discrete locations in reciprocal space with little intensity visible between Bragg spots in most cases. For crystallography, this causes the total scattering of the illuminated volume of the crystal to be concentrated in small volumes of reciprocal space, resulting in very high intensity measurements. These high-intensity measurements, when merged with symmetry-related reflections, result in large signal-to-noise ratios, often out to high resolution. For SAXS, however, the scattering profile is continuous, resulting from the underlying continuous molecular transform of single particles being sampled at multiple orientations and summed together. This continuity is, in fact, quite advantageous for analyzing scattering data in analyzing the context of information content of the SAXS experiment. Information content in SAXS can be assessed using the Shannon sampling theorem. The Shannon sampling theorem (Shannon, 1948a, b) states that any continuous signal can be exactly reconstructed from a discrete sampling of that signal if the sampling rate is more than two times the highest frequency present in the signal. The highest frequency in reciprocal space corresponds to the largest pair distance present in the object; this is defined as the maximum particle dimension, or D_{max}. The width of a so-called Shannon channel can then be calculated as $\Delta q_{Shannon} = \pi/D_{max}$. Therefore, according to the Shannon sampling theorem, the continuous SAXS profile can be reconstructed, provided we sample the continuous reciprocal space scattering profile at intervals less than or equal to π/D_{max}. In practice, the experimental sampling rate is determined by several factors such as X-ray wavelength, sample-to-detector distance, the number and size of pixels in the detector, and the binning procedure used when radially averaging measurements from two-dimensional detectors. Typical SAXS profiles for biological samples contain on the order of several hundred data points. One can estimate the number of Shannon channels (i.e., the number of unique and independent pieces of information in the profile) as the total q range of the data divided by the width of the Shannon channels, as shown by equation 2.22:

$$N_{Shannon} = \frac{q_{max} - q_{min}}{\Delta q_{Shannon}}. \tag{2.22}$$

For SAXS data from a particle that is 100 Å in diameter, collected to $q < 0.50$ Å^{-1}, the number of Shannon channels is ~15. Thus, not only is SAXS a low-resolution technique, it is also a low-information technique. However, since there are 500 data points (i.e., measurements) in total, each unique piece of information in the scattering profile will have been measured approximately 33 times, a phenomenon commonly known as "oversampling." In essence, this means that while there are very few independent pieces of information in the scattering profile, each one of those pieces of information is known to very high precision.

Resolution is also difficult to describe when discussing models created to fit SAXS data. Unfortunately, ab initio envelopes are often viewed as the equivalent of electron density maps in crystallography, albeit at much lower resolution. Electron density maps are calculated from measured crystallographic data after determining phases and thus are the real-space representations of the data to which atomic models are fit. Crystallographic resolution manifests in an electron density map as the ability to discriminate between two adjacent peaks corresponding to the electron cloud centered at atomic coordinates; thus, higher resolution allows one to better estimate the atomic center, based on the peak position. Ab initio envelopes, however, are themselves models of the data and thus are not equivalent to low-resolution electron density maps. Therefore, a comparison of resolution as the ability to distinguish features in an envelope is misleading. The resolution of SAXS data is, in reality, the ability of the data to distinguish between models. Such models may include ab initio envelopes, and while many ab initio envelopes will fit the data equally well, gross structural features not reminiscent of the particle shape can be reliably discriminated against. Additionally, high-resolution models obtained via other methods, such as crystallography or NMR, can be compared against SAXS data, and differences, such as small conformational changes in a fold, can be discerned by comparing calculated scattering profiles.

One of the most commonly used methods to generate ab initio envelopes in SAXS is bead modeling, described in more detail in Chapter 4. There has recently been an important development in ab initio bead modeling: the ability to quantify resolution (Tuukkanen et al., 2016). As there are many bead models that will describe the SAXS curve equally well, the convention has been to calculate multiple envelopes (~10), each starting with a different random seed, and to align and "average" the set of bead models, generating a consensus bead model. The advantage of this approach lies in the ability to assess the variability within the set of bead models and to determine which features of the envelopes are consistent with each other and which are specific to the random seed. Features that are consistent across all or most of the bead models can be considered true aspects of the particle shape, whereas features not consistent across the envelopes are considered to be artifacts. The variability within the ensemble of bead models can be quantified by taking a page from the book of cryo-electron microscopy. In cryo-electron microscopy experiments, resolution is estimated by splitting the images into two independent sets, generating three-dimensional volumes from the two sets of images, and calculating the correlation between the two volumes to quantify their similarity. This correlation is done in reciprocal space, using a process known as Fourier shell correlation (FSC), which ranges from 0 (no correlation) to 1 (perfect correlation) and is plotted as a function of resolution. To quantify resolution of bead models, a similar approach is taken: the variability within the ensemble of bead models is determined by first calculating the FSC curves for each envelope to every other envelope, and then averaging the set of FSC curves. The variability within the ensemble is estimated by assuming a cutoff of the averaged FSC curve, typically taken to be where the FSC falls below 0.5. However, the variability of the ensemble is not a direct estimate of resolution. Nevertheless, it was found that the variability of the ensemble correlates well with the estimated resolution of the bead models when comparing them against known structures through a simple linear relationship. Thus, one can estimate the resolution of the consensus bead model with knowledge of the ensemble variability. This ability to quantify resolution from bead models will help us to understand the limitations of determining particle structure from SAXS. It has already been discovered that current state-of-the-art algorithms for envelope reconstruction from SAXS are fundamentally limited to worse than 10 Å by the assumptions of bead modeling, depending on the size and complexity of the particle. These developments are likely to spur advances in the field to push the resolution capabilities of SAXS to ever-greater values.

2.7 Summary

An X-ray small angle scattering experiment is conceptually simple. A parallel, monochromatic beam of X-rays illuminates a volume of solvent containing many identical particles in random orientations. Because the solution is isotropic, the scattered X-rays form a circularly symmetric pattern on the detector. A radial trace through this intensity pattern represents the full experimental information available from the system. This intensity trace is conventionally called $I(q)$, where q is the magnitude of the momentum change undergone by a scattered X-ray photon: $q = 4\pi\sin(\theta)/\lambda$, where 2θ is the scattering angle, and λ is the wavelength of the X-rays.

The intensity $I(q)$ can be calculated from a model of the particle structure by using the Debye equation:

$$I(q) = \sum_{i=1}^{N}\sum_{j=1}^{N} f_i f_j \frac{\sin(qr_{ij})}{qr_{ij}}.$$

Here, f_i is the scattering factor for the scattering element—usually an atom—located at point $\boldsymbol{r}_i$, and r_{ij} is the distance between $\boldsymbol{r}_i$ and $\boldsymbol{r}_j$. The $\sin(x)/x$ factor, called $\text{sinc}(x)$, arises from the rotational averaging.

Initially, however, one rarely has a model. In these cases, $I(q)$ can be transformed into a direct space function that provides clearer insight into various characteristics of the structure. In the same way that the Fourier transform of a crystal diffraction pattern gives a direct space vector map called the Patterson function, the Fourier transform of $I(q)$ yields the Patterson function of a single particle, but with the function rotationally averaged to reflect the rotational average in $I(q)$. This function is also called the characteristic function $\gamma(r)$:

$$I(q) = \int_0^{D_{\max}} 4\pi r^2 \gamma(r) \frac{\sin(qr)}{qr}\, dr,$$

$$r^2\gamma(r) = \frac{r}{2\pi^2}\int_0^{\infty} qI(q)\sin(qr)dq.$$

The function $r^2\gamma(r)$ is given its own name: the pair distance distribution function, or $P(r)$; $P(r)dr$ is the number of scattering centers in the particle, with the centers separated by distances between r and $r + dr$. The value of r at which $P(r)$ falls to zero represents the maximum distance between scattering centers—effectively, the maximum dimension of the particle. Bimodal curves for $P(r)$ usually indicate a particle with two distinct domains.

For small angles $q \leq \pi/D_{max}$, the particle dimension is smaller than the most rapidly changing component in its Fourier expansion. Thus, the density within the particle cannot vary appreciably and is effectively constant. In this angular range, the entire scattering signal from the particle arises from the *contrast* between it and the surrounding solvent, and if the solvent and particle electron densities are equal, the small angle scattering vanishes.

Fourier ideas are also useful in defining the information content of solution scattering. The largest pair distance r_{ij} present in the object corresponds to the maximum particle dimension, or D_{max}, and can be determined by the point at which the pair distribution function falls to zero. This allows the calculation of the width of a Shannon channel as $\Delta q_{Shannon} = \pi/D_{max}$. Therefore, according to the Shannon sampling theorem, the continuous SAXS profile can be reconstructed, provided we sample the continuous reciprocal space scattering profile at intervals less than or equal to π/D_{max}. A typical $I(q)$ curve may contain no more than 15 such channels, so that the information content is quite low. However, the oversampling implicit in measuring a continuous curve means that the values at the Shannon points are determined very precisely.

3

Quantities Directly Measurable by Scattering

3.1 Introduction

In Chapter 2, we described some of the theory for small angle scattering. We now discuss how we gain information and understanding from small angle scattering experiments. Small angle scattering provides direct determination of "invariants" associated with the system under study. These are called invariants because they represent quantities that any model must agree with. We discuss these invariants here, coupling them to a qualitative understanding of how features of different scales within the structure contribute to the scattering. The regions in q-space corresponding to these features are illustrated in Figure 3.1. This shows a schematic of the scattering from a macromolecular sample broken down into scattering in four regions: (i) scattering at zero angle, (ii) scattering primarily determined by the radius of gyration, (iii) scattering associated with shape, and (iv) scattering due to the interface of the particle and solution. In panel (iii), showing scattering associated with shape, the influence of spherical particles, disk-like particles, and rods are shown.

3.2 Invariants

We begin with a discussion of the invariants.

The principal invariants are as follows:

- the molecular mass M
- the radius of gyration R_g
- the Porod invariant Q
- the particle volume V
- the maximum particle dimension D_{max}
- the particle surface area S
- the correlation length l_c
- the volume of correlation V_c

Biological Small Angle Scattering: Theory and Practice. Eaton E. Lattman, Thomas D. Grant, and Edward H. Snell.

Published in 2018 by Oxford University Press.
DOI: 10.1093/oso/9780199670871.001.0001

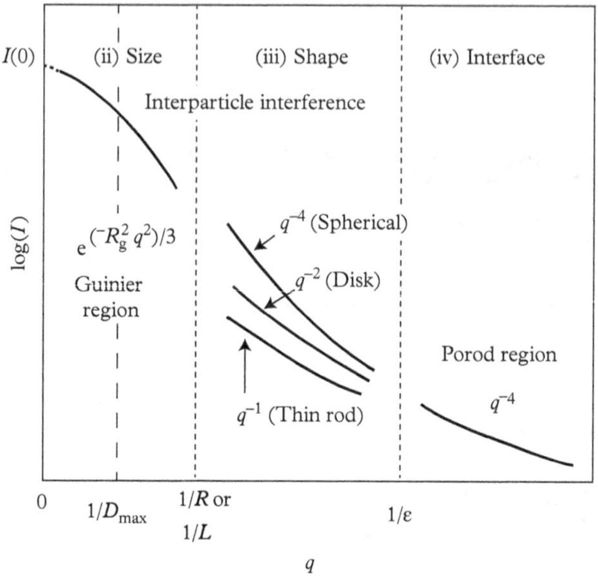

Figure 3.1 A schematic of a typical experimental scattering curve adapted from Matsuoka et al. (1999) with the help of Dr. Tsutomo Matsui. Four regions of interest are illustrated: (i) the intensity at zero angle, (ii) scattering primarily influenced by the radius of gyration, (iii) scattering associated with shape, and (iv) scattering associated with the interface of the particle and solution; D_{max} is defined in the text, R and L are radius (sphere and disk) and length (rod), and ε is the thickness of a disk (also notated as T in the text) or the diameter of a thin rod, depending on particle shape.

We discuss the first six of these quantities in some detail but spend less time on the last two, which have less application to biological material than to some other classes of specimens.

3.2.1 Molecular Mass from Intensity at the Origin

Experimentally, the intensity at the origin, (i) in Figure 3.1, can provide an estimate of the molecular mass to within an error of 10% for globular proteins, provided the particle concentration is measured with an accuracy of 5%–10% (Mylonas & Svergun, 2007). To understand how this is accomplished, perhaps the most straightforward place to start is with the value of $I(0)$, the forward scattering or, more accurately, the value $I(q)$ as $q \to 0$. At very small angles, all electrons or scattering centers in the particle scatter in phase, so that $I(0)$ is proportional to n_e^2, the square of the number of electrons in the molecule. At the same time $I(0)$ is proportional to N_m, the number of molecules in the illuminated volume. Therefore, ignoring solvent contrast,

$$I(0) \propto N_m n_e^2. \tag{3.1}$$

The value of N_m can be obtained from knowledge of the molar concentration of particles in the solution, and of the illuminated volume. We can thus calculate n_e^2. We can go from this quantity to the *molecular weight* of the particle. The following slight modification adds solvent contrast to the argument. If we assume a 1 cm^3 sample then we can equate this to

$$I(0) = N_m \Delta\rho^2 V^2, \tag{3.2}$$

where $\Delta\rho$ is the scattering density difference between the particle and the solvent (proportional to the electron density difference), and V is the volume of the particle. If the particle is mass M and the particle concentration is c, the particle number density is $n = (c/M) \cdot N_A$, where N_A is Avogadro's number. We can also express the electron density difference per unit mass of the particles: then $\Delta\rho V = \Delta\rho_m M/N_A$, where $\Delta\rho_m$ is the scattering density difference per unit mass. This leads to

$$I(0) = \frac{c\Delta\rho_m^2 M^2}{MN_A} \tag{3.3}$$

and

$$M = \frac{I(0)N_A}{c\Delta\rho_m^2}. \tag{3.4}$$

If the concentration of particles in the sample, the contrast factor, and the intensity at $q = 0$ are known, the molecular weight can be calculated. Experimentally, water can be used as a calibration source for the intensity and is typically run as a control before collecting data on any sample.

The same approach can also be used to calculate the mass of an unknown sample using that from a standard applying:

$$M_s = \frac{I_s(0)}{c_s} \cdot \frac{M_c}{I_c(0)/c_c}, \tag{3.5}$$

where M_s is the mass of the unknown sample, M_c is the mass of the control, I_s and I_c are the scattering of the sample and the control at $q = 0$, respectively, and c_s and c_s are the concentrations of the sample and the control, respectively.

Complementary methods, using the peak in the Kratky plot and the integral of $I(q)q^2$ (Fischer et al., 2010), will be described later.

3.2.2 Radius of Gyration R_g: Low Angular (Guinier) Region

The low angular region, (ii) in Figure 3.1, provides information on the radius of gyration and the maximum particle dimension. Because the determination of R_g is so important, we present two derivations of the Guinier relation, both of which provide insight.

For SAXS, the square of the radius of gyration, R_g^2, is defined as the mean-square distance of scattering centers from the center of mass of the particle:

$$R_g^2 = \int |\mathbf{r}|^2 \rho(\mathbf{r}) dV \Big/ \int \rho(\mathbf{r}) dV. \tag{3.6}$$

Scattering centers are electrons for X-rays, and nuclei for neutrons. In mechanics, the definition often refers to the mean-square distance from a chosen axis, but since particles are rotationally averaged in SAXS, all axes are often considered equivalent, and the distance from the center of mass is a more useful definition.

The maximum particle dimension, D_{max}, is defined as the maximum intra-particle distance between scattering centers. At the very lowest angles, when $q < 2\pi/D_{max}$, the image of a particle will appear blurry because there are no short wavelength Fourier components to delineate the sharp boundary at the surface. This blurriness only increases when one looks at the pair distance distribution function $P(r)$. One way to think about the scattering in this region is to construct a crude model of the pair distance distribution function using a Monte Carlo procedure. The $P(r)$ tells us how many vectors of a given length are present in the particle. We can simulate this by taking a vector, putting one end down at a random position and in random orientation, and asking if the other end lies within the object. Repeating this many times for each vector length lets us gather statistics about how frequently a vector of a given length will terminate in the particle. A vector of zero length always terminates within the structure, while a vector of length greater than D_{max} always falls outside. For values between zero and D_{max}, we can estimate the frequency by realizing that the placement of vectors represents a random sampling of the space of possible inter-scatterer vectors in the particle. Such sampling results, to a first approximation, in a Gaussian distribution whose variance σ^2 is the mean-square distance between scattering elements $\langle r_{ij} \rangle^2$. More formally,

$$P(r) = N \cdot \exp\left(-\frac{x^2}{2\sigma^2/3}\right) \cdot \exp\left(-\frac{y^2}{2\sigma^2/3}\right) \cdot \exp\left(-\frac{z^2}{2\sigma^2/3}\right).$$

Recognizing that $x^2 + y^2 + z^2 = r^2$, where r is the distance from the origin, we have

$$P(r) = N \cdot \exp\left(-\frac{3r^2}{2\sigma^2}\right) = N \cdot \exp\left(-\frac{3r^2}{R_g^2}\right). \tag{3.7}$$

Here, N is a normalization factor. The factor of 1/3 arises because one-third of the variance of the distribution occurs along each of the three independent Cartesian coordinates.

Equation 3.7 makes use of the relation from Cantor and Schimmel (1980) that $2\sigma^2 = R_g^2$, where R_g is the radius of gyration of the object. Recall that R_g^2 is the mean-square distance of the scattering elements from the center of mass of the particle. We do not reproduce this derivation, but the insight required is to realize that any vector running from a scattering element A to a scattering element B can be written as the sum of two vectors, with the first running from A to the center of mass M, and the second running from M to B. This relation links σ^2 and R_g^2.

Recall from Chapter 2 (equation 2.8) that $P(r)$ is the Fourier transform of $I(q)$. This transform is readily available from tables, yielding

$$I(q) = I_0 \exp\left(-\frac{q^2 R_g^2}{3}\right). \tag{3.8}$$

The use of this equation in determining R_g is discussed in the next section.

3.2.3 The Guinier Approximation

The Guinier approximation (Guinier, 1939; Guinier & Fournet, 1955) is the classical approach used to obtain R_g. A plot of $\ln(I)$ versus q^2, termed the Guinier plot, is a straight line with a slope $-R_g^2/3$, from which R_g can be easily derived. For a monodisperse solution, the Guinier plot is linear in the low-resolution regime, where $q < 1.3/R_g$ (Guinier & Fournet, 1955). The Guinier approximation is derived as follows. The Maclaurin series of $\sin(qr)/qr$ is given by

$$\frac{\sin qr}{qr} = 1 - \frac{(qr)^2}{3!} + \frac{(qr)^4}{5!} - \cdots. \tag{3.9}$$

Considering only very small angles, we will approximate by ignoring terms of order greater than 2. Equation 3.9 can then be inserted into equation 2.15 to yield

$$I(q) = 4\pi \int_0^{D_{max}} P(r)dr - \frac{4\pi}{6} \int_0^{D_{max}} P(r)\cdot q^2 r^2 dr. \tag{3.10}$$

By taking the limit $q \to 0$ in equation 2.15 for $I(q)$, it is easy to show that the last term on the right of 3.10 equals $I(0)$:

$$I(0) = 4\pi \int_0^{D_{max}} P(r)dr. \tag{3.11}$$

The integral in the second term is proportional to $q^2 R_g^2$ (as will be shown in equation 3.17). We anticipate that equation here for clarity:

$$R_g^2 = \frac{\int_0^{D_{max}} P(r)\cdot r^2 dr}{2\int_0^{D_{max}} P(r)dr}. \tag{3.12}$$

Equation 3.10 thus becomes

$$I(q) = I(0)\left(1 - \frac{q^2 R_g^2}{3}\right). \tag{3.13}$$

Guinier recognized this as the first two terms in the Maclaurin series of e^x such that

$$I(q) = I_0 e^{-\frac{q^2 R_g^2}{3}} \tag{3.14}$$

for sufficiently small q. One recognizes this as the same as equation (3.8), which was derived in section 3.2.3, from the statistical development of $P(r)$.

Thus, by taking the natural logarithm of both sides, we obtain

$$\ln I(q) = \ln I(0) - \frac{q^2 R_g^2}{3}, \tag{3.15}$$

which is in the familiar form of the linear equation $y = mx+b$. Therefore, by plotting the natural log of intensity versus the square of the momentum transfer q, we can obtain R_g via the slope of the line, such that

$$R_g = \sqrt{-\frac{slope}{3}}. \tag{3.16}$$

It is important to note that while R_g is calculated from the slope of the line through the data points in the Guinier region, the Guinier region is dependent upon R_g. Typically, R_g and the Guinier region are determined through an iterative cycle of calculating R_g and adjusting the Guinier region accordingly, followed by recalculating the R_g, and so on. The final determination of the accepted Guinier region and calculated R_g is then arrived at through a somewhat subjective interpretation of an acceptable linear region. This becomes particularly difficult when particles are large, resulting in very few data points and a heightened sensitivity to the estimation of R_g when varying the Guinier region by as little as one data point.

The term R_g is a measure of the size of a particle, but its meaning takes a little contemplation, as R_g is usually smaller than our picture of the particle. For a solid ball of radius R, for example, $R_g^2 = (3/5)R^2$. For a spherical shell of radius R (such as a table tennis ball), $R_g^2 = R^2$.

The Guinier formula is valid for particle shapes that are not too eccentric, but if something is known about the shape, treatments that are more elaborate can be brought into play. A powerful approach involves the definition of R_g^2 as the second moment of the mass distribution about the center of mass. This quantity can be readily calculated from the pair distance distribution function $P(r)$, which gives the relative number of vectors of length r in the particle; thus,

$$R_g^2 = \frac{1}{2} \frac{\int_0^{D_{\max}} r^2 P(r) dr}{\int_0^{D_{\max}} P(r) dr}, \tag{3.17}$$

where $D_{\max}$ is the maximum dimension of the particle. This also provides a second check of the R_g derived from the Guinier formula. The factor of 2 in the denominator arises because equation 3.17 calculates the mean-square value of the inter-scatterer distances, not R_g^2 directly.

3.2.4 The Porod Invariant, *Q*

The Porod invariant Q, like R_g, is independent of particle shape. The Porod invariant comes from relating two expressions for $\gamma(0)$. As mentioned in Chapter 2, section 2.7, the chapter summary,

$$\gamma(r) = \frac{1}{2\pi^2} \int_0^{\infty} q^2 I(q) \frac{\sin(qr)}{qr} dq \tag{3.18}$$

and

$$\gamma(\boldsymbol{r}) = \int \Delta\rho(\boldsymbol{u}) \Delta\rho(\boldsymbol{r} + \boldsymbol{u}) dV = \Delta\rho(\boldsymbol{r}) \otimes \Delta\rho(-\boldsymbol{r}). \tag{3.19}$$

The *Porod invariant* Q is defined as

$$Q = \int_0^{\infty} q^2 I(q) dq. \tag{3.20}$$

Setting $r = 0$ in equations 3.18 and 3.19 we find

$$\gamma(0) = \frac{1}{2\pi^2} \int_0^{\infty} q^2 I(q) dq = \int \Delta\rho^2 dV = V \Delta\bar{\rho}^2, \tag{3.21}$$

$$V \Delta\bar{\rho}^2 = \frac{Q}{2\pi^2}. \tag{3.22}$$

This result makes use of the limit $\sin ax/x \rightarrow a$ as $x \rightarrow 0$.

Thus the Porod invariant, the integral of the q^2-weighted intensity curve over all reciprocal space, is proportional to the mean-square fluctuation of the electron density in the particle. This relation is an application of Parseval's theorem in Fourier analysis. The mean-square fluctuation of the electron density does not change if components of the structure are moved, so long as their overlap with other components of the structure is not altered. Thus, if a structure is composed of two subunits with a linker, all conformations of the pair of subunits will have the same value for mean-square fluctuation of the electron density, hence the label "invariant."

3.2.5 Particle Volume, V or V_p

We now have the tools to estimate the particle volume, using quantities that we have already calculated. Recall that

$$I(0) = N_m n_e^2. \tag{3.23}$$

The number of electrons, n_e, is equal to the mean density times the volume of the particle, that is

$$I(0) = N_m(\Delta\bar{\rho}V)^2. \tag{3.24}$$

If the data is on an absolute scale, then

$$I(0) = \Delta\bar{\rho}^2 V^2. \tag{3.25}$$

Recall from equation 3.22 that

$$V\Delta\bar{\rho}^2 = \frac{Q}{2\pi^2},$$

and thus,

$$V = \frac{2\pi^2 I(0)}{Q}. \tag{3.26}$$

Here, V (also commonly written V_p) is the Porod volume, the volume of the hydrated particle.

3.2.6 Maximum Particle Dimension, D_{max}

The maximum length of the vectors that can be drawn within the particle is termed D_{max}. In some circumstances, this is not a truly well-defined quantity, as when one end of the longest vector lies in a flexible region. Nevertheless, knowing D_{max} can be a powerful piece of information in establishing particle shape or polymerization state. The most direct way of determining D_{max} is to identify it with the point where $P(r)$ falls to zero. This apparently straightforward task is, however, fraught with difficulty; $P(r)$ is the Fourier transform of $I(q)$, so that accurate representation of $P(r)$ requires that we know $I(q)$ over the full range of q. If $I(q)$ is truncated, this introduces ripples into $P(r)$, introducing uncertainty into the zero point. Noise in $I(q)$ introduces additional uncertainty.

The method of indirect Fourier transformation introduced by Glatter (1975, 1977) deals well with these issues. Glatter assumed that an initial estimate of $D_{\max}$ can be obtained (which can later be refined). He then approximated $P(r)$ by an expansion in a set of orthogonal cubic spline functions Φ:

$$P_{\mathrm{A}}(r) = \sum_{s=1}^{N} C_s \phi_s. \tag{3.27}$$

The Fourier transforms of Φ_s are termed ψ and can be used to expand $I(q)$:

$$\psi_s(q) = \int_0^{\infty} \phi_s(r) \frac{\sin(qr)}{qr} dr. \tag{3.28}$$

Recall that

$$I(q) = 4\pi \int_0^{\infty} P(r) \frac{\sin(qr)}{qr} \tag{3.29}$$

Combining equations 3.27, 3.28 and 3.29 gives

$$I(q) = 4\pi \sum_{s=1}^{N} C_s \psi_s(q_i), \tag{3.30}$$

where the values q are the points at which the curve has been measured. The values of C_s are unaltered by this linear Fourier transformation. This treatment has omitted various data processing protocols that may need to be incorporated, such as steps, such as desmearing (see Chapter 8), that relate the experimental curve to a theoretical one. The values of C_s are determined by a constrained least-squares protocol that incorporates a stabilization term. One minimizes N_c, the point-to-point variation of C_s, to produce the smoothest possible curve for $P_{\mathrm{A}}(r)$, subject to the constraint that the same set of C_s values fit the $I(q)$ curve well through the target function L:

$$L = \int_{S_{\min}}^{S_{\max}} \left(\frac{1}{\sigma^2(q)} \right) \left(I(q) - \sum_{p=1}^{N} C_s \psi_s(q) \right)^2 dq, \tag{3.31}$$

$$N_c = \sum_{s=1}^{N-1} (C_{s+1} - C_s)^2. \tag{3.32}$$

The final values of C_s are used to generate a new $P(r)$ from which a much more reliable value of $D_{\max}$ can be extracted.

3.2.7 Correlation Length

The correlation length l_c is defined through the relation

$$l_c = 2 \int_0^{D_{max}} \gamma_0(r) dr. \tag{3.33}$$

The meaning behind this definition is not superficially clear. To gain insight, recall that $\gamma_0(r)$ is proportional to the volume of overlap between the original object and a copy translated by r, averaged over all directions of r. The integral of $\gamma_0(r)$ thus represents total overlap, summed over all r. If the object is a rod, for example, then for r greater than the width of the rod, there are many orientations that do not overlap at all, so that $\gamma_0(r)$ falls rapidly with r, and the integral is relatively small. If the object is a sphere, then overlap is the same for all orientations, $\gamma_0(r)$ falls slowly with r, and the integral is larger. Indeed, in the case of a homogeneous sphere, it can be shown that $l_c = 3D_{max}/4$ or $6R/4$, where R is the radius of the sphere. Figure 3.2 shows $\gamma_0(r)$ for simple geometric shapes: a sphere, a disk, and a rod, all with the same R_g. The correlation length is defined as twice the area under this graph. The correlation length, along with other invariants, is described in Table 3.1 and is 75.00 Å, 29.29 Å, and 30.19 Å for the sphere, disk, and rod, respectively.

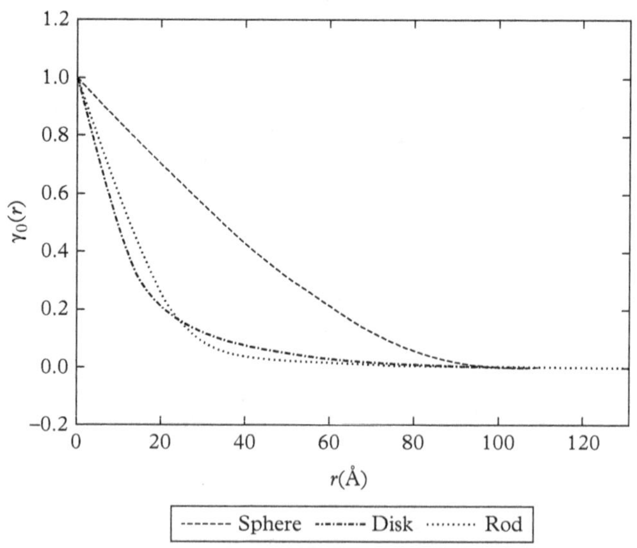

Figure 3.2 This figure shows $\gamma_0(r)$ for three hypothetical curves: from a sphere, a disk, and a rod. The sphere has a radius of 50 Å, and the disk and the rod have a 5:1 ratio of radius to thickness and a 1:10 ratio of radius to length, respectively.

Table 3.1 *Invariants calculated for the objects described in Figure 3.6*

Invariants	Sphere	Disk (5:1)	Rod (1:10)
Radius of gyration (Å)	38.73	38.73	38.73
Radius (Å)	50.00	54.59	13.03
Thickness/length (Å)	n/a	10.92	130.31
Maximum dimension (Å)	100.00	109.18	130.31
Volume (Å^3)	5.24×10^6	1.02×10^6	0.69×10^6
Volume of correlation (Å^2)	1110.72	555.53	366.45
Length of correlation (Å)	75.00	29.29	30.19

By using the equation for the Fourier transform of $\gamma_0(r)$, it can be shown that

$$l_c = \frac{\pi}{Q} \int_0^\infty qI(q)dq \tag{3.34}$$

and from that,

$$l_c = \frac{\pi \int_0^\infty qI(q)dq}{\int_0^\infty q^2 I(q)dq}. \tag{3.35}$$

Thus, l_c is also related to the invariant Q and to the integrated scattered intensity.

3.2.8 Remaining Invariants

We postpone discussion of the particle surface and the volume of correlation until later in this chapter.

3.3 Global Characteristics of the *I*(*q*) Curve

3.3.1 Intermediate Angular Region (or Shape Region)

The intermediate angular region (remembering that this is still a very small angle compared to crystallographic scattering) contains information on the shape of the particle. In this region, the Guinier plot is still valid but may have to be modified to take eccentric shapes into account (Glatter & Kratky, 1982). For instance, given a cylinder-like particle of length L and radius R, the low-q Guinier approximation remains as given in equation 3.11, where

$$R_g^2 = \frac{L^2}{12} + \frac{R^2}{2}$$

is the true radius of gyration for a cylinder. However, at intermediate q, the Guinier approximation becomes

$$I(q) = \frac{I(0)}{q} \exp\left(-\frac{q^2 R_g^2}{2}\right) \tag{3.36}$$

and the apparent $R_g^2 = R^2$.

The intermediate-q Guinier plot becomes $\ln(q\, I(q))$ versus q^2.

Similarly, for a lamella or disk of thickness T, the intermediate-q Guinier approximation becomes

$$I(q) = \frac{I(0)}{q^2} \exp\left(-\frac{q^2 R_g^2}{1}\right), \tag{3.37}$$

where $R_g^2 = T^2/12$.

The intermediate-q Guinier plot becomes $\ln(q^2 I(q))$ versus q^2.

In each of these cases, the contribution of the larger dimension of the object to $I(q)$ becomes negligible as q increases.

While the low-q region (regions (i) and (ii) in Figure 3.1) reveals information about the size and molecular weight of a particle, the intermediate-q region (region iii) describes the shape of a particle. The specific rate of decay of the scattering intensity in this region directly corresponds to the particle's gross structural contour. If the particle is globular, the intensity will decay proportional to q^{-4}. If it is

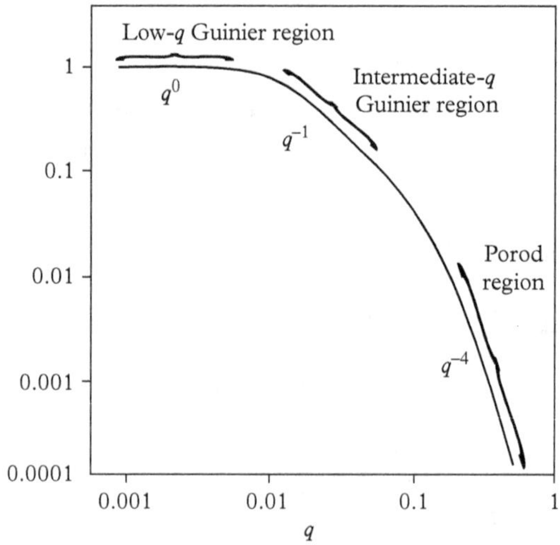

Figure 3.3 Form factor for a cylinder showing the low-q Guinier region, the intermediate-q Guinier region, and the high-q Porod region.
(Hammouda, 2016)

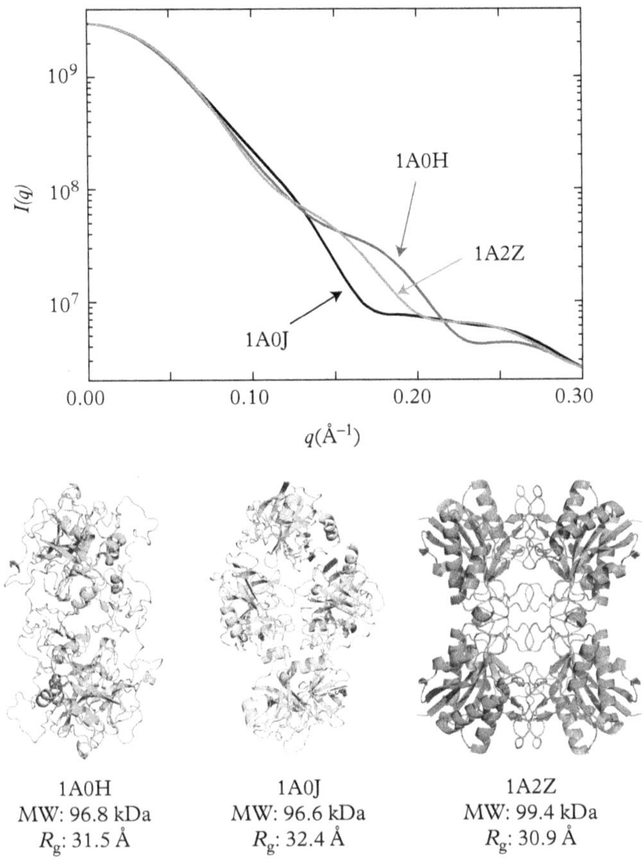

Figure 3.4 The top panel shows scattering profiles calculated from crystal structures of three different proteins with similar molecular weights and radii of gyration: meizothrombin (1A0H; Martin et al., 1997), non-psychrophilic trypsin (1A0J; Schroder et al., 1998), and pyrrolidone carboxyl peptidase (1A2Z; Singleton et al., 1999). The bottom panel shows the corresponding crystallographic derived structures; the structures are identified by their unique four-letter Protein Data Bank accession code.

disk-like, the intensity will decay proportional to q^{-2}. If the particle is rod shaped, then the intensity will decay as q^{-1}. We will return to a more complete discussion of these q-power dependences shortly.

It is the data in this region that can distinguish between different geometries of particles that may be of similar molecular weight and overall size. Figure 3.3 shows the calculated $I(q)$ for a cylinder. Note the distinct regions, and the differing q dependencies in them.

Figure 3.4 shows the shapes and associated scattering profiles of three different protein structures that have been deposited into the Protein Data Bank and have similar molecular weights and R_g values. As can clearly be seen, while the scattering profiles are similar in the low-q region, corresponding to the molecular weights and R_g values, deviations are clear beyond $q \gtrsim 0.07$ Å^{-1}. Such differences justify the idea that crude models of proteins can be obtained from small angle scattering data.

3.3.2 The Porod Region

Net scattering comes primarily from surfaces

In the following sections, we explore two important and related ideas from Porod. The first is that beyond the Guinier region, most of the *net* scattering from a globular particle (having a well-defined surface) occurs at interfaces, either between the particle and the solvent, or between particle components that have significantly different scattering densities, such as protein components and lipid components. The second important idea is that the high-q scattering from most compact particles approaches asymptotic dependence of q^{-4}. Rigorous treatment of these matters is quite mathematically intensive. In the spirit of this book, to keep it short and simple, we take an intuitive and heuristic approach to these ideas.

A useful way to think about the sources of signal at various values of q is to visualize a solution scattering pattern developing from a single molecule rotating in time in front of the X-ray beam. Previously, we described the calculation of $I(q)$ via the Debye equation, which uses the set of inter-scatterer vectors r_{ij}. In fact, the Debye equation is a very cumbersome tool for actually doing this calculation. The number of terms goes up as the square of the number of scatterers: a protein with 10^4 atoms requires 10^8 terms. The calculation is more efficiently done using the structure factor calculation, which leads to the values of $\boldsymbol{F}(\boldsymbol{q})$ for a given orientation of the molecule, and then averaging this over all angles either digitally or, more conveniently, analytically, by expanding the equation in spherical harmonics. For convenience, we repeat this equation:

$$\boldsymbol{F}(\boldsymbol{q}) = \sum_{j=1}^{N} f_j \exp(i\boldsymbol{q}\cdot\boldsymbol{r}_j). \tag{3.38}$$

It is important to realize that the global properties of the scattering curve have to be true for most individual orientations of the particle. For example, the idea that for certain ranges of q, most of the scattering takes place from the surface of the particle implies that for most orientations of the particle, scattering from volume elements in the interior of the particle largely cancels.

At the very low angles, where q is less than $1/D_{\max}$, $\boldsymbol{q}\cdot\boldsymbol{r}$ is small, and $\exp(i\boldsymbol{qr})$ will be nearly equal to 1. Thus, all of the volume elements in the particle scatter nearly in phase, so that the total scattered intensity is very large, proportional to the square of the number of electrons in the particle. This is another way of understanding the very strong scattering in the forward direction.

At larger values of q, the values of $\exp(i\boldsymbol{q}\boldsymbol{r}_j)$ will sample equally all angular values between 0 and 2π. Thus, the summed scattering from these volume elements would tend to be quite small. This cancellation through random phases will fail when the scattering power—that is, electron density in the case of X-ray scattering—of nearby volume elements is quite different. Such abrupt changes in scattering strength occur at boundaries, such as the interface between the particle and the solvent, or between materials of different average electron densities, such as lipids and proteins.

This argument makes plausible the demonstration by Porod that for larger values of q, most of the *net* scattering from a particle takes places from surfaces or interfaces.

Asymptotic scattering varies as $\mathbf{q}^{-4}$

A general simple protocol for the semiquantitative analysis of scattering at larger q is to do as follows:

- divide the object up into blocks of size $(1/q)$, with each containing n_e electrons, so that there are N_S blocks in the whole object
- calculate $I(q) = N_q \cdot n_e^2$

Let us apply this idea to a solid spherical particle of radius R (Figure 3.5). We know that net scattering occurs primarily at the surface. We can therefore ignore the interior, and cover the surface of the sphere by smaller spheres of diameter $(1/q)$, each half inside and half outside the specimen:

$$N \propto \frac{\textit{area big sphere}}{\textit{area of small spheres}} \propto \frac{4\pi R^2}{\pi\left(\frac{1}{q^2}\right)} \propto 4R^2q^2, \tag{3.39}$$

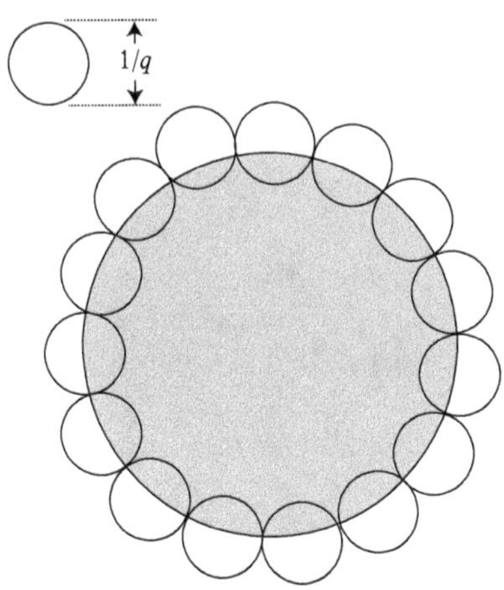

Figure 3.5 A solid specimen of radius R, with the surface covered by N smaller spheres of diameter $1/q$.

$$n_e \propto \textit{volume small sphere} \propto \frac{2\pi}{3q^3}, \tag{3.40}$$

$$I(q) \propto N n_e^2 \propto R^2 q^{-4}. \tag{3.41}$$

Another way to see the same result is to extrapolate from the result of a truly spherical specimen. The scattering from a sphere of radius R is given by

$$F_s(q) = \frac{3(\sin Rq - Rq\cos Rq)}{(Rq)^3}. \tag{3.42}$$

As q becomes large, the sine term becomes negligible. If we set the cosine term equal to 1 to illustrate the envelope of the scattering, we find that

$$F_s(q) \propto Rq/(Rq)^3, \tag{3.43}$$

$$F_s(q) \propto 1/(Rq)^2, \tag{3.44}$$

$$I(q) = F_s^2(q) \propto 1/(Rq)^4, \tag{3.45}$$

which again falls off as $(1/q^4)$. An ellipsoidal object would have the same q^{-4} dependence, but the ripples that characterize $F(q)$ would have different periods along different axes of the diffraction pattern and would wash out during rotational averaging.

Particle surface area, S

The derivation above makes clear that at large q, the scattering is proportional to the particle surface area, even as it falls off as q^{-4}. The heuristic nature of the presentation did not develop the exact values of this constant of proportionality, but it was shown by Feigen and Svergun (1987) that in the limit of large q,

$$\lim_{q\to\infty} I(q) \approx \frac{2\pi}{q^4}\Delta\bar{\rho}^2 S, \tag{3.46}$$

where S is the particle surface area. By combining this equation with equation (3.22) for Q, we can eliminate $\Delta\bar{\rho}^2$ and arrive at an equation for the specific surface area of the particle, independent of intensity scale:

$$\frac{\pi}{Q}\lim_{q\to\infty} q^4 I(q) = \frac{S}{V}. \tag{3.47}$$

Volume of correlation, V_c

The volume of correlation V_c is a newly developed SAXS invariant (Rambo & Tainer, 2013), derived from the scattered intensities that are specific to the structural state of the particle, but independent of concentration and the requirements of a compact, folded particle; V_c also provides a new route to the determination of molecular mass. As defined in the paper,

$$V_c = \frac{I(0)}{\int qI(q)dq} = \frac{c \cdot V_p^2 \cdot (\Delta\rho)^2}{c \cdot V_p \cdot (\Delta\rho)^2 \cdot 2\pi l_c} = \frac{V_p}{2\pi l_c}. \tag{3.48}$$

The paper provides experimental examples validating the proposed concentration independence. In addition, the authors provide several examples of systems in which ligand-dependent conformational changes produce corresponding change in V_c, suggesting that V_c is an informative descriptor of the macromolecular state.

The authors discovered that a parameter Q_r, the ratio of V_c^2 to R_g, expressed in cubic angstroms, is linear, versus molecular mass in a log–log plot, with the relationship given by

$$M = \left(\frac{Q_r}{e^c}\right)^{1/k}, \tag{3.49}$$

where k and c are empirically determined and specific to the class of the particle (e.g., 1.00 and 0.1231 for protein, and 0.81 and 0.009 for RNA (Rambo & Tainer, 2013). An advantage of this new invariant is its usefulness for estimating molecular weight for both globular and intrinsically disordered particles, in contrast to the invariant Q, which is only well defined for globular particles.

3.4 Comparison of Invariants

It is useful to understand the relationship between invariants for simple geometric objects. Figure 3.6 shows the $P(r)$ for three simple objects, all of identical R_g. The sphere has a radius of 50 Å; the disk and the rod have a 5:1 ratio of radius to thickness and a 1:10 ratio of radius to length, respectively. Table 3.1 lists some of the resulting invariants.

3.5 The Kratky Plot Distinguishes Globularity and Flexibility

The Kratky plot, a plot of $q^2I(q)$ versus q, is useful for distinguishing specimens that have a defined surface, such as a folded protein molecule, from those that do not have well-defined surfaces or interiors, such as polymer coils or denatured proteins. Clearly, this plot will distinguish $I(q)$ curves by the power dependence of their falloff at high q. If I falls as q^{-4}, $q^2I(q)$ will fall as q becomes large, while if I falls

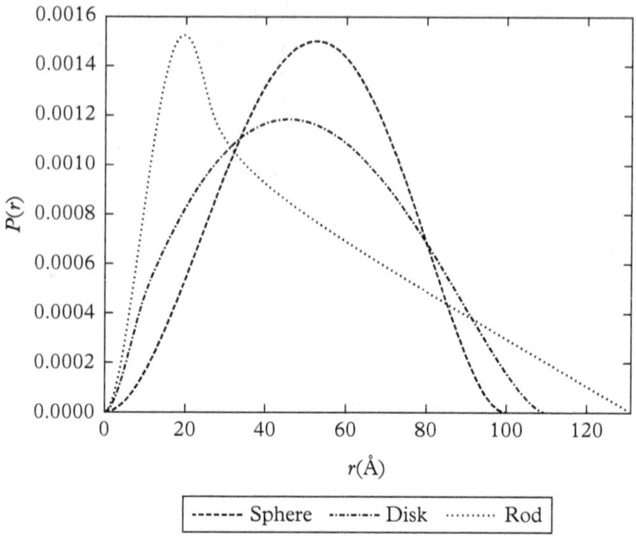

Figure 3.6 Calculated $P(r)$ for three simple objects, all of identical R_g, 38.73 Å. The sphere has a radius of 50 Å; the disk and the rod have a 5:1 ratio of radius to thickness and a 1:10 ratio of radius to length, respectively.

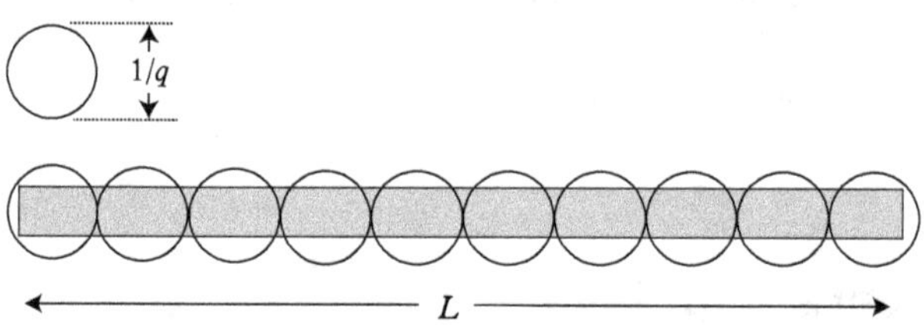

Figure 3.7 A rod-shaped particle of length L with diameter $D = 1/q$.

as q^{-1}, $q^2I(q)$ will rise. Full understanding of this plot requires that we explore a bit further how the $I(q)$ curves for various model systems behave.

Consider a rod of length L and diameter D. Using the arguments made for a sphere above, the rod can be covered by a series of spheres of diameter $(1/q)$ (see Figure 3.7). The number of spheres is given by

$$N \propto \frac{L}{1/q} \propto Lq, \tag{3.50}$$

and the number of electrons in each sphere is proportional to the volume of the rod contained within a sphere:

$$n_e \propto \frac{D^2}{q} \tag{3.51}$$

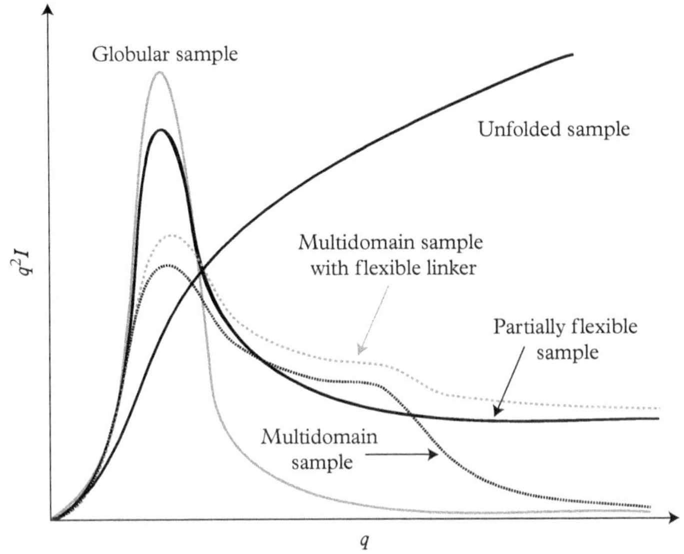

Figure 3.8 Kratky plots for a range of idealized samples including a globular sample, an unfolded sample, a multidomain sample, and samples consisting of flexible regions. Kratky plots offer a rapid characterization of the sample.

and

$$I(q) \propto \frac{D^4 L}{q}. \tag{3.52}$$

Thus, the $I(q)$ curve falls off as $1/q$. An unfolded protein molecule can be regarded schematically as a series of rods freely joined together, where the length of the rod is the persistence length of the polypeptide chain: 5–10 amino acids. Since the relation of the rods is not constant, the scattering curve is closely similar to that of a single rod. A different way to see this result is to realize that the diffraction from a rod is a disk-like function with relatively uniform intensity. When this two-dimensional pattern is rotationally averaged, it is smeared out over a three-dimensional sphere, so that the intensity in the sphere falls off as 1/radius or $1/q$.

Applying the same argument to a disk-like object shows a q^{-2} falloff.

Figure 3.8 show Kratky plots for a number of different samples. The globular sample displays a rapid falloff, consistent with a q^{-4} dependence. At the other extreme, the unfolded protein in black continues to rise at large q, fully consistent with a q^{-1} dependence. Kratky plots have been used to follow protein unfolding upon the addition of denaturant and to identify the scattering curves arising from folding intermediates present at low concentration. The multidomain sample gives a curve that is roughly the sum of the curves from two independent globular units.

Often, fluctuations in the intermediate-q region can be useful in identifying conformational changes in a protein under varying solution conditions or in distinguishing between multiple proposed models of a single particle assembly. This is a particularly powerful application of SAXS that can help to unambiguously determine particle conformation or assembly that may be elusive to other structural techniques such as X-ray crystallography, NMR, or electron microscopy. In the case of X-ray crystallography, crystal contacts may trap a conformation or assembly that is nonbiological. NMR, while a solution technique, has limitations in protein size and in determining long-range interactions. In many cases, these long-range distances may be required to discriminate between models that are similar in terms of local atomic environments and secondary structure elements but different in terms of tertiary or quaternary assemblies.

3.6 Summary

Solution scattering data can be divided into four regions. Scattering at zero angle provides information on the molecular weight of the particle in solution. Beyond that, the scattering is influenced by the radius of gyration. As the scattering angle increases, the scattering is influenced by the particle shape, and, finally, by the interface with the particle and the solution.

There are a number of important invariants that can be calculated directly from the data:

1. The molecular mass M
2. The radius of gyration R_g
3. The Porod invariant Q
4. The particle volume V
5. The maximum particle dimension D_{max}
6. The particle surface area S
7. The correlation length l_c
8. The volume of correlation V_c

The *molecular mass* is basically determined by $I(0)$, the scattering in the forward direction:

$$M = \frac{I(0)}{c\Delta\rho_m^2 N_A}.$$

The *radius of gyration*, R_g, measures the root-mean-square distance of scattering centers from the particle center of mass:

$$R_g^2 = \int |\boldsymbol{r}|^2 \rho(\boldsymbol{r}) dV \Big/ \int \rho(\boldsymbol{r}) dV.$$

It can also be defined in terms of *P*(*r*):

$$R_g^2 = \frac{1}{2} \frac{\int_0^{D_{max}} r^2 P(r) dr}{\int_0^{D_{max}} P(r) dr}.$$

Since *P*(*r*) is directly available from *I*(*q*), R_g can be calculated directly from this equation. Alternately, it can be determined from a Guinier plot by using the relation

$$\ln I(q) = \ln I(0) - \frac{q^2 R_g^2}{3}.$$

Ideally, a plot of ln(*I*(*q*)) versus q^2 gives a straight line from which R_g can be determined:

$$R_g = \sqrt{-\frac{slope}{3}}.$$

However, the Guinier plot is valid only at very small scattering angles, while the definition in equation (3.12) uses all the data.

The *Porod invariant Q* is a global quantity in *q*-space that is simply related to a key global particle quantity:

$$Q = \int_0^{\infty} q^2 I(q) dq,$$

$$V \Delta \bar{\rho}^2 = \frac{Q}{2\pi^2}.$$

However, *Q* may not be defined if $q^2 I(q)$ does not fall to zero within the range of observation.

The particle volume *V* or V_p was shown to be

$$V = \frac{2\pi^2 I(0)}{Q}.$$

The *maximum dimension of the particle* D_{max} is a useful number, but it may not be precisely defined for particles in which flexibility causes variation over time in the maximum dimension. It can be determined by the value of *r* at which *P*(*r*) falls to zero. This value can difficult to determine in cases where truncation of the *I*(*q*) curve introduces ripples into *P*(*r*). The recovery of *P*(*r*) by the method of indirect Fourier transformation can ameliorate this difficulty.

The particle surface area S is most readily recovered from

$$\frac{S}{V} = \frac{\pi}{Q} \lim_{q\to\infty} q^4 I(q),$$

where V is the particle volume. This relation requires that $I(q)$ fall off as q^{-4}, which arises from the idea that the particle is compact and has a surface.

The *correlation length* l_c is defined as the mean width of the *correlation function* and is given by

$$l_c = 2 \int_0^{D_{max}} \gamma_0(r) dr.$$

The *volume of correlation* V_c is a newly defined invariant that is independent of concentration:

$$V_c \equiv \frac{I(0)}{\int qI(q)dq} = \frac{V_p}{2\pi l_c}.$$

The asymptotic behavior of $I(q)$ at large q provides important information about the compactness of a particle: whether it is a folded protein or a random coil, for example. It is probed by the *Kratky plot*, in which $q^2 I(q)$ is displayed versus q. For compact particles with well-defined surfaces, $I(q)$ falls off as q^{-4} for large q. The Kratky plot for such a protein tends to zero for large q. For expanded, highly solvated coils, $I(q)$ falls off slowly, as q^{-1} and the Kratky plot will continue to slope upward for large q.

4

Shape Reconstructions from Small Angle Scattering Data

Recent advances in computational speed and software have brought about a renaissance in biological small angle scattering through advanced three-dimensional modeling. There is a dramatic loss of information in small angle scattering experiments because of spherical averaging from molecules tumbling in solution and because detectors sense only intensity information and provide no phase information. This has made it challenging to calculate three-dimensional electron density maps directly from scattering profiles. However, by generating three-dimensional shapes and calculating scattering profiles from these shapes, it is possible to create three-dimensional models that agree well with small angle scattering data. Specialized algorithms for this purpose have enabled a variety of different types of three-dimensional modeling, depending on the amount of a priori information available. If no information is available, ab initio modeling can be used to reconstruct envelopes with shapes closely agreeing with known atomic-resolution structures. If the high-resolution structure is known, accurate profiles can be calculated and fitted to experimental small angle scattering data to test agreement between the solution structure and the experimentally determined high-resolution structure. The latter may be different, particularly if it was determined via crystallography, where the crystal lattice may restrict motions available in solution. If full or partial high-resolution structures are available, rigid body modeling may be used to determine the orientation of domains or subunits in biomolecular complexes, or ensemble modeling may be used if conformational dynamics or mixtures are suspected. If multiple sources of additional structural information are available, hybrid modeling algorithms can be employed to fit all data simultaneously, resulting in both rigid high-resolution models and ensembles of high-resolution models. In this chapter, we describe the theoretical foundations and applications of a variety of three-dimensional modeling algorithms, with examples of applications presented in Chapter 12.

Biological Small Angle Scattering: Theory and Practice. Eaton E. Lattman, Thomas D. Grant, and Edward H. Snell.

DOI: 10.1093/oso/9780199670871.001.0001

4.1 Calculating Scattering Profiles from Three-Dimensional Models

All modeling algorithms rely on the ability to test the accuracy of the model against experimental data. The method of choice for calculating model agreement with experimental data is by using the chi-squared (χ^2) test, given by the following equation:

$$\chi^2 = \frac{1}{(N-1)} \sum_i^N \left(\frac{I_{\text{exp}}(q_i) - cI_{\text{calc}}(q_i)}{\sigma_i} \right)^2, \tag{4.1}$$

where $I_{\text{exp}}(q_i)$ and $I_{\text{calc}}(q_i)$ are the experimental and calculated scattering profiles, σ_i is the standard error in the measurement of intensity for each data point i, and c is a scaling factor. The summation is taken over all N data points. Theory provides a direct means of calculating scattering profiles from three-dimensional models via the well-known Debye equation, which reproduced from equation 2.11 below:

$$I(q) = \sum_{i=1}^N \sum_{j=1}^N f_i(q) f_j(q) \frac{\sin(qr_{ij})}{qr_{ij}}. \tag{4.2}$$

The double sum over all N atoms in the Debye equation makes this equation a costly computation, scaling as $O(N^2)$. When modeling SAXS data, iterative modeling procedures are often employed, which often require hundreds of thousands or millions of evaluations of models before an accurate model is selected. Therefore, computational efficiency is highly advantageous. Approximations to the Debye formalism using methods such as spherical harmonics and coarse graining are often used to speed up this calculation significantly, with little trade-off in accuracy at small angles. Additional considerations must be taken into account for nontrivial contributions from solvent scattering. In this section, we briefly describe some of the common methods of calculating scattering profiles from three-dimensional models at various levels of resolution.

4.1.1 Debye Equation

In equation 4.2, $f_i(q)$ represents the form factor for each volume of space located at position $\mathbf{r}_i$ that makes up the entire volume occupied by the particle when summed over all N volumes. Often, this form factor is taken to be the atomic form factor defined by an atom located at $\mathbf{r}_i$ and is assumed to be spherical. However, the form factors do not necessarily need to be defined by the conventional atoms found in biological molecules. They may also be defined by arbitrary volumes of space filled with electrons. This representation is advantageous for three-dimensional modeling of shapes, with beads representing volumes of space uniformly filled with electrons, as are often used in the ab initio modeling methods described in section 4.2, or to represent dummy residues whose form factors can be accurately approximated.

A particularly simple case arises for low-resolution models for which there is no appreciable density variation within the molecular surface or envelope.

4.1.2 Gaining a Feel for the Spherical Harmonic Function

In the representation of single particles, functions called spherical harmonics play a role analogous to those of trigonometric functions in crystallography. As with trigonometric functions, low values of indices m and n label functions with large spacing and low spatial resolution, while increasing values of these indices label harmonics with high resolution. Figure 4.1 shows the first few spherical harmonics represented on the surface of a sphere and suggests how they can combine, much as Fourier terms do, to synthesize any desired function. Three-dimensional objects can be represented by stacking spherical shells like an onion, with each shell being represented by different combinations of the harmonics. The coefficients used in a spherical harmonic expansion can be mathematically transformed to yield the corresponding coefficients of the Fourier transform (diffraction pattern) of the object. This provides a path to the spherical harmonic–based calculation of $I(q)$ from a particle model.

Spherical harmonics approximations can be made more or less accurate by using more or fewer harmonics, but typically, these approximations have difficulties when

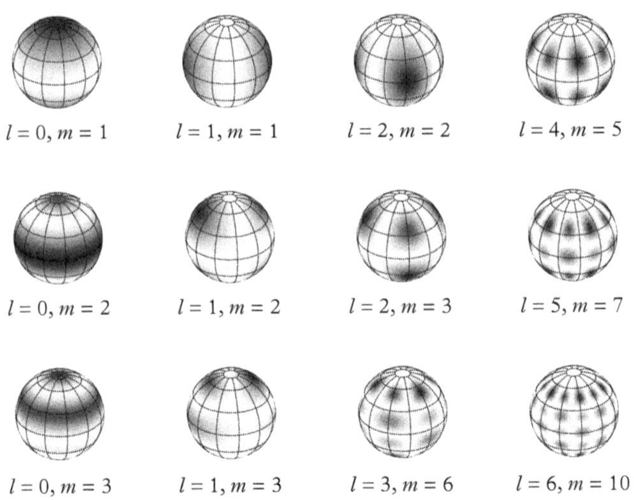

Figure 4.1 The normalized plots of the real component of several spherical harmonics of various order and degree. The order l corresponds to the number of null circles (white) that pass through the poles while m corresponds to the number of null circles that are parallel to the latitude lines.

(DeCaria & Van Knowe, 2014)

modeling highly complex shapes with features such as cavities and shells. The major advantage of the spherical harmonics approximation is a reduction in computational complexity from $O(N^2)$ to $O(N)$, and many modeling programs enlist spherical harmonics for this reason.

4.1.3 Spherical Harmonic Representation of Envelopes

One way of speeding the calculation of $I(q)$ arises from an insight by Stuhrmann (1970b) that reduces the dimensionality of the particle representation from three dimensions to two dimensions, with a corresponding reduction in the computation of $I(q)$. At low resolution, the particle can be represented as a surface within which the density is unity, and outside of which the density is zero; $F(\omega)$ (equation 4.3) maps the distance from the particle center to the molecular surface in terms of two angles that identify points on the surface of a sphere. The expansion uses a series of spherical harmonics that simplifies the rotational averaging by the orthogonal properties of the harmonics.

Note a possible confusion in notation here. In naming $F(\omega)$, we are following well-established conventions in the literature. However, the symbol F is also used to represent scattering in other equations.

As an introduction, consider the problem in two dimensions. Instead of fitting a three-dimensional shape, let us imagine that we want to fit a two-dimensional closed curve surrounding the origin of the coordinate system. Such a curve can be characterized in polar coordinates, where a radial coordinate running from the origin to a point on the curve is written as a function of the angle between the radius and the horizontal axis. This function is periodic because it repeats every 360°. Thus, it can be fit using a Fourier series, using as many terms as our knowledge of the resolution permits. In the same manner, a three-dimensional surface can be fit by a function that describes a radial coordinate as a function of two angles. The appropriate expansion of such a function involves spherical harmonics.

The spherical harmonic expansion for $F(\omega)$ is summarized in equations 4.3–4.5. Two basic issues arise in fitting a shape. First, we have to evaluate how many coefficients we have the data to fit. This is determined by the amount of detail available in the experimental intensity profile. Second, we have to connect the expansion coefficients to the observed intensity profile, so that they can be refined. Let us look first at the number of terms.

The number of coefficients (equation 4.3) is equal to $(L+1)^2$. However, not all of these are independent, since there are constraints among them. To find the appropriate value of L for a given problem, we need to know how many independent parameters it takes to represent the intensity profile. This is discussed in section 2.6.

A surface can be represented as a set of peaks and valleys imposed on the unit sphere, where the value of the function is the radius of the surface. Such an expansion assumes the form

$$F(\omega) \equiv F(\theta, \varphi) = \sum_{l=0}^{L} \sum_{m=-1}^{l} f_{l,m} Y_{l,m}(\omega), \tag{4.3}$$

where the $Y_{l,m}$ are spherical harmonics given by

$$Y_{l,m}(\theta, \varphi) = N \exp(im\varphi) P_l^m(\cos\theta) \tag{4.4}$$

and the Ps are the Legendre polynomials. Then, because the $Y_{l,m}$ are a complete, orthonormal set, we can recover the coefficients f_{lm} from the inverse relationship

$$f_{l,m} = \int F(\omega) Y_{l,m}(\omega) d\omega. \tag{4.5}$$

Stuhrmann (1970b) also developed rapid analytical algorithms, alluded to previously, to calculate the scattering curve $I(q)$ from $f_{l.m}$. This provides the basics of a refinement cycle in which deviations between I_{obs} and I_{calc} can be minimized by adjusting $f_{l.m}$.

4.1.4 Expansion of the Structure Factor Equation in Spherical Harmonics

It is often important to be able to calculate $I(q)$ from a set of atomic coordinates. This application is useful, for example, to confirm that a solution structure is essentially identical to one found in the crystal. This confirmation is provided by close similarity between the $I(q)$ curve calculated from the set of atomic coordinates in the crystal structure, and the experimental $I(q)$ curve. It is equally useful in the case where a dummy atom model (section 4.2.4) is being studied. The formalism of spherical harmonics, which has just been discussed in terms of shape modeling, can also be used to derive a computationally rapid formula for the $I(q)$ curve in terms of the atomic coordinates for the structure.

We begin with the structure factor equation (see appendix A1), which is repeated here for convenience, and the corresponding equation for the intensity:

$$\boldsymbol{F}(\boldsymbol{q}) = \sum_{j=1}^{N} f_j \exp(i\boldsymbol{q}\boldsymbol{r}_j), \tag{4.6}$$

$$\boldsymbol{I}(\boldsymbol{q}) = \boldsymbol{F}(\boldsymbol{q})\boldsymbol{F}^{\star}(\boldsymbol{q}) = \sum_{j=1}^{N} f_j \exp(i\boldsymbol{q}\boldsymbol{r}_j) \sum_{k=1}^{N} f_j \exp(-i\boldsymbol{q}\boldsymbol{r}_k). \tag{4.7}$$

After expressing $\boldsymbol{r}$ in spherical rather than Cartesian coordinates, the exponential factor $\exp(i\boldsymbol{q}\boldsymbol{r})$ can be expanded in terms of spherical harmonics through a relationship developed in the book by Morse and Feschbach (1953). We do not reproduce the details here, but the orthogonal relationships among the spherical harmonics mean that, when I is rotationally averaged, many cross-terms vanish, and one is left with a sum over a limited number of spherical harmonics, rather than the double sum over all atoms in the Debye equation. Then,

$$I(q) = \frac{1}{2} \sum_{n=0}^{n_{\max}} \sum_{m=0}^{n} \varepsilon_m N_{m,n} |\boldsymbol{G}_{m,n}(q)|^2, \tag{4.8}$$

where

$$G_{m,n}(q) = \sum\nolimits_{\text{all atoms}} f_j Y_{m,n}(\theta_j,\ \varphi_j) j_n(qr_j). \tag{4.9}$$

The index j runs over all atoms; r, θ, φ are the atomic coordinates in spherical coordinates, and ε and N are constants. The $Y_{m,n}$ are complex spherical harmonics similar to those illustrated above, and the j_n are spherical Bessel functions. The f_j are the atomic scattering factors for the jth atom.

These equations look quite daunting, but the key point to take home is that this method of finding $I(q)$ is *much* faster than the Debye equation. First, the double sum in equation 4.8 is quick to evaluate, since it goes over the spherical harmonic indices, which are much smaller than the number of atoms. Second, the sum required to create $\boldsymbol{G}$ is a one-dimensional sum over the number of atoms at limited number of values of q. The related method for envelope models, mentioned in section 4.1.3, is faster still. This method has been implemented in various software packages.

Virtually any three-dimensional shape can be represented by these functions, provided enough terms are included. Spherical harmonics are ideal for modeling in the case of SAXS because, as shown in equation 4.8, these functions can also be used to expand the calculated $I(q)$ curve—the experimental data serves as a convenient restraint on any model.

Envelope modeling is the most common application of spherical harmonics. The method defines a three-dimensional envelope function, $F(\omega)$, such that for $r < F(\omega)$, that is, inside the envelope, the electron density function $\rho(r) = 1$, and outside the envelope, $\rho(r) = 0$. By fitting the spherical harmonic coefficients to the SAXS profile, accurate low-resolution envelope functions can often be found for moderately complex shapes.

4.1.5 Zernicke Polynomial Expansion

Another approach to fitting a shape uses three-dimensional Zernike polynomials. Using this method, after scaling down to fit in a unit sphere, the electron density of a three-dimensional model can be represented by

$$\rho(\boldsymbol{r}) = \sum_{n=0}^{\infty} \sum_{l=0}^{n} \sum_{m=-1}^{l} c_{\text{nlm}} Z_{\text{nlm}}(\boldsymbol{r}), \tag{4.10}$$

where $\boldsymbol{r}$ is a three-dimensional vector, n is the number of expansion orders, c_{nlm} is a complex expansion coefficient (known as three-dimensional Zernike moments) and $Z_{\text{nlm}}(r)$ is a three-dimensional Zernike polynomial. The reader is referred to a full mathematical treatment given elsewhere (Liu, Morris, et al., 2012; Liu, Poon, et al., 2012). The Fourier transform of $\rho(\boldsymbol{r})$ in this form can be achieved very economically to yield $I(q)$. In the Debye equation, for a molecule with N atoms and M desired intensity points, the computational complexity is $O(MN^2)$. Spherical harmonics improves the computational intensity to $O(MN)$, but the computation time still linearly depends on the number of data points. Using the Zernicke polynomial expansion approach, the atom positions and momentum transfer q are decoupled, yielding a computational intensity of $O(N)$. A test with ten real data sets showed that the Zernicke method was comparable to spherical harmonic approaches, with fits to the experimental data being very similar to each other (Liu, Morris, et al., 2012). The Zernike method is also more sensitive to cavities in macromolecules (such cavities are not well modeled by other methods).

4.1.6 Solvent Considerations

The Debye equation as written above implicitly assumes that the molecule is in vacuo. As shown in equations 2.17 and 2.20, at low resolution (small angles), submersion of the molecule in solvent reduces the scattered intensity by the factor $(\Delta\rho)^2 = (\rho_{\text{particle}} - \rho_{\text{solvent}})^2$, representing the contrast between the electron density of the atoms and that of the solvent. If one considers an idealized continuum solvent with no fluctuations, then $I(q)$ for such a solvent would be a delta function at $q = 0$, and the Fourier transform of this is a constant density everywhere. To a first approximation, $I_{\text{sample}} - I_{\text{solvent}}$ is therefore the curve obtained for a particle of density $\rho_{\text{particle}} - \rho_{\text{solvent}}$ surrounded by a zero-density medium. This correction is made in practice by measuring the scattering curve for the buffer alone and then subtracting it from the curve from the solution. This nontrivial task requires an exact match of the bulk solvent to that in the protein solution, which is often obtained by dialysis or collected from purification procedures.

If the scattering from the solvent blank is not identical to the background scattering from the protein solution, errors in background subtraction will occur. This often manifests most noticeably at higher angles, where finer features of the particle shape are dominant. Many algorithms that calculate scattering profiles from atomic models attempt to correct for this error by subtracting a constant term from the entire data set to force the Porod law decay of q^{-4} at high angles. As this term is usually small compared to the total scattering intensity, the lower resolution part of the curve, a region which is much higher in intensity than that at higher angles and which contains most of the information about size and shape, is generally unaffected, while the higher resolution data is corrected, often leading to more accurate scattering profiles. However, care must be taken to prevent overcorrection for background subtraction errors, since not all samples will obey the Porod law, particularly non-globular, elongated, or flexible molecules. In cases such as these, where background subtraction may be inaccurate, it is suggested to recollect data with a more accurate buffer blank and remove any constant subtraction corrections.

Since the experimental intensity is determined after subtraction of a sample of bulk solvent from an equivalent volume of particle immersed in solvent, the solvent volume displaced by the volume of the particle must be subtracted from the scattering equation. This displaced solvent term is known as the excluded volume and contributes to the overall scattering intensity, so it must be accounted for.

Additionally, experiments and simulations have shown that the solvent immediately surrounding the particle often associates with the surface of the particle and becomes more ordered on average than the bulk solvent between molecules. This increase in order of solvent molecules increases the average electron density in this region by approximately 5%–10%. This region is commonly referred to as the hydration shell.

Most of the algorithms to calculate scattering profiles agree well with each other at SAXS resolutions; however, discrepancies at higher resolutions in the wide angle X-ray scattering (WAXS) regime are readily apparent when compared with known experimental WAXS data (Schneidman-Duhovny, Kim, et al., 2012; Putnam et al., 2013). The WAXS regime begins at values of q where the approximation on constant density within the particle begins to break down. It is not surprising that the approximation of constant density solvent fails at the same point. Explicit solvent models therefore often provide better fits than implicit solvent models in the WAXS regime, although the calculations are substantially more complex. In some cases, explicit solvent modeling does not require fitting solvent terms such as contrast and electron density values; molecular dynamics simulations directly calculate these values from the model, leading to less biased profiles at high resolution (Putnam et al., 2013).

Various methods of accounting for each one of these solvent terms have been developed, using both implicit and explicit solvent models. Procedures for implicitly fitting the hydration shell typically surround the envelope of a particle with a thin shell of uniform electron density, where the electron density inside this region is a fitted parameter. Explicit solvent algorithms surround the particle with water molecules to calculate the scattering profiles directly. Explicit methods are generally more accurate than implicit methods. However, they incur a substantial computational cost when solvating the molecule, as they often involves complex procedures such as molecular dynamics. Implicit methods are much faster computationally, but suffer from an inability to properly model cavities and complex shapes, although recent advances in heterogeneous modeling of the hydration shell have proven highly accurate, even for complex shapes.

4.2 Ab Initio Modeling

The ability to reconstruct low-resolution molecular envelopes from the $I(q)$ scattering profile without any a priori knowledge has been a major advance made possible by algorithmic development and sheer computational power. As mentioned in section 4.1, by the term "envelope," we mean a surface that encloses a constant density structure and conforms closely to the atomic or molecular surface. At first glance, this capability is remarkable; how can a one-dimensional curve, quite smooth and featureless, contain enough information to determine the structure of a molecule, even at quite low resolution? There are several components to this question:

- First, how significantly do $I(q)$ curves differ from one protein to the other? (If all the curves were the same when scaled by the radius of gyration it would be impossible to find a unique envelope.)
- Second, even if $I(q)$ curves are distinct, can a unique envelope really be extracted from the data?
- Finally, how do we convince ourselves that any particular solution is correct?

This lays out a productive viewpoint to detail the theory behind molecular envelope reconstruction.

4.2.1 Modeling by Simple Shapes

One of the first methods of modeling three-dimensional shapes using SAXS data was through simple bodies such as ellipsoids, cylinders, and shells. Such bodies can be fitted to SAXS data by altering shape parameters such as axial radii. While too simple to reconstruct accurate envelopes effectively for even modestly complex shapes, these algorithms are able to yield gross shape and size characteristics at very low resolution. Simple shape models will usually agree with the observed R_g, D_{max}, and other invariants (Kratky & Pilz, 1972).

4.2.2 Simple Model Examples

Envelope reconstruction has proven to be an excellent method for modeling scattering data with molecular envelopes with little or no bias. To understand the general process of envelope reconstruction, let us start by considering a simple special case in which the possibility of developing a molecular envelope from a solution scattering curve is relatively clear. In this case, the density is assumed to equal one inside the surface, and zero outside it, so that the model assumes that the particle displays no internal fluctuations in density. This assumption is a consequence of a low-resolution data set and we will expand this later to allow for some internal density fluctuations. Figure 4.2 taken from Kuzmanovic et al. (2003) shows the SANS scattering curve from a solution of bacteriophage MS2, a 275 Å RNA virus that infects male *Escherichia coli* (Stockley et al., 1994). Overlaid on this is the theoretical scattering curve from a ball of identical radius to the virus. It is also known from genetic and biochemical studies analysis that the viral capsid is composed of 180 copies of a single coat protein molecule. This information strongly suggests that the capsid and the isometric virus have a $T = 3$ triangulation number, and that the resulting structure can be regarded a perturbation of a sphere.

The process can be improved by replacing the single hollow sphere model with smaller components. To develop an envelope, we create an initial model of the virus by arranging a set of balls on the surface of a sphere to give an object with $T = 3$ icosahedral symmetry on the surface. From this model, we can calculate an $I(q)$ curve, which will, to some extent, agree with the observed curve $I_o(q)$. We then proceed to make small alternations in the model, moving balls on the surface

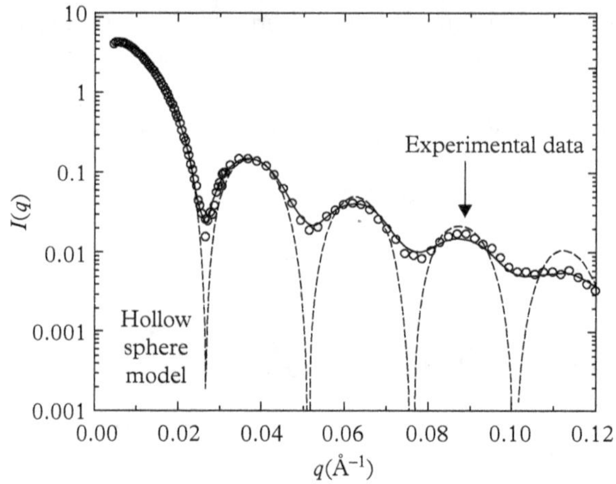

Figure 4.2 SANS data from MS2 fitted to a core-shell sphere model (Guinier & Fournet, 1955). The neutron scattering length density of the RNA core was an extra fitting parameter.

(Kuzmanovic et al., 2003)

or normal to it. After each adjustment, we calculate an index of model quality, such as the χ^2 metric shown by equation 4.1, where we seek to minimize χ^2 to the lowest value possible.

In this case, the model of the virus has relatively few parameters. At each vertex of the icosahedron, there is a pentagon of identical subunits, while in the center of each face, there is a hexagon with identical subunits. There are various ways to describe this model, but none of them requires many variables. Thus, reducing the value of χ^2 can be achieved by conventional least-squares analysis. In fact, quite good models of virus structure can be achieved in this way.

The high symmetry that makes the case of reconstruction for isometric viruses fairly straightforward is not a common feature among protein assemblies. More general methods of creating such reconstructions are required. There are two broad classes of methods. One class fits the envelope of the particle using analytic functions. The other set of methods fits the envelope as the sum of simple shapes such as closely packed spheres. Let us begin the discussion with analytic fitting.

4.2.3 Envelope Fitting Using Analytical Functions

From the discussion of spherical harmonics above, virtually any three-dimensional shape can be represented by these functions, provided enough terms are included.

Envelope modeling is the most common application of spherical harmonics. The method defines a three-dimensional envelope function, $F(\omega)$, such that for

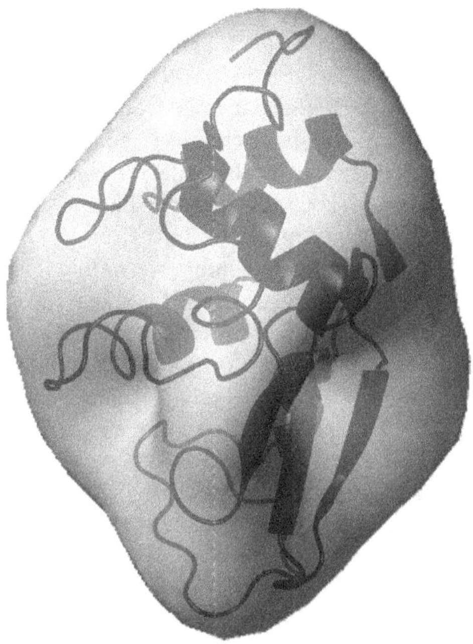

Figure 4.3 Ab initio spherical harmonics envelope of lysozyme. The program SASHA was used to calculate spherical harmonics coefficients fitting the SAXS profile of lysozyme (Svergun et al., 1996). The spherical harmonics envelope is shown in transparent gray, and the lysozyme crystal structure is shown in black cartoon representation (Konarev et al., 2001).

$r < F(\omega)$, that is, inside the envelope, the electron density function $\rho(r) = 1$ and outside the envelope $\rho(r) = 0$. By fitting the spherical harmonic coefficients to the SAXS profile, accurate low-resolution envelope functions can often be found for moderately complex shapes (Stuhrmann, 1970a). An example of such a reconstruction of lysozyme is shown in Figure 4.3. While many diverse envelope shapes can be reconstructed in this manner, highly complex particle shapes, such as those with cavities or extended regions, will prove difficult to reconstruct accurately.

4.2.4 Envelope Fitting Using the Sum of Simple Volumes

Bead modeling is an alternative avenue for creating envelopes and has gained widespread popularity in the community for its versatility. Bead modeling algorithms attempt to arrange small beads, typically spheres, in a volume of space such that the calculated scattering profile from the collection of beads matches experimental data. Such methods, often termed dummy atom modeling, have proven very useful at reconstructing envelopes from SAXS data when no a priori information is available.

Several programs exist which perform bead modeling via a variety of algorithms (Chacon et al., 1998; Zhao et al., 1998; Svergun, 1999; Walther et al., 2000; Svergun et al., 2001; Franke & Svergun, 2009) but the general outline of each remains similar. To begin, beads are arranged (either randomly or in defined starting locations) in a volume of space chosen to be large enough to enclose the particle. Often, identical spherical beads are arranged in a hexagonally close-packed lattice. The radius of the beads is set such that the search volume contains some number of beads, typically on the order of ten thousand. In other algorithms, beads are allowed to float with some restrictions, for example remaining within some distance of adjacent beads. Each bead can have a density of either zero (solvent) or unity (particle). The refinement process switches the density of beads to arrive at an arrangement for which the calculated scattering profile matches the experimental one. The method of selection varies between algorithms. For a hexagonally packed array of 10,000 beads, there are some $2^{10,000}$ possible combinations, making it impossible to search computationally to find the best arrangement of beads. All bead modeling algorithms suffer from this combinatorial explosion, but researchers have devised methods to approximate solutions while searching only a small fraction of all possible configurations. These methods range from simulated annealing protocols to genetic algorithms, all with the same basic procedure: rearranging a random collection of beads, testing agreement with the experimental SAXS data, adjusting the arrangement of beads, retesting, and gradually converging to a solution. These methods still require some hundreds of thousands or millions of model evaluations to reach convergence. While several algorithms use the Debye equation for calculating profiles from candidate bead models, others use expansion methods mentioned in section 4.1.4, and all require adjustments accounting for solvent scattering. In addition, criteria that the particle model be compact and comprise contiguous pieces are often invoked.

To assist in the prevention of overfitting SAXS data and in creating realistic models, many bead modeling algorithms utilize additional restraints and penalties during the bead selection process. In addition to characterizing models based on χ^2 fitting, physical characteristics of models, such as how globular, connected, or loose collections of beads are, become parameters within more generalized scoring functions. These additional characterizations penalize non-protein-like shapes (e.g., models that have disconnected beads), since proteins and other biomolecules are known to be connected, and act similarly to chemical restraints used during restrained refinement of crystallographic or NMR structures to weight the observations-to-parameters ratio favorably. Without these restraints, bead models that fit experimental data often suffer from non-physical characteristics. However, when restraints are combined with accurate fits, plausible and realistic models that agree with known structures are often obtained. Figure 4.4 shows examples of bead modeling for the protein lysozyme, with and without using appropriate real space restraints. It is clear that even though the χ^2 value is slightly lower for the model created without restraints, the model created with restraints has a much better fit to the actual structure of the protein.

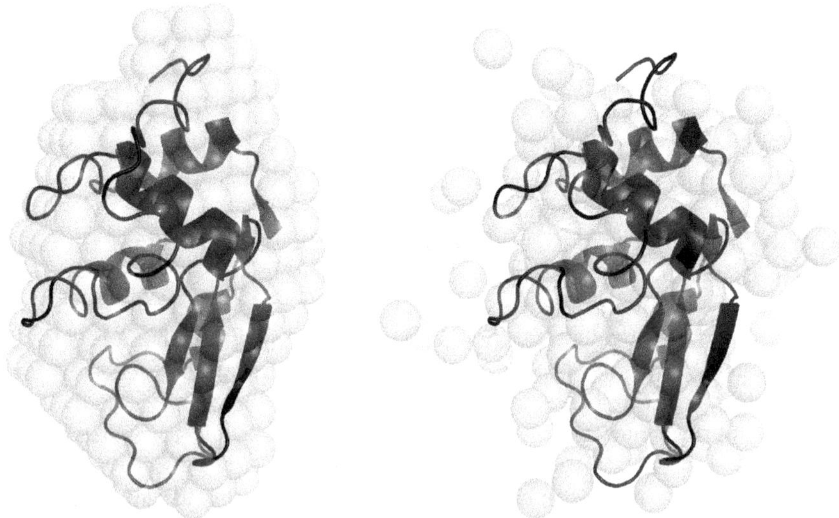

Figure 4.4 Ab initio bead modeling of lysozyme. The left panel shows a bead model of lysozyme created using the default parameters in DAMMIN (Svergun, 1999). The right panel shows a bead model of lysozyme created while disabling real space penalties such as connectivity and looseness. The bead models are shown as transparent gray spheres aligned to the crystal structure of lysozyme, which is shown in black cartoon representation (PDB ID: 6LYZ).

4.3 Flexible Fitting

In cases where molecules may have alternate conformations, direct comparison of calculated scattering profiles from high-resolution structures may fail to provide acceptable fits to experimental scattering data. Discrepancies may be the result of differences between the conformational states obtained in solution and those obtained in environments found during high-resolution data collection (e.g., packed in a crystal lattice). In these instances, flexible fitting may provide a path to determining a solution state model that closely resembles the solution scattering profile. In flexible fitting, techniques such as normal mode analysis identify ways in which the molecule may move and flex to access available conformations consistent with known biological motions. These alternate states are then fitted to the experimental scattering data and optimized, resulting in a conformation that matches the scattering profile yet is different from that which has been determined via high-resolution structural methods. Algorithms also exist to flexibly fit high-resolution models to ab initio envelope reconstructions, rather than scattering profiles, as is often done in cryo-electron microscopy (Zheng & Tekpinar, 2011; Panjkovich & Svergun, 2016).

4.4 Rigid Body Modeling

In cases where the high-resolution structures are known for components of a particle, a common method for utilizing this information is rigid body modeling (Petoukhov & Svergun, 2005). In these cases, the structures often comprise domains or fragments of full-length proteins, or subunits of macromolecular complexes. Rigid body modeling attempts to rotate and translate structures relative to one another, such that the resulting scattering profile calculated from the model agrees with the experimental SAXS data. A six-dimensional search is thus required for each independent rigid body. Simulated annealing is often employed to make the search computationally efficient. The large number of restraints provided by the known high-resolution structures greatly increases the number of observations to parameters, resulting in reliable, pseudo-high-resolution models in many cases.

Recent improvements to rigid body algorithms enable one to model missing fragments such as linkers between domains (Petoukhov et al., 2012). If available, SAXS profiles for various combinations of fragments can provide important additional information for modeling, provided that the solution structures of these fragments reflect those found in the solution structure of the complete molecule.

However, limitations due to treating each molecule as a rigid body may result in inaccurate models or a failure to converge entirely if the rigid body structures do not reflect the solution structures of each individual component. Additionally, in cases where flexibility or ensembles of conformations may be present in solution, attempts at modeling the structure with one single body will likely fail.

4.5 Docking Algorithms

Rigid body modeling has the advantage of combining known high-resolution structural models with low-resolution solution structure information from SAXS to create a pseudo-high-resolution model. However, the large six-dimensional conformational landscape that must be searched during rigid body modeling makes finding the most precise positions and orientations of subunits difficult. Molecular docking algorithms attempt to use energetics calculations to determine the lowest energy surface interactions between molecules and then pose molecules relative to each other to minimize these surface energies. Incorporating SAXS restraints into docking algorithms helps to combine low-resolution shape information from SAXS with high-resolution interface energy calculations from docking to determine the most likely complex interfaces between molecules. SAXS docking algorithms are thus an extension to conventional rigid body modeling and help to overcome the precision lost to low-resolution SAXS data.

Conventional docking procedures vary between software implementations, but most begin by searching molecular surfaces to create thousands of candidate conformations (Jimenez-Garcia et al., 2015; Schindler et al., 2016; Schneidman-Duhovny et al., 2016). The energy for each conformation is calculated and the results clustered in groups based on scoring functions. SAXS data is often introduced in two ways. First, the initial candidate conformations are filtered coarsely for gross size characteristics obtainable directly from the SAXS profile, such as R_g and D_{max}. Next, SAXS profiles are calculated for each of the remaining candidate conformations and fit to the experimental SAXS data to identify the most likely candidates. For the remaining candidate conformations, the docking algorithm calculates the energy scores from molecular surface interactions and then proceeds to cluster the candidates into groups based on scoring functions, followed by refinement to yield the most likely conformation.

4.6 Mixtures

Most modeling algorithms discussed so far are predicated on the assumption that one molecular shape exists in solution and that this shape can accurately describe the total solution structure and experimental scattering profile. However, in many cases, multiple oligomeric states or conformations may exist in solution, leading to polydispersity and an inability to model the scattering profile with one unique shape. However, polydisperse mixtures can often be modeled if full or partial structures are known. When complete structures of multiple oligomeric species are known or can be predicted, for example, from crystallographic packing arrangements, it is possible to perform a linear combination of scattering profiles to determine volume fractions of oligomeric states (Konarev et al., 2003). Even in cases where high-resolution structural models do not exist, experimental SAXS profiles of individual components can be used to determine these volume fractions. Additionally, recent improvements to rigid body and ab initio modeling programs include extensions for modeling fractional quantities of subspecies of models, allowing for improved fitting of polydisperse mixtures (Petoukhov et al., 2012).

4.6.1 Principal Component Analysis and SVD

Principal component analysis (PCA) in the form of singular value decomposition (SVD) has been successfully applied to SAXS data with mixtures of species or conformations in a number of studies, including observation of the early compaction step in horse cytochrome c (Kathuria et al., 2014), the unfolding of necocarzinostatin (Perez et al., 2001), and the structural transition of aspartate transcarbamylase binding an analog (Fetler et al., 1995).

PCA is a central tool for data analysis in a wide variety of disciplines. It is used to reduce dimensionality and provide insight into individual components making up complex data sets while minimizing information loss (Jolliffe, 2002). As an example, consider a time-resolved study measuring a solution scattering profile as a function of time after some triggering event. Many scattering profiles will result that collectively contain data across a spectrum of potential states. In many examples, only two or three species give rise to this spectrum (e.g., mixtures of folded protein, unfolded protein, and folding intermediates). The profiles collected are actually sums of the profiles from these three states, where only the weights vary from one time point to the next. PCA is a tool to reveal these three basis profiles, thus reducing the dimensionality of the data set from tens of spectra to three and, in the process, revealing the distinct states.

SVD is a powerful technique for the analysis of SAXS data where multiple species are involved (Williamson et al., 2008) and is used to calculate a principal components analysis from a data matrix. SVD is a matrix algebra method that can be used to determine the minimum number of components required to accurately represent data sets having high redundancy (Malaby et al., 2015). SAXS (or SANS) data sets can be represented as an $M \times N$ matrix $\boldsymbol{A}$, with N columns corresponding to individual scattering curves sampled at Mq values; $\boldsymbol{A}$ is decomposed (Golub & Van Loan, 1996) as a product of orthonormal basis vectors (columns of a $M \times N$ matrix $\boldsymbol{U}$), singular values (elements of an $N \times N$ diagonal matrix $\boldsymbol{S}$), and orthonormal coefficients (columns of a transposed $N \times N$ matrix $\boldsymbol{V}$):

$$A = USV^T. \tag{4.11}$$

The singular values applied to the coefficients in $\boldsymbol{V}^T(\boldsymbol{S}\boldsymbol{V}^T)$ specify the linear combination of basis components in $\boldsymbol{U}$ required to exactly reconstruct both the signal and the noise in each scattering curve making up $\boldsymbol{A}$. SVD can rank the contribution of each column of $\boldsymbol{U}$ such that the number of components making up the system can be determined and described.

4.7 Ensembles

Not only can oligomeric polydispersity exist in solution, but conformational polydispersity can also exist, leading to an inability to describe the solution structure of a molecule with one unique shape. This has led to the advent of ensemble modeling of SAXS data, and this has proven to be one of its most powerful and commonly used features (Bernado et al., 2007; Pelikan et al., 2009; Yang et al., 2010; Rozycki et al., 2011; Tria et al., 2015). Ensemble modeling begins by creating thousands of candidate models, often including known structural information from high-resolution models, by generating new conformational states derived from sampling available conformations of flexible linkers between domains, either randomly or using known sequence information, such as through molecular dynamics. From these thousands of candidate conformations, groups of candidates are selected, and the scattering profiles of all candidate conformations in the group are averaged together. Sampling procedures such as genetic algorithms attempt to create and select the group of candidate conformations whose average scattering profile best matches the experimental data. Ensembles can be created for flexible, dynamic particles, including intrinsically disordered molecules. SAXS is one of the few experimental approaches that can access large-scale conformational dynamics in solution.

While ensemble methods are particularly powerful for extracting dynamic information from SAXS data, they also suffer from a potential to overfit SAXS data, given the large number of parameters used in fitting multiple candidate conformations to a single SAXS profile. Care must be taken to prevent overinterpretation of these models. Rather than accepting each of the individual candidate conformations as accurate models in solution, the creators of ensemble modeling algorithms often recommend analyzing the results in terms of only gross size characteristics. Therefore, distributions of size characteristics such as R_g and D_{max} are created to demonstrate the propensity for the algorithm to choose various sizes of conformations and may lead to identifying multiple distinct sizes, which would suggest the presence of discrete populations in solution, or broad distributions, which would suggest largely random conformational dynamics.

4.8 Hybrid Modeling

Hybrid modeling approaches attempt to combine many of the modeling methods described here into one large fitting procedure. These methods can often incorporate nearly as much a priori information as is available, including NMR, residual dipolar couplings, X-ray structures, fluorescence resonance energy transfer (FRET), circular dichroism, and bioinformatics, as well as multiple scattering methods, including SAXS, WAXS, and SANS. These hybrid methods have proven to be extraordinarily powerful at solving the solution structures of large, dynamic complexes that would ordinarily be impossible to uncover using any single method alone (Russel et al., 2012; Schwieters et al., 2018).

4.9 Summary

This chapter discussed a number of methods of recovering shape or structural information from SAXS data. Key to any such process is the ability to generate a calculated intensity $I_{calc}(q)$ from a model and then compare this curve with the experimental one, $I_{exp}(q)$. The method of choice for performing the comparison is via the χ^2 test, using equation 4.1:

$$\chi^2 = \frac{1}{(N-1)} \sum_{i}^{N} \left(\frac{I_{exp}(q_i) - cI_{calc}(q_i)}{\sigma_i} \right)^2 .$$

The Debye equation (equation 4.2),

$$I_{calc}(q) = \sum_{i=1}^{N} \sum_{j=1}^{N} f_i(q) f_j(q) \frac{\sin(qr_{ij})}{qr_{ij}},$$

remains the preferred way of calculating scattering curves, but it is very computationally demanding, with a computational intensity of $O(N^2)$, where N is the number of atoms. It is more tractable when the structure is being modeled with a larger number of dummy atoms, since N becomes much smaller.

For computationally intensive applications, where, for example, thousands of $I_{calc}(q)$ curves are calculated during iterative modeling, the scattering density can be represented by Zernicke polynomials (section 4.1.4), which are readily Fourier transformed. Alternately, the structure factor equation

$$\boldsymbol{F}(\boldsymbol{q}) = \sum_{j=1}^{N} f_j \exp(i\boldsymbol{q}\boldsymbol{r}_j)$$

can be expanded in spherical harmonics to give a direct equation, whose computational intensity is $O(N)$, for $I_{calc}(q)$.

In the favorable case in which internal density fluctuations can be ignored, the particle can be represented by a two-dimensional equation that defines its surface and separates the particle interior ($\rho = 1$) from the exterior ($\rho = 0$):

$$F(\omega) \equiv F(\theta, \varphi) = \sum_{l=0}^{L} \sum_{m=-1}^{l} f_{lm} Y_{l,m}(\omega)$$

Here, F is the distance from the particle center to the point on the surface defined by the angles θ and φ. Rapid algorithms for obtaining $I_{exp}(q)$ from F have also been developed.

Proper treatment of solvent is critical. The Debye equation as written assumes that the molecule is in vacuo. To account for solvent, at low resolution (small angle), the contrast between the electron density of the atoms and that of the solvent, $(\Delta\rho)^2 = (\rho_{particle} - \rho_{solvent})^2$, must be included as a scaling factor in front of the double sum. This approximate treatment is only valid when the density within the particle can be assumed to be constant. Treatments that are more detailed may include a separate category for "bound" solvent molecules adjacent to the particle surface.

Models for the particle scattering density can be as pure geometric shapes such as ellipsoids or, in the case of viruses, icosahedra. These representations contain very few free parameters. More complex particle surfaces can be represented by the spherical harmonic representation above or by a set of close-packed beads, each of which can have density 0 or 1. In bead modeling, $I_{calc}(q)$ can be calculated using the spherically averaged structure factor equation, where f then represents the form factor for the bead.

Sometimes, high-resolution structures are known for components of a particle. These elements often comprise domains or fragments of full-length proteins, or subunits of macromolecular complexes. Rigid body modeling attempts to rotate and translate structures relative to one another, such that the resulting scattering profile calculated from the model agrees with the experimental SAXS data. A two-body structure requires a six-dimensional search. Simulated annealing is often employed to make the search computationally efficient.

Solutions may not always contain a homogeneous set of particles. A common case is the presence of two or more conformations of a single particle. The SVD method can extract $I(q)$ for conformationally distinct species. SVD requires that a number of $I_{exp}(q)$ curves be measured under conditions that alter the proportion of different subspecies. Studying protein folding as a function of denaturant concentration is an example.

Part 3

Practical Aspects of Small Angle Scattering

5

Before the Beamtime

For the novice user performing the first experiment, step one is learning as much about the technique as possible. In addition to this work, many excellent texts are available that cover both theoretical and practical aspects (Trewhella et al., 1998; Putnam et al., 2007; Hura et al., 2009; Jacques & Trewhella, 2010; Jacques et al., 2012; Svergun et al., 2013; Dyer et al., 2014; Jeffries & Svergun, 2015; Trewhella, 2016). There is also a wealth of expertise that can be gained from beamline personnel and from those offering laboratory systems. This expertise can make the difference between success and failure and should be taken advantage of before performing any experiment.

5.1 Sample Production

SAXS requires a sample concentration similar to that typically used for crystallization (e.g., a few milligrams per milliliter) and a volume on the order of tens of microliters. SANS can require an order of magnitude more than this.

For SAXS, a rule of thumb described by Dyer et al. (2014) is that the concentration in mg/mL multiplied by the molecular weight in kDa should be greater than 100. In the case of individual concentration points, at least three concentrations are used to detect potential concentration-dependent aggregation or oligomerization processes—more concentrations are preferable. SANS, as noted, requires an order of magnitude more sample, and if X-ray scattering has been used to detect concentration-dependent effects beforehand, it does not usually require a concentration series. Normally several different contrast points are recorded by appropriately varying the ratio of D_2O to H_2O. Jeffries et al. (2016) provide a comprehensive primer on the preparation of samples for both X-ray and neutron scattering.

The sample should be well characterized and free from impurities. For example, Jeffries et al. (2016) note that a 15 kDa protein sample purified to 98% that contains a 2% 100 kDa impurity will generate a forward scattering intensity that is almost twice that expected from the pure 15 kDa sample. The sample should be verified but there have been cases in crystallography where the protein being expressed and studied was not the protein that was required (Niedzialkowska

Biological Small Angle Scattering: Theory and Practice. Eaton E. Lattman, Thomas D. Grant, and Edward H. Snell.

DOI: 10.1093/oso/9780199670871.001.0001

et al., 2016). It is good practice to be sure of the sample produced, for example, via a suitable assay, mass spectrometry, or other technique.

All laboratories should have access to native and sodium dodecyl sulfate (SDS)–polyacrylamide gel electrophoresis (PAGE) analysis. SDS–PAGE acts as a check on the approximate molecular weight and determines if contaminants may be present. Native PAGE can be used to determine if a sample is homogeneous or may consist of a range of structural species. For protein purification, size-exclusion chromatography (SEC) is also a standard laboratory technique and it should be used after SDS–PAGE to characterize the sample purity further, preferably through a dilution series. UV spectrophotometry of the elution fractions can reveal the presence of oligomeric species or aggregation. Oligomerization or aggregation can be a time dependent process and some beamlines have inline UV detection capabilities to monitor this; for an SEC SAXS experiment, this is a useful diagnostic. For a preprepared concentration series, it is useful to determine if oligomerization or aggregation occurs over time and proceed with that knowledge. Concentration-dependent aggregation does not preclude successful SAXS data collection; it just limits the concentration range that should be studied. Similarly, the same analysis techniques can be used to study the effects of sample storage before data collection.

Dynamic light scattering (DLS) and static light scattering (SLS) are useful techniques to evaluate polydispersity and measure the hydrodynamic radius of the sample. With accurate concentration measurements, SLS can be used as a check on the molecular weight of the sample in solution, and how this may change with concentration or buffer choice. Both DLS and SLS should be conducted as a function of concentration. Jeffries et al. (2016) note specifically that these techniques are extremely sensitive to aggregation and if no aggregation is detected, the sample is probably of sufficient quality to obtain useful SAXS data. DLS and SLS are not as common in laboratories as the other techniques mentioned but are often found where crystallographic research is conducted.

5.2 Buffer Choice and Matching

The SAXS signal itself is small, given the dependence on the small density difference between the sample and solubilizing solution absent the sample. The solution without the sample is often referred to as the buffer blank. The signal itself is proportional to the sample concentration and the difference in density between the sample and the buffer blank. For that reason, the buffer blank itself should be designed to minimize its own scattering component, for example, by avoiding high atomic number elements. However, the buffer should also stabilize the sample. Components such as glycerol, high salt concentrations to reduce interparticle interactions, detergents in the case of membrane systems, lipids and carbohydrates (e.g., for contrast matching in the X-ray case) may all be needed. There can be a subtle balance between optimum conditions for the sample and the signal-to-noise in the resulting data.

Buffer components can also play a role in helping to reduce radiation damage through scavenging free radical species. This has been explored in the X-ray case where radical electrons are considered, but the solution state is a more complex system and flow or oscillation methods offer better protection from this artifact.

The first rule of a buffer for SAXS is that the sample must be stable in the buffer. In order to minimize X-ray absorption and subsequent production of the free radicals (primarily, $OH^{\bullet}$), the buffers should be as free from high atomic number elements as possible. Detergents should be kept to a minimum, unless they are absolutely necessary to solubilize the protein. Detergents form assemblies at concentrations exceeding the critical micelle concentration (CMC), and these produce strong scattering signals themselves. The staff at the SIBYLS beamline have established a list of preferred detergents that minimize scattering and can be used at concentrations below the CMC (see table 10.1 in section 10.1.2). These are 0.1% 1-s-Nonyl-β-D-thioglucoside, 0.2% *n*-Decanoylsucrose, 0.3% *n*-Nonyl-β-D-maltoside, 0.4% DDAO, 0.5% C_8E_5, 0.8% FOS-choline®-10 and 1.1% FOS-choline®-9. These are described as detergents that can be used below the CMC and, with a strong protein signal, above the CMC also. Other detergents that can be used below the CMC but not above are also described. Detergents help solubilize the protein by absorbing on the surface but in doing so change that surface. This may have to be taken into account during the subsequent modeling.

Salts can be used to decrease the long-range electrostatic repulsion between the macromolecules at the measurement conditions. Adjusting the pH of the buffer toward the protein's isoelectric point can also decrease long-range electrostatic repulsion but can also lead to increased aggregation (Grishaev, 2012).

SAXS causes radiation damage to both the sample and the buffer. The field of X-ray crystallography has made extensive studies on radiation chemistry effects in crystalline samples and some of this work is applicable to SAXS. X-rays produce free radicals in crystals and in solution through their interaction with water. Chemical scavengers such as glycerol, ascorbate, ethylene glycol, or sucrose can increase the longevity of the sample (Kuwamoto et al., 2004; Grishaev, 2012), but they can also affect the signal-to-noise ratio (Jeffries et al., 2015). Quantitative measurement have been carried out on radiation dose and damage in SAXS revealing that the efficacy of additives is somewhat correlated to their scattering signal and is concentration dependent (Brooks-Bartlett et al., 2017). For an experiment using a concentration series, this would suggest trialing several scavengers if necessary.

Once a buffer is chosen, a good match between the sample and its solubilizing solution and the buffer blank is critical. There are a number of ways of ensuring this. If the sample is produced by gel filtration, the sizing buffer is usually a good match. Other favored ways are to concentrate the sample and use the flow through as the buffer blank or use dialysis methods with the dialysis solution consisting of the blank. In all cases, good practice is to prepare a fresh, concentrated, sterilized buffer stock solution from which the working buffer is prepared (Skou, Gillilan, et al., 2014). Grishaev (2012) provides good advice on buffer preparation, by ensuring the exact sample/buffer match with a multistage dialysis through a semi-permeable membrane with the molecular weight cutoff smaller than that of the macromolecule of interest. A total dialysis time of at least 16–48 h should be sufficient to achieve the exact sample/buffer match. The buffer should be stirred throughout the process and the membrane area large enough to ensure complete solvent equilibration.

5.3 Sample Optimization

For most experimental studies, the sample must have chemical and physical monodispersity. Chromotography techniques are recommended to purify the sample with characterization by DLS or multi-angle laser light scattering (MALS). As noted above, a single band on an SDS or native gel is not a sufficient indicator of monodispersity and DLS may not be sensitive to small oligomers. As a practical consideration, samples should show no signs of precipitate and should be able to be concentrated without significant loss of protein.

Some beamlines offer the ability to use SEC experiments in line with SAXS data collection. This may be coupled with a UV or a MALS instrument to monitor the sample. SEC (inline or performed in advance) is arguably the most effective means to ensure data is as high a quality as possible and can have a considerable positive impact on outcome (Rambo & Tainer, 2010; Yang et al., 2011). In this case, it is not only the sample that needs optimizing but also the choice of SEC column. The sample preparation and column need to be matched so that it is possible to resolve the sample from any possible contaminants. This requires characterization and preparation in the home laboratory. Buffers should be filtered (0.22–0.44 μm) to remove any particulate matter and preferably degassed to minimize potential bubble formation (also a good strategy for regular data collection if the sample is not damaged by the process).

The molecular weight of the sample should be known, and/or checked by mass spectrometry. This information enables characterization of oligomeric state. Analytical ultracentrifugation can also be used to supplement this and confirm the oligomeric state observed in the scattering experiments.

5.4 Transport of the Sample

For a home laboratory system, transport of a sample is not a significant issue. For synchrotron or neutron facilities, transport of liquid samples has proven increasingly difficult unless ground transportation can be used. Many beamlines offer at least basic wet laboratory facilities. In these cases, the sample can be freshly prepared at the beamline either by thawing and diluting the flash frozen sample or by concentrating a dilute solution of the sample. Absent these possibilities, another shipping option is the freshly prepared sample with ice packs. In this case, the sample can be shipped with a concentration series preprepared. It is not recommended to ship samples at non-controlled temperature conditions or as a prefrozen concentration series. In the former, sample denaturation can occur, and for the latter, aggregation can result. Sample lifetime is also a consideration. Some samples will degrade or aggregate over time so it is always preferable to use as fresh a sample preparation as possible.

5.5 Practical Preparations before Data Collection

Skou et al. (2014) also provide extensive details on preparation for a SAXS experiment, either in the laboratory or at the beamline. Access to a centrifuge is recommended for spinning samples down before data collection. This can sometimes be an immediately effective technique to eliminate aggregation problems. A UV-visible spectrophotometer provides a good check on protein concentration. If SEC is to be employed, it is better to use a column that has already proved successful in sample preparation rather than a generic one. Useful disposable equipment includes filters for large volumes (e.g., for buffer blank preparation), centrifugal filters for small volumes at appropriate molecular weights (keeping in mind the possibility of oligomer formation in solution) and potentially dialysis membranes, although the dialysis is slow. Feedback during the data collection can indicate that the sample or buffer may need to be changed. For this reason, a desalting column and centrifugal concentrators with appropriate molecular weight cutoffs should be available. For those uncertain which materials will be available at the beamline, many of these items and standard laboratory equipment (e.g., pipets, tips, and appropriate tubes and stands for manipulating the sample) should be brought to the beamline. Proper protective equipment (e.g., safety glasses and gloves) are usually available but it is always best to talk to the beamline staff before making any assumptions.

Many experiments work well the second or third time simply because a problem seen the first time was not noticed or could not be addressed at the beamline. Standard samples are recommended to check on the performance of the instrument, to ensure that the first use provides good data, and to act as a calibration standard for the sample of interest. They also serve to identify any potential operational problems before they impact the sample to be studied. There are many standards established including chicken egg white lysozyme, or xylose isomerase although the latter is no longer commercially available. Any well-characterized, stable, globular protein that is not prone to oligomerization is suitable for use as a standard. It is also recommended to have access to buffer components and potentially additives that can reduce radiation damage or minimize aggregation.

Most home sources and beamlines have adequate protocols for cleaning the sample holder that usually involve water and ethanol, bleach, detergent, or various mixtures of those components. A good experimental practice is to take buffer blank shots before and after a sample series and cleaning, and then compare the scattering from the two buffers. It should be emphasized that cleaning time is never wasted time; if in doubt, clean.

For experiments that make use of a concentration series, serial dilution using matched solvent is recommended. For inline SEC experiments, Jeffries et al. (2016) provide a useful table of column properties, advantages, and limitations, although at the stage of the experiment, the performance of the sample on the column of choice should already be well defined.

5.6 Quality Control Checks

In addition to studying the sample of interest there are a number of quality control checks that can be carried out to ensure successful data collection and subsequent analysis. An important step is to measure the concentration of the sample being studied accurately. This can be done with UV spectrophotometry and is a standard laboratory technique with many synchrotron sources offering this facility at or near the beamline. The molecular weight of the sample should be well characterized—this can be used as a useful check of potential oligomer formation in solution. It is also recommended to record a scattering pattern from water and the blank capillary before starting an experiment. This can be useful for putting the results on an absolute scale, or for potential future developments in data analysis.

5.7 Contrast Matching for X-Ray and Neutron Cases

Jeffries et al. (2016) provide a detailed primer on contrast matching in both X-ray and neutron cases. Contrast, as discussed in Chapter 2, section 2.5, and Chapter 11, section 11.3, is the difference in scattering length density between the macromolecule and the solvent. For SAXS, the chemical environment can be changed to alter the electron density. Sucrose has been used as a contrast agent to locate heavy atoms within a protein (Grishaev et al., 2012) and evaluate distances between atoms with an uncertainty of ~1 Å. In this case, 65% sucrose was used to mask the protein contribution and thus reveal the heavy atom scattering component. SAXS studies of protein–DNA complexes have also used sucrose as a contrast agent. Chen et al. (2014) performed a challenging time-resolved study to follow the structural process of protein and DNA unwrapping in the nucleosome core particle. A solution containing 50% sucrose matched the protein scattering, leaving only the contribution from the DNA. There are disadvantages; the signal is reduced and the viscosity of the solution can cause sample-handling issues. Similarly, a rigorous analysis needs to be conducted to ensure sample function is maintained. Contrast matching has also proved useful in studying transmembrane proteins. Unlike soluble proteins with a two-state system (i.e., buffer and sample), studies with membrane proteins can involve a third state, the protein–detergent complex, buffer, and free detergent micelles. Bu and Engelman (1999) used sucrose to match the scattering from the micelles and study the α-helical transmembrane domain of the human erythrocyte glycophorin.

SANS makes use of the scattering differences between hydrogen and its isotope deuterium. This is addressed in more detail in section 11.3 but it makes use of the different and opposite signed neutron scattering lengths of H and D. The density difference between the buffer and sample can be manipulated by using a mixture of H_2O and D_2O, as shown in Figure 11.3 in section 11.3. Sample preparation is relatively straightforward, and this ability provides one of the great strengths of SANS, as it is thus possible to study the protein, nucleic acid, and detergent structures of a complex, individually or in combination. Examples of this are given in section 12.1.1. It is also possible to alter the H-to-D ratio within the protein itself by introducing D recombinantly at non-exchangable hydrogen positions. This is a more specialized approach used for very specific structural studies. Note that pH and pD are not equivalent, with pD being equal to the measured pH +0.4.

Svergun et al. (2013) discuss both the X-ray and the neutron case, with a useful table of scattering lengths for water, heavy water, sucrose, sodium chloride, and biological components of macromolecules. Salts can be used to contrast mask lipids but require high molar concentrations to mask other components and may be damaging at those concentrations. Contrast matching may not be exact; the resultant scattering may have a contribution from the sample, the buffer, and the cross-term. For SANS, with data from an incremental set of H_2O-to-D_2O ratios, the contribution from each individual component of a complex can be evaluated (Jeffries et al., 2016).

5.8 Summary

SAXS requires a sample concentration similar to that typically used for crystallization, and a volume on the order of tens of microliters. SANS can require an order of magnitude more than this. The sample should be well characterized, and free from impurities. Standard laboratory techniques are suitable for characterization. Buffer choice and accurate matching is a component of experimental success and after sample preparation is the next most critical step. The sample should be monodisperse, stable, and well characterized.

6

Making the Best Use of Beamtime

Synchrotron beamlines provide the highest-quality SAXS data. They offer high flux with low beam divergence, and as a shared resource, access to detectors that represent the state of the art. In some cases, robotic sample handling has made the collection of high numbers of samples and concentrations efficient and reproducible, to the point that some beamlines now accept shipped samples for remote data collection. However, as a shared facility, synchrotron time is limited and knowledge of the basic considerations during the beamtime can help to make the most efficient use of this scarce resource. In this chapter, we will discuss some of the practical considerations during beamtime; however, the same considerations apply equally to a laboratory source but the time constraints are less pressing.

Each experiment has unique considerations for instrumentation setup and includes a variety of potential modifications to maximize data quality and information content in the SAXS profile, which will each be described in turn below. The most important preparation is to contact the beamline staff to understand the strengths and limitations of the beamline, including beam fluence, size, divergence, energies, detector size, detector sensitivity, sample-to-detector distance limitations, temperature control, robotic sample handling, and so on, as well as the availability of sample preparation facilities.

6.1 Sample-to-Detector Distance

Unlike the case in crystallographic studies, where changing the sample-to-detector distance is straightforward, the process can take some time on a SAXS beamline. Therefore, an important consideration prior the experiment is the appropriate choice of the sample-to-detector distance. As larger samples scatter at lower angles it is important to have a long enough sample-to-detector distance to ensure the Guinier region is adequately sampled and data points lower than the first Shannon channel, π/D_{max}, are collected. Thus, having a rough estimate of the size of the particle is particularly important when designing the experimental setup. Estimates of particle size can be obtained from other solution methods such as SLS or DLS and will yield either the radius of gyration or the hydrodynamic radius, R_g or R_h, respectively. Rough estimates of D_{max} can be estimated using the rule of thumb of

Biological Small Angle Scattering: Theory and Practice. Eaton E. Lattman, Thomas D. Grant, and Edward H. Snell.

DOI: 10.1093/oso/9780199670871.001.0001

$3.5 \times R_g$, and the ratio of R_g/R_h can yield an estimate of particle shape. Circular dichroism, NMR, or disorder prediction algorithms may also be useful to identify the likelihood of unfolded regions of proteins that may lead to extended conformations being present, longer estimates of D_{max}, and hence longer sample-to-detector distances being needed.

Some useful relationships are the maximum and minimum θ angles. The maximum θ, θ_{max}, is given by

$$\theta_{max} = \frac{1}{2}\tan^{-1}\left(\frac{l_{dect}}{d}\right), \tag{6.1}$$

where l_{dect} is the maximum usable dimension of the detector from the beam center, and d is the sample-to-detector distance. The minimum θ, θ_{min}, is given by

$$\theta_{min} = \frac{1}{2}\tan^{-1}\left(\frac{b}{d}\right), \tag{6.2}$$

where b is the diameter of the beamstop from the beam center. The maximum and minimum q, q_{max} and q_{min}, are given by

$$q_{max} = 4\pi\sin(\theta_{max})/\lambda, \tag{6.3}$$

$$q_{min} = 4\pi\sin(\theta_{min})/\lambda, \tag{6.4}$$

where λ is the incident wavelength.

The maximum and minimum d spacing that can be sampled, d_{max} and d_{min}, are given by

$$d_{max} = \left(2\sin(\theta_{min})/\lambda\right)^{-1} = 2\pi/q_{min}, \tag{6.5}$$

$$d_{min} = \left(2\sin(\theta_{max})/\lambda\right)^{-1} = 2\pi/q_{max}. \tag{6.6}$$

In addition to ensuring the collection of low-resolution data points below π/D_{max}, large sample-to-detector distances also provide a denser sampling rate of reciprocal space, which effectively leads to greater signal-to-noise at the expense of higher resolution information, as described in Chapter 2.

Depending on the information sought from the experiment, an appropriate balance must be struck between collection of low-resolution information and high-resolution information. Signal-to-noise decreases rapidly with resolution, and most modeling algorithms have difficulty calculating accurate scattering profiles at large scattering angles; thus, the ability to interpret high-resolution scattering information is limited. However, small conformational changes or differences in shape may only be visible at high resolution; thus, a priori information about the system is useful for designing the experiment and selecting an appropriate sample-to-detector distance to optimize the quality of the data at the desired resolution. It may even be appropriate to make use of a number of sample-to-detector distances.

If the sample-to-detector distance is limited by the beamline configuration, another option for adjusting the resolution limits of the data is by selecting longer or shorter wavelengths using a monochromator (see equation 6.3 and 6.4). The characteristic synchrotron radiation flux can decrease rapidly as wavelength decreases; thus, the signal-to-noise ratio of the resultant data may be affected if the wavelength is used to vary the accessible resolution.

6.2 Instrument Calibration

Before data collection, the instrument is usually calibrated by the beamline staff. This is an important step to ensure accurate photon counts and pixel resolution are recorded. This includes wavelength selection, beam alignment, collimation, dark current correction, and slit width adjustments to ameliorate parasitic scattering. Additionally, a well-defined conversion from pixel value to reciprocal space q vector is obtained by calibration with a known scatterer, such as silver behenate crystallites, a sample that provides powder diffraction rings at known resolutions. This allows the accurate determination of reciprocal space coordinates of the detector pixels.

Intensity varies from beamline to beamline due to variations in source brightness, detector response, sample cell transmission, and so on. While many small angle scattering parameters can be calculated independent of the absolute scattering intensity, such as R_g and volume, other parameters, such as molecular weight, can be calculated with better accuracy if the intensity has been calibrated. Intensity calibration can be obtained using solutions of known scattering strength, such as water, or protein standards.

Water is well suited as a calibration standard because the scattering is predominately angle independent and depends only on isothermal compressibility (the relative volume change as a response to pressure). The differential scattering cross section, $d\Sigma/d\Omega$, is proportional to the mean-square fluctuation density. The scattering intensity of water, I_w, can be calculated from

$$I_w = \frac{d\Sigma}{d\Omega} = \rho^2 kT\chi_T \tag{6.7}$$

where ρ is the scattering length density (9.388×10^{10} cm^{-2} for water), k is the Boltzman constant (1.38×10^{-23} m^2 kg s^{-2} K^{-1}), T is the temperature, and χ_T is the isothermal compressibility (10^{-10} Pa^{-1} for water at 293 K).

To convert data to an absolute intensity, the measurement of the empty holder is subtracted from that of a water sample. The scattering of water is weak, so careful consideration needs to be given to measuring the scattering accurately. The resulting water signal can be used to calibrate the intensity by fitting the scattering and extrapolation to $q = 0$. Water gives an intercept of 1.632×10^{-2} cm^{-1} at 293 K (Orthaber et al., 2000). The calculation of absolute intensity, and from that the molecular weight is illustrated by Orthaber et al. (2000). For lysozyme, a molecular weight of 14.6 kDa was calculated compares to the known 14.3 kDa. The uncertainty for determining molecular weight using this method is estimated as 5%.

Well-characterized protein samples can also be used to calibrate intensity provided sample concentration and molecular weight are known accurately. Xylose isomerase was often chosen as a protein calibrant due to its reliability to remain monodisperse even at high concentrations and its low cost; however, it is now commercially unavailable. Molecular weight of a protein sample can be calculated according to the following equation:

$$\frac{MW_{\text{sample}} \cdot C_{\text{sample}}}{I_{0,\text{sample}}} = \frac{MW_{\text{standard}} \cdot C_{\text{standard}}}{I_{0,\text{standard}}}, \tag{6.8}$$

where MW is the molecular weight of the sample in daltons, C is the concentration of the sample in milligrams per milliliter, and I_0 is the forward scattering intensity of the sample. This method requires accurate knowledge about the concentration of both the protein sample and the protein standard.

6.3 Sample Concentration

Scattering by a solution of molecules is an additive process, meaning that the scattered intensity of the solution is directly proportional to the number of molecules in the irradiated volume. By increasing the concentration of the sample, and thus the number of molecules in the irradiated volume, the signal-to-noise ratio of the scattered intensity increases. However, greater concentrations also lead to a greater likelihood of interactions between neighboring molecules in solution, and therefore a greater tendency for repulsion, attraction, or aggregation. Interparticle effects such as these can ruin the ability to accurately interpret size and shape information from scattering profiles as they violate the assumption of monodispersity in many basic scattering equations. Thus an appropriate balance must be struck between signal-to-noise (higher concentration) and monodispersity (lower concentration), with caution leaning toward monodispersity, as noisy profiles are more useful than aggregated samples in most cases. Larger molecules scatter more strongly and thus need less concentration to achieve usable signal-to-noise, as do electron-rich molecules such as RNA and DNA. Typical concentrations used for scattering experiments for protein molecules ranges from 1–10 mg/mL to obtain usable signal from solutions dilute enough to avoid interparticle interactions, although these guidelines are highly protein and instrument specific. A useful rule of thumb for estimating the concentration in milligrams per milliliter needed to achieve high signal-to-noise is to divide 100 by the molecular weight of the protein in kilodaltons (Dyer et al., 2014). A minimum of three concentrations should be used to check for concentration-dependent effects and to allow extrapolation to zero concentration.

Monodispersity is often difficult to assess accurately directly from the scattering profile alone. Typically, this is done using analytical tools such as the Guinier plot. For monodisperse samples, the Guinier plot is linear at low q values. However, molecular envelopes that deviate largely from simple globular shapes often deviate from linearity in the Guinier plot as well, potentially showing nonlinear behavior despite monodispersity. Additionally, subtle interparticle effects or aggregation may not be detectable in Guinier plots, either due to low signal-to-noise or due to too few data points on the region. Thus, linearity in the Guinier plot is not a sufficient condition to prove monodispersity.

The need for highly monodisperse solutions requires one to obtain a scattering profile from more than one concentration to assess any potential interparticle interactions. This allows one to evaluate changes in the scattering profile as a function of concentration, which, if present, indicate subtle interparticle interactions that must be dealt with prior to data analysis and interpretation. If the degree of interparticle interactions is moderate or severe (i.e., changes in particle size are greater than a few percentage points over the concentration range), it is recommended to recollect the data using alternative solution conditions promoting greater solubility and monodispersity, such as altering variables such as temperature; pH; available ligands or cosolutes; salt type and concentration; and preservatives and surfactants. However, if the interparticle interactions are minor, it is possible to merge the information from all concentrations by performing zero extrapolation of the data, described in Chapter 7, section 7.10.

The concentration of macromolecules should be estimated prior to collecting scattering data, for example, by UV fluorescence measurements. Collecting scattering data from multiple concentrations is done often by first performing serial dilutions of the highest concentration sample available with the buffer blank.

The best dilution strategy will be dependent upon the specific sample to be studied. For most well-behaved, soluble proteins, a range of concentrations from 1 to 10 mg/mL is suitable to ensure adequate signal-to-noise at the highest concentrations and limited or no interparticle interactions at the lowest concentrations. The minimum number of concentrations in the series should be at least three. Five concentrations helps to insure against instrumentation or sample problems that may invalidate data (e.g., air bubbles) and provide enough sampling for adequate assessment of concentration-dependent effects, with more concentrations being useful for dealing with such effects. Note that without at least three concentrations, SAXS data is not typically publishable.

When diluting to low concentrations, pipetting error of small volumes may contribute to inaccurate estimations of concentration. Combined with lower signal-to-noise, small angle scattering parameters calculated using these lower concentration scattering profiles are often subject to large degrees of error. One potential means of decreasing the error in the estimation of concentration is by utilizing inline UV spectroscopy to monitor the concentration of the sample while scattering data is being collected. This ensures that the concentration of the actual sample solution being collected is accurate. However, this feature is not available on all beamlines, and fluorescence is dependent on the presence of tryptophan in the protein sequence. Accurate knowledge of concentration can help improve the accuracy of the estimation of parameters such as molecular weight, using the above equation.

Dilution series are not typically employed in SANS due to limited beamtime availability. They are replaced by precharacterization using X-ray methods.

6.4 Number of Exposures/Exposure Time

For many samples, higher concentrations often result in aggregation of the sample despite much effort in searching for soluble chemical conditions. In such cases, it is possible to increase the signal-to-noise by increasing either the number of X-ray exposures collected or the exposure time of each image. Increasing the total number of X-rays incident on the sample increases the likelihood of radiation damage. In SAXS, this is often manifested by aggregation or repulsion evident in the low-angle region of a Guinier plot. By utilizing more exposures rather than increased exposure time per image, one can evaluate changes as a function of total dose and simply remove images that appear to exhibit radiation damage. However, using more exposures also increases the amount of detector readout noise, as the data from the detector must be read more often. Low-noise detectors or noise-less detectors such as pixel-array detectors do not suffer from this and are becoming more common on beamlines. Utilizing longer exposures can also increase signal-to-noise without increasing detector readout noise, however longer exposures run the risk of damaging the sample without the ability to remove the damaged signal. Many beamlines oscillate the sample in the cell during data collection to mitigate radiation damage. Some use continuous flow that, while desirable, requires large amounts of sample volume to feed fresh solution to the X-ray beam.

While signal-to-noise increases proportionally with concentration of the sample, increasing the total exposure time does not increase the signal-to-noise of the data linearly. Ideally, signal-to-noise scales as the square root of the intensity, thus four times the intensity only yields twice the signal (ignoring other sources of noise). However, for proteins that do not tolerate high concentrations, or in situations where high signal-to-noise is desirable and large quantities of protein solution are available, long total exposure times may provide the necessary intensity. Typical exposure times on a source like the SSRL are on the order of a second for the concentrations described above.

6.5 Image Integration

Since particles tumble in solution in random orientations over the course of the exposure, the resulting scattering is isotropic and circularly symmetric. Scattering data on beamlines are often collected on a two-dimensional area detector. During the instrument calibration step (e.g., using silver behenate crystallites whose spacing between diffraction rings is well-known), the center of the image is defined, and each pixel on the detector is assigned a q value corresponding to its radial distance from the center. As detector images are usually in the mega-pixel range, millions of data points are collected. However, to limit file sizes, provide more manageable data quantities, and increase signal-to-noise, q values are binned to provide a limited number of data points, typically around 500 points per image (see Figure 6.1). The intensity values for each pixel are averaged with other pixel values in the same q bin to increase the signal-to-noise. This is particularly useful for higher resolution data as there are many more pixels in high-q bins than there are in low-q bins. Oftentimes, the detector will be offset from the center to maximize the q range of data collected at the expense of fewer pixels for averaging. This is also a consideration in making effective use of the detector-to-sample distance. More detail is given on image integration in Chapter 7.

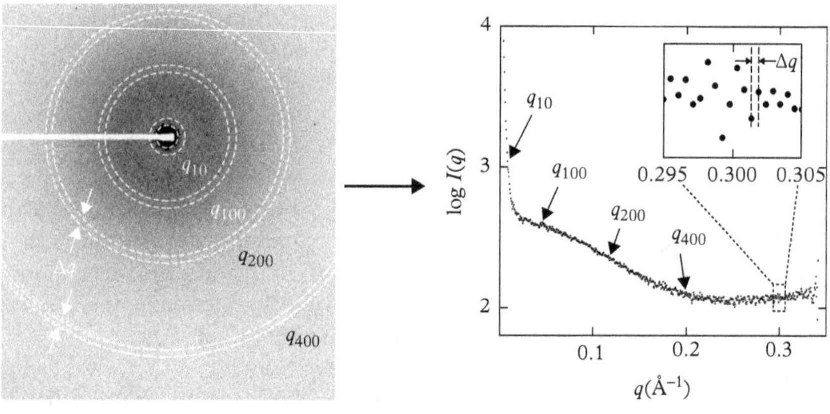

Figure 6.1 Circular integration and averaging of lysozyme SAXS data. The left panel shows the raw detector image read from a CCD detector. Black pixels correspond to higher intensity whereas white pixels correspond to lower intensity. The beamstop can be seen as the white horizontal bar extending from the beam center to the left of the image. White dashed circles show the inner and outer rings defining a selection of four bins (q_{10}, q_{100}, q_{200}, q_{400} for data points 10, 100, 200, and 400, respectively). Intensities of each pixel within each bin are averaged, resulting in the final intensity for the corresponding q value. White arrows illustrate the width of each bin, Δq. For illustration purposes, Δq is exaggerated. The panel on the right shows the resulting one-dimensional scattering profile. The data points corresponding to each of the four bins are identified by black arrows. The inset shows a zoomed-in region around $q \approx 0.3$ Å^{-1}, showing the individual data points and the actual bin size, Δq. Note that the data shown in this figure has not been buffer subtracted. The sharp increase in intensity at very low q is primarily due to the scattering from the sample cell, which gets subtracted during buffer subtraction.

6.6 Buffer Collection and Subtraction

Particles in solution are not the only source of scattered radiation in solution scattering experiments. Bulk solvent scatters, as does the sample cell or any other electrons between the source and the detector. To mitigate such contaminating sources of scatter, solution containing only buffer is also shot during an experiment, and the signal is subtracted from the sample solution signal to yield the signal from just the macromolecule of interest.

The signal from buffer blank is often much less than that from the protein solution at low resolution. However, at higher angles, the differences become very small and the scattering profile becomes particularly sensitive to inaccurate buffer subtraction, resulting in poor fits of accurate models to the data. Therefore, it is extremely important that buffer blank solutions precisely match the buffer in macromolecular solutions to ensure accurate buffer subtraction.

Experimental data for a protein sample (bovine serum albumin [BSA]) and buffer blank is shown in Figure 6.2. The buffer blank contains signal from the buffer and the sample cell. While the sample cell scattering dominates at low q, subtracting the buffer blank scattering reveals the signal attributable to the protein only.

Serial dilutions prepared with the buffer blank can help identify buffer subtraction errors as they provide a systematic increase in the volume of the matching buffer to compare dilutions. As the only difference between serial dilutions should be the number of molecules in solution, the intensity over the entire q range should

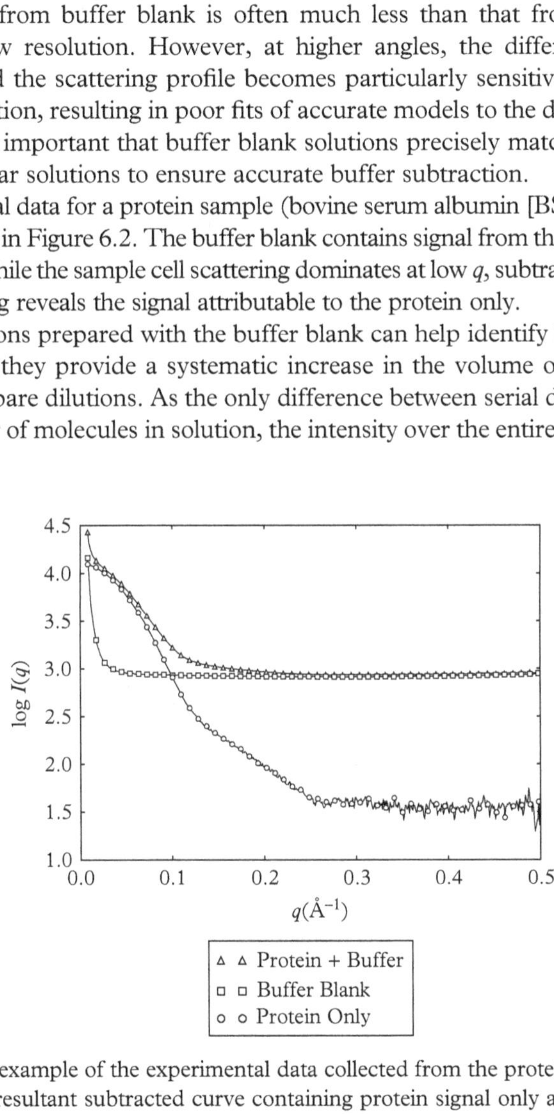

Figure 6.2 An example of the experimental data collected from the protein sample, buffer blank, and the resultant subtracted curve containing protein signal only and which is then used for the subsequent steps of analysis.

scale linearly with concentration. Nonlinearity at high q values indicates inaccurate buffer subtraction, particularly if it is persistent throughout all serial dilutions, whereas nonlinearity at low q values is more likely due to the presence of interparticle interactions, although inaccurate buffer subtraction can also present at low resolutions as well. More details for assessing buffer subtraction errors are given in section 6.15.

Subtraction of the buffer blank eliminates signal from the bulk solvent; however, solvent surrounding the immediate region of the protein, termed the hydration shell, typically associates slightly more strongly with the macromolecule surface and thus becomes more ordered than the bulk solvent resulting in an increase in the average electron density of this region by approximately 10%–15% (Svergun et al., 1998). This signal from the hydration shell is not subtracted during buffer subtraction and must be accounted for during modeling. Additionally, subtracting the buffer signal also assumes that the entire irradiated volume of solution contains only bulk solvent, whereas in an aqueous solution containing protein, some of the bulk solvent volume is replaced by protein molecules. Hence, this excluded volume of buffer must also be accounted for during modeling.

6.7 Initial Parameter Evaluation

During the data collection, it is important to ensure that any data being collected is of good quality and to adjust sample conditions or instrumentation setup if necessary. Several beamlines have computational processing pipelines for automatically integrating and circularly averaging detector images, subtracting buffer blanks, and performing quick data analysis to retrieve estimates of R_g, $I(0)$, D_{max}, molecular weight, low-resolution envelopes, and algorithms for detecting radiation damage and concentration dependence (Knaapila et al., 2009; Nielsen et al., 2009; Franke et al., 2012; Martel et al., 2012). Here, we will describe avenues for assessing data quality and flagging problematic data either manually or through automated pipelines.

6.8 The Guinier Plot

Often one of the first methods used to evaluate data quality is the Guinier plot, described in Chapter 3, section 3.2.3. The Guinier plot shows the natural log of the intensity, $\ln(I(q))$, versus q^2 in the low-q part of the scattering profile. For monodisperse data, the Guinier plot should be linear, ideally below $q < 1.3/R_g$. Due to shape effects, particularly elongated particles may only be linear below $0.8/R_g$. Nonlinearity in the Guinier plot suggests interparticle interactions or aggregation in the sample. Positive curvature (a "smile") suggests aggregation is present. Negative curvature in the Guinier plot (a "frown") suggests repulsion of particles and can artificially decrease the estimates of particle size. In such cases, it is important to alter the solution conditions by diluting the protein concentration, altering chemical conditions such as salt concentration or pH, or passing the solution through a filter to get rid of aggregates prior to continuing to collect data. In most cases, aggregation precludes the ability to interpret SAXS data accurately. The curvature in the Guinier plot due to attractive or repulsive interparticle interactions can often be described in terms of the so-called structure factor.

Figure 6.3 illustrates the concept of the "structure factor" in SAXS described in Chapter 2. The term "structure factor" here should be distinguished from the same term applied in crystallography. In crystallography, the "structure factor" refers to the Fourier transform of the electron density function; this is a complex wave function containing amplitudes and phases at each point in reciprocal space. In SAXS, the (unfortunately) named "structure factor" is instead a one-dimensional function of q that reflects the interparticle interactions present in the solution. The intensity profile collected in a scattering experiment is thus a product of the form factor of the particle (which contains the desired size and shape information) and the structure factor. For most SAXS analysis, monodispersity and a lack of interparticle interactions are required (i.e., a structure factor equal to 1 throughout the entire q range). In such cases, the intensity profile will thus be equal to the form factor of the particle, and information regarding size and shape can thus be extracted directly from $I(q)$. Attractive interactions result in increased intensity at low q values, whereas repulsive interactions result in decreased intensity at low q values (see Figure 6.3).

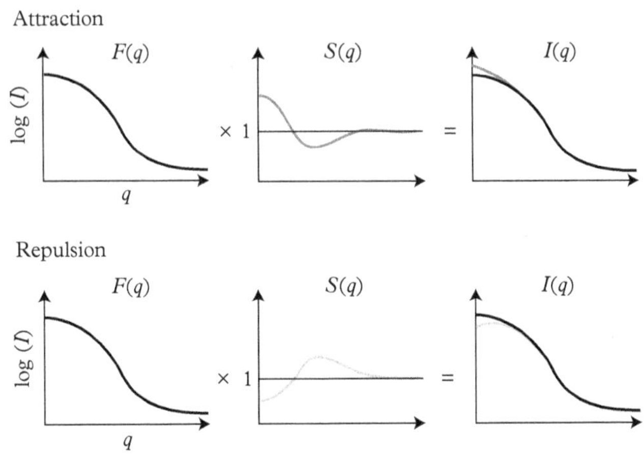

Figure 6.3 An illustration of the addition of an attractive component to the structure factor, $S(q)$, and a repulsive component, showing the net effect on the resultant scattering curve. For a monodisperse, noninteracting solution, $S(q)$ should equal 1. For attractive interactions, $S(q)$ will be greater than one at low q, and for repulsive interactions, $S(q)$ will be <1 at low q.

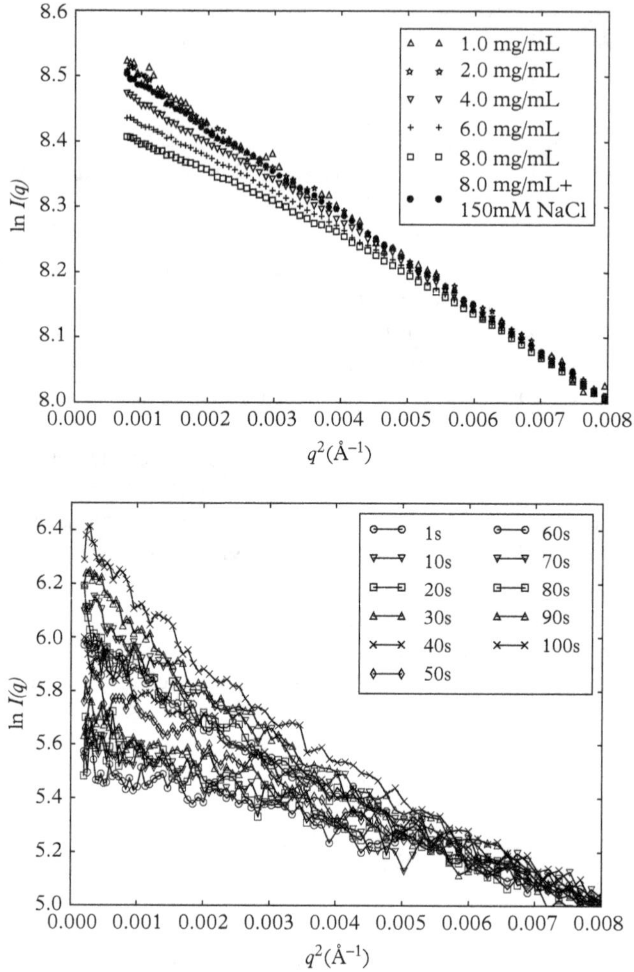

Figure 6.4 Guinier plots for repulsive and aggregated lysozyme. The q range shown here corresponds to the typical Guinier range (i.e., $q < 1.3/R_g$; note the abscissa for a Guinier plot is q^2). The top exhibits repulsive interactions, shown by the decrease in intensity at low q as a function of increasing concentration (the intensities have been scaled to the high concentration for illustration). The bottom shows lysozyme (at 1 mg/mL) in aggregating conditions. Here, the aggregation is caused by progressive radiation damage from the X-ray beam. The legend shows the accumulated exposure time for the sample.

An example of a real experimental case is shown in Figure 6.4. Guinier plots are shown for lysozyme under attractive and repulsive conditions. Repulsive interactions (top panel) are distinguishable by the nonlinear downturn in intensity at low q values (the "frown"). Adding salt to the highest concentration helps to restore linearity in the Guinier plot while retaining high signal-to-noise desired from high concentrations. The panel on the bottom shows lysozyme in attractive or aggregating conditions (here a product of progressive radiation damage). The upward curvature is indicative of aggregation resulting from an increasing amount of large aggregates (steeper Guinier plots at lower q) mixing with the smaller monomers (shallower Guinier plots at slightly higher q), which average to result in a nonlinear upward trend.

The terms attraction and aggregation are often used almost interchangeably to describe the positive curvature in a Guinier plot. However, there is a subtle difference. As the opposite of repulsion, attraction is the transient, self-associations of particles in solution that do not result in the formation of large oligomers. These interactions occur on average more often than in non-interacting solutions, thus resulting in typically long-distance interparticle scattering events that are prominent at low q values. Aggregation on the other hand refers to the random, non-specific self-association of particles that do result in the formation of a variety of large oligomers in solution. The subtle differences between attraction and aggregation are indistinguishable in SAXS profiles; both resulting in the upward curvature of intensity at low q, although aggregation tends to be more severe. Aggregated data is most often unusable as it contains a complex mixture of a variety of aggregates. This may be distinguished from oligomeric mixtures containing large volume fractions of a few oligomeric states, which oftentimes may be useful for further modeling and analysis, provided careful consideration is given to account for the mixture.

6.9 Estimating the Radius of Gyration, R_g

The radius of gyration, R_g, is often the first parameter determined from SAXS data as it is one of the most easily calculated parameters and can provide a quick assessment of whether the experiment was successful, agrees with expectations, or gives insight into potentially new, unexpected science. One of the most common ways to calculate R_g from SAXS data is through the Guinier plot. As noted in section 6.8, the Guinier plot should be linear at low q for monodisperse solutions. The slope of the linear region of the Guinier plot can be used to calculate R_g as described in section 3.2.3.

The range of low-q data points where the assumption of linearity is valid is dependent on the size of the particle and is often limited to $q < 1.3/R_g$. As R_g itself is determined from the slope of the Guinier plot, R_g must be estimated in an iterative fashion. For example, one may start with the first 50 data points, fit a line to the Guinier plot, calculate R_g, revise the selection of data points to $q < 1.3/R_g$, and continue this process until a stable R_g value is found. Algorithms such as those contained in the software AUTORG exist to automate this process (Petoukhov et al., 2007).

An alternative to calculating R_g via the Guinier plot is to use the $P(r)$ curve; R_g can be calculated from the $P(r)$ curve as

$$R_g^2 = \frac{\int_0^{D_{max}} r^2 P(r) dr}{2\int_0^{D_{max}} P(r) dr}. \qquad (6.9)$$

This method requires the *P*(*r*) to be calculated first (described in section 6.10). Using the *P*(*r*) curve to calculate R_g has the advantage of being less sensitive to interparticle interactions than the Guinier estimate of R_g, as the *P*(*r*) curve takes into account data from all resolutions of the profile, whereas the Guinier estimate is only valid at low *q*, where interparticle interactions are most pronounced. Good agreement between both measures of R_g is an indication of good quality sample, whereas poor agreement is indicative of potential interparticle interactions being present in the sample. Plotting R_g as a function of multiple concentrations using both methods (Guinier and *P*(*r*)) should ideally yield a flat line with a slope of zero, indicating no influence of concentration on apparent particle size.

6.10 Calculating the Pair Distance Distribution Function, *P*(*r*)

The *P*(*r*) function yields the most immediate, gross shape characteristics of the particle (see Chapter 2, Figure 2.9). The *P*(*r*) function, as described in Chapter 2, can be calculated directly from the scattering profile as

$$P(r) = \frac{r}{2\pi^2} \int_0^{\infty} qI(q) \sin(qr) \qquad (6.10)$$

One can calculate *P*(*r*) using this equation with the raw *I*(*q*) profile by numerically estimating the area under *I*(*q*) × *qr* × sin(*qr*), using the trapezoid formula for each *r*. This method is subject to large errors and oscillations in the *P*(*r*) profile and should not be used in practice. However, as equation 6.9 includes division by the total area under *P*(*r*), the calculation of R_g using this method tends to be robust to such errors. To mitigate any such errors introduced by direct numerical calculation of *P*(*r*), the indirect Fourier transform utilizing *I*(*q*) fitted to the raw scattering profile and an estimate of D_{max} can also be used (discussed in more detail in Chapter 3, section 3.2.6). This method also suffers from the potential for statistical errors and oscillations in the *P*(*r*) plot, yet can be coupled with regularization to smooth the *P*(*r*) profile while still fitting *I*(*q*). This method requires much more sophisticated algorithms, but can also be performed automatically using software such as DATGNOM (Petoukhov et al., 2007).

The *P*(*r*) function is sensitive to sample quality conditions such as repulsion and aggregation. Aggregated samples will exhibit larger R_g and D_{max} values than the well-behaved sample. Extended tails at large values of *r* are often reflective of aggregated samples; however, they may also be due to true elongated particle characteristics. Samples containing repulsive interactions will often show smaller R_g and D_{max} values than expected. Additionally, repulsive samples may cause the $P(r)$ function to go negative at larger r values, indicating the presence of these interparticle interactions. However, in rare cases, negative regions of the $P(r)$ plot are permissible. Such cases include particles containing low-density components, such as lipids, which have lower density than the surrounding bulk solvent and thus appear as negative on the $P(r)$ plot after the bulk solvent has been subtracted from the scattering profile. For typical protein or nucleic acid samples, however, the $P(r)$ plot should not be negative in any region. For both aggregated and repulsive samples, plotting $P(r)$ curves (normalized by concentration) from multiple concentrations on the same plot can help to determine which of these features are concentration-dependent artifacts, and which are true features of the particle shape. For ideal, monodisperse samples, normalized $P(r)$ curves should overlap nearly perfectly. Differences in the normalized $P(r)$ profiles trending as a function of concentration can thus be attributed to concentration-dependent artifacts and are not representative of the true particle shape.

6.11 Estimating Forward Scattering, *I*(0), and Maximum Dimension, D_{max}

Forward scattering, $I(0)$, can be used for estimating the molecular weight of the particle in solution. This is useful as a check on potential oligomeric state and guidance for experimental parameters that may need to be adjusted; $I(0)$ can be estimated from the Guinier plot as the y-intercept of the line fitted to the data in equation 3.15. It can also be calculated from the $P(r)$ curve using equation 6.11:

$$I(0) = 4\pi \int_0^{D_{max}} P(r)dr. \qquad (6.11)$$

Like R_g, $I(0)$ calculated with the $P(r)$ curve is typically less sensitive to interparticle interactions than the Guinier method, due to the use of the entire scattering profile to calculate $P(r)$, compared to only using the low-q region of the curve for the Guinier plot, the area which is most severely affected by interparticle interactions; $I(0)$ can then be used with equation 6.1 to estimate molecular weight from a known protein standard.

To calculate the $P(r)$ profile using the indirect Fourier transform, one must first have an estimate of D_{max}; D_{max} can be estimated from the $P(r)$ curve by examining where $P(r)$ falls to zero. However, as one needs D_{max} to estimate $P(r)$, this is an iterative process. As noted in the start of this chapter, an initial estimate can be made as $3.5 \times R_g$ for relatively globular particles, and further refinements made.

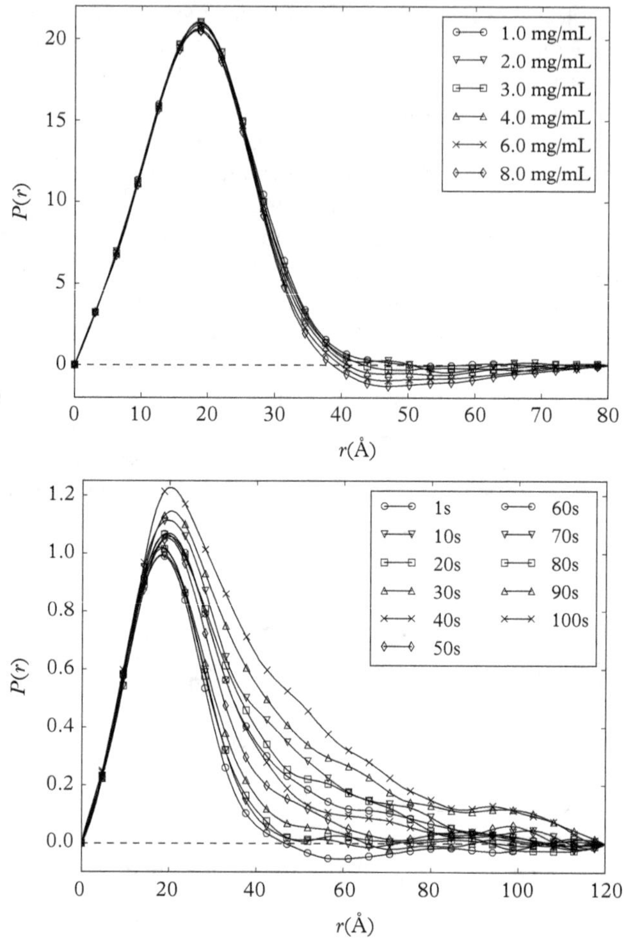

Figure 6.5 Pair distribution function of lysozyme in repulsive and aggregating conditions. The top panel shows the $P(r)$ function of lysozyme from multiple concentrations in repulsive conditions. The $P(r)$ functions have been scaled by their concentration for comparison. Note that at higher concentrations, D_{max} (where $P(r)$ crosses zero) decreases, and the $P(r)$ function goes negative. The bottom panel shows the $P(r)$ function for lysozyme in aggregating conditions. Here, aggregation is caused by progressive radiation damage from the X-ray beam. In this case, the $P(r)$ functions have not been scaled, since the concentration of lysozyme is the same for each curve (1 mg/mL). Note that the overall area under the $P(r)$ function, corresponding to $I(0)$ (see section 6.10), increases with increasing exposure, demonstrating larger particle size on average due to aggregation. Also note the elongated tail at high r values, indicative of increasing aggregation.

Additionally, one can start by using a large D_{max}, seeing where the $P(r)$ curve naturally falls to zero, and then recalculating $P(r)$ with the revised D_{max}.

As mentioned above, for aggregated and repulsive samples, $I(0)$ and D_{max} can be greater or less than expected, respectively. In particular, D_{max} can be difficult to assess for aggregated samples that exhibit elongated tails at large values of r. In such cases, increasing D_{max} to determine its correct value may only result in longer tails extending to even larger r values. Such instances are indicative of data that should not be used for further processing, as the aggregation may result in false interpretations and incorrect conclusions being drawn. For repulsive cases, $P(r)$ will fall below zero at some value of r and continue to go negative, suggesting that value of r as D_{max} (see Figure 6.5). However, that value will most often underestimate the true D_{max}. For mild cases of repulsion, zero extrapolation can be performed from a dilution series (described in section 6.14, Figure 6.9), which may help mitigate these interparticle effects. For severe repulsion, even zero extrapolation may not be effective, and the data may need to be recollected using reformulated sample conditions, such as higher salt concentrations, that do not exhibit interparticle interactions.

As is the case for R_g described above, plotting D_{max} versus concentration can alert one to concentration-dependent artifacts affecting particle size (see Figure 6.6, top). Similarly, $I(0)$ can also be used to indicate these concentration-dependent effects; however, since $I(0)$ scales linearly with concentration, one must first divide $I(0)$ by the concentration before plotting, yielding a plot of $I(0)$/concentration versus concentration. For R_g, D_{max}, and $I(0)$/concentration plots, the well-behaved sample will exhibit a flat line with a slope of zero as a function of concentration.

6.12 Assessing Flexibility

Qualitative estimations of flexibility can be obtained quickly using the Kratky plot, $q^2 \times I(q)$ versus q (see Chapter 3, section 3.5). This quick qualitative assessment of flexibility can reveal if the protein is unfolded, and if this is not expected, can alert the researcher to potential issues with sample preparation, such as conditions causing unfoldedness, or buffer mismatch. For globular molecules, the Kratky plot rises to a peak at low q values (with the peak position roughly corresponding to R_g) and falls back to the baseline at high q values (Figure 6.7, top panel). For intrinsically disordered molecules, the Kratky plot will not exhibit such a peak but will continue to rise and likely plateau at high q values (Figure 6.7, bottom panel). For molecules that are partially ordered or for molecules that have domains with flexible linkers, multiple peaks in the Kratky plot may be present, and the curve may not fall to the baseline at high q values.

Quantitative estimations of flexibility can be obtained by plotting $q^4 \times I(q)$ versus q^4; this is the Porod–Debye plot. For globular molecules, this plot will plateau beginning at moderate q values; for flexible molecules, this plot will instead

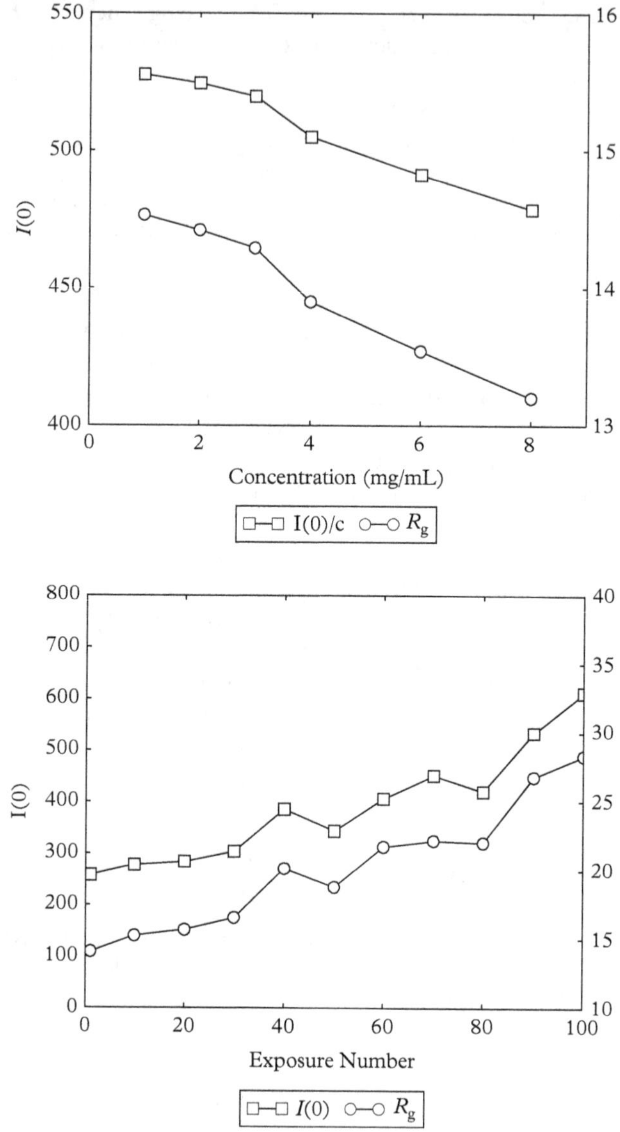

Figure 6.6 SAXS parameters of lysozyme as a function of concentration and X-ray exposure. The top panel shows $I(0)$ divided by concentration ($I(0)/c$) and R_g for a series of concentrations of lysozyme. The decreasing values of $I(0)/c$ and R_g as concentration increases indicate that the solution conditions result in repulsive interparticle interactions. The bottom panel shows lysozyme in aggregating conditions caused by increasing radiation damage from the X-ray beam. The rise in both $I(0)$ and R_g indicate that the lysozyme is aggregating with greater dose.

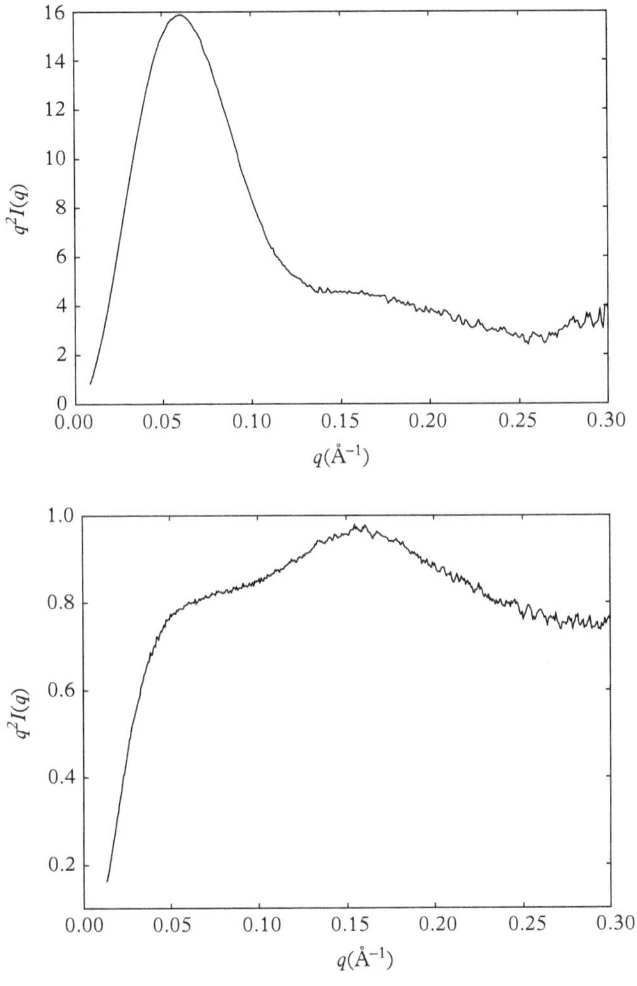

Figure 6.7 Experimentally determined Kratky plot of globular and unfolded particles. The top panel shows the Kratky plot from BSA. The characteristic peak at $q \approx 0.06$ followed by the fall in intensity to near baseline demonstrates a well-folded, globular protein in solution. The bottom panel shows an RNA riboswitch that is unfolded (BIOISIS ID: 2LYSRR). The random coil nature of the particle is shown by the fact that the curve does not fall back to baseline after peaking.

rise to high q values. The Porod exponent can be calculated from this plot using the program ScÅtter, which can be used as a quantitative estimate for flexibility (Rambo & Tainer, 2013). Globular molecules have a Porod exponent approaching 4 (following the Porod law decay for globular molecules of q^{-4}), intrinsically disordered molecules resembling a random chain will have a Porod exponent of 2,

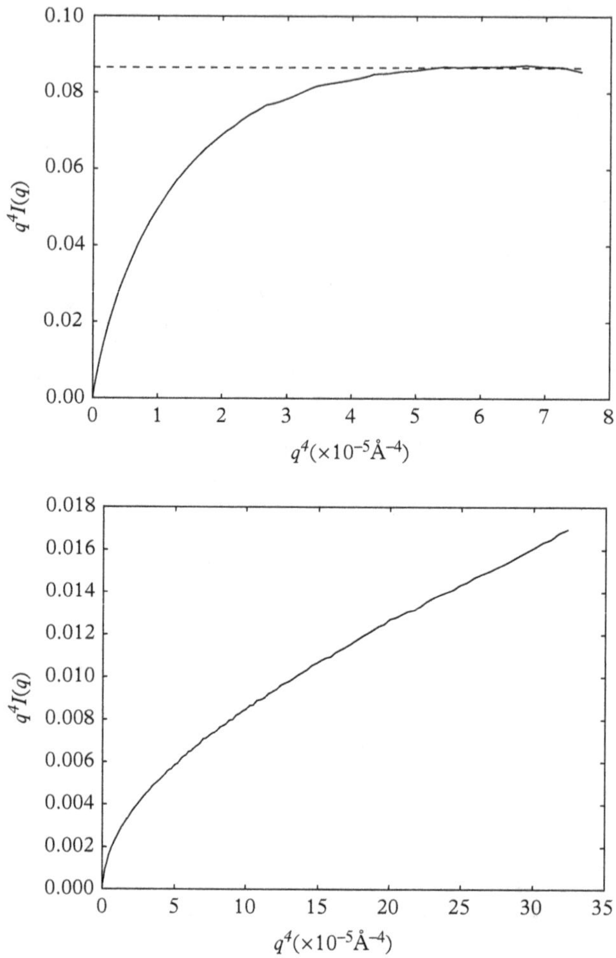

Figure 6.8 Porod–Debye plot for globular and unfolded particles. Globular proteins, such as BSA (top panel), display a plateau (dashed line) in a Porod–Debye plot, reflecting the q^{-4} decay in intensity, here occurring at approximately $q^4 = 0.00005$ Å^{-4}. The bottom panel shows the Porod–Debye plot for an unfolded riboswitch (BIOISIS database reference: 2LYSRR). The flexibility of the particle shows a characteristic rise in the Porod–Debye plot, never plateauing as for globular particles, reflecting the random coil nature of the unfolded riboswitch.

and partially folded or flexible molecules will exhibit a Porod exponent between 2 and 4 (Figure 6.8). Globular particles showing a plateau in the Porod–Debye plot may also exhibit slight nonlinearity in the curve immediately after the plateau (such as a dip in the plot); this is often due to the contributions of the shape of the particle. For the purpose of assessing flexibility, these effects are still consistent with globularity.

6.13 Estimating Volume and Molecular Weight

In addition to using equation 6.1 to estimate molecular weight, one can also estimate molecular weight independent of protein standards or sample concentration. The particle volume can be estimated from the Porod law according to equation 3.26:

$$V = \frac{2\pi^2 I(0)}{Q}, \tag{6.12}$$

where V is volume, and Q is the Porod invariant defined as

$$Q = \int_0^\infty [I(q) - k]\, q^2 dq, \tag{6.13}$$

where k is a constant subtracted to ensure the Porod law decay of q^{-4}. To determine the constant k, $I(q)$ is defined such that it obeys the Porod law (only valid in the Porod region):

$$I(q) = Aq^{-4}, \tag{6.14}$$

$$q^4 I(q) = A, \tag{6.15}$$

where A is a scaling constant. Create a new $I'(q)$ from which the constant k will be subtracted:

$$I'(q) = I(q) - k, \tag{6.16}$$

$$q^4 I'(q) = A,$$

$$q^4 \left(I(q) - k\right) = A,$$

$$q^4 I(q) = kq^4 + A, \tag{6.17}$$

$$y = mx + b.$$

Thus, one can determine the constant k by taking the slope of the line of the Porod–Debye plot ($q^4 \times I(q)$ vs. q^4) in the Porod region. The Porod region, like the Guinier region, has limited extent, and as the Porod law states that the scattering intensity decays as q approaches larger values, the Porod law is only valid at higher q values. The beginning of the Porod region can be estimated visually as the region of the Porod–Debye plot that begins to plateau. In Figure 6.8 for example, the Porod region begins at approximately $q^4 = 0.00005$ and extends to higher q values. While subtracting the constant k modifies the experimental data, the constant to subtract is often very small, as it is defined by the data at high q values, where the intensity is weak. Thus, subtracting such a small constant has little effect on the overall shape of the profile, particularly at low resolution, where the intensity is often orders of magnitude stronger and information corresponding to size and shape is found. Constant subtraction thus helps to mitigate minor buffer subtraction errors to force the Porod law decay while having little impact on size and shape estimates in many cases. However, as the Porod law is only applicable to globular molecules with a sharp boundary between the particle and the solvent, data from flexible molecules should not be treated this way.

Once the constant k has been determined, the protein volume can be calculated from equation 6.12, and the molecular weight can be estimated by dividing the Porod volume by ~1.7 (Petoukhov et al., 2012). The program DATPOROD can automatically estimate Porod volume. For flexible molecules, the plot of $q^2 \times I(q)$ versus q does not converge, and therefore the Porod invariant is inaccurate in such cases. Indeed, the volume of correlation, V_c, can be calculated from the plot of $q \times I(q)$ versus q, which converges for both globular and flexible molecules. The ratio of V_c^2/R_g has been shown to correlate well with molecular weight and can be used as an alternative to Porod volume estimations of molecular weight in cases of flexibility; V_c can be calculated using the programs ScÅtter and DATVC (Petoukhov et al., 2012; Rambo & Tainer, 2013).

6.14 Evaluating Radiation Damage and Concentration Dependence

Overexposing samples to ionizing radiation generally results in radiation damage that presents as aggregation or dose-dependent changes in particle size. To monitor radiation damage, one can calculate each of the above mentioned parameters to examine them as a function of exposure (Figure 6.6, bottom). Any statistically significant dose-dependent changes suggest radiation damage is present, and only non-damaged images should be kept and averaged for further analysis.

Concentration-dependent changes can also be analyzed in a similar manner to assess whether interparticle interactions are present that may be too small to be detectable in noisy Guinier plots (Figure 6.6, top). If concentration-dependent changes are identified, adjustments should be made to the sample solution, such as filtering aggregates or altering physical or chemical conditions such as temperature, pH, ionic strength, and so on. The statistical significance of both dose-dependent and concentration-dependent changes can be estimated quickly using Student's T-test (Grant et al., 2015). If minor repulsive interparticle effects are present, zero extrapolation may be used to mitigate these effects in data analysis. Zero extrapolation involves scaling a series of data from a range of concentrations, extrapolating to a theoretical infinite dilution at low q values (where interparticle interactions are most prominent), and merging with high-q data from the highest concentration. This ensures minimal contamination of the data from interparticle effects at low q while retaining high signal-to-noise from higher concentrations at

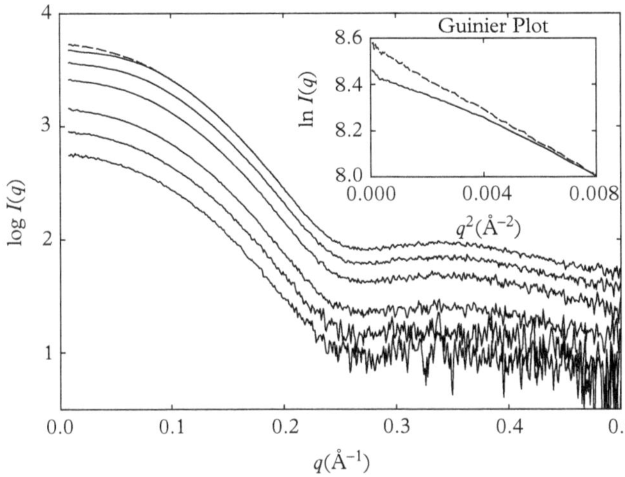

Figure 6.9 Extrapolation of repulsive lysozyme data to infinite dilution. A dilution series of different concentrations of lysozyme from 1–8 mg/mL is shown (thin solid lines). The highest concentrations show a slight downturn in intensity at the lowest q values, due to interparticle repulsion. Zero extrapolation is used to mitigate these repulsive effects. Overlaid on the highest concentration is the zero extrapolated curve (thick, dashed line), showing high signal-to-noise at high q, and lack of repulsion at low q. Inset shows the Guinier plot of the highest concentration (solid line) and the zero extrapolated curve (dashed line). Note the downturn in the highest concentration Guinier plot is absent in the zero extrapolated curve.

high q. An example of zero extrapolation for lysozyme in repulsive conditions is shown in Figure 6.9.

6.15 Assessing Accuracy of Buffer Subtraction

Identification of buffer mismatch is important to ensure accurate serial dilutions and buffer subtraction. Accurate buffer subtraction will be independent of sample concentration. Performing serial dilutions using the same buffer enables monitoring of concentration-dependent changes at high q where buffer subtraction errors are most prominent. By analyzing scattering profiles of subsequent concentrations, a well-matched buffer should exhibit linearly dependent changes only (i.e., the intensities should simply scale as a factor of concentration). For well-folded proteins, the difference between the protein solution scattering and buffer scattering should approach zero at high q (i.e., they should be nearly identical). If the buffer scattering is greater than the protein scattering at high q (i.e., the subtracted profile is negative at high q), this suggests buffer mismatch. If the protein solution scattering is significantly greater than the buffer scattering at high q, this may also

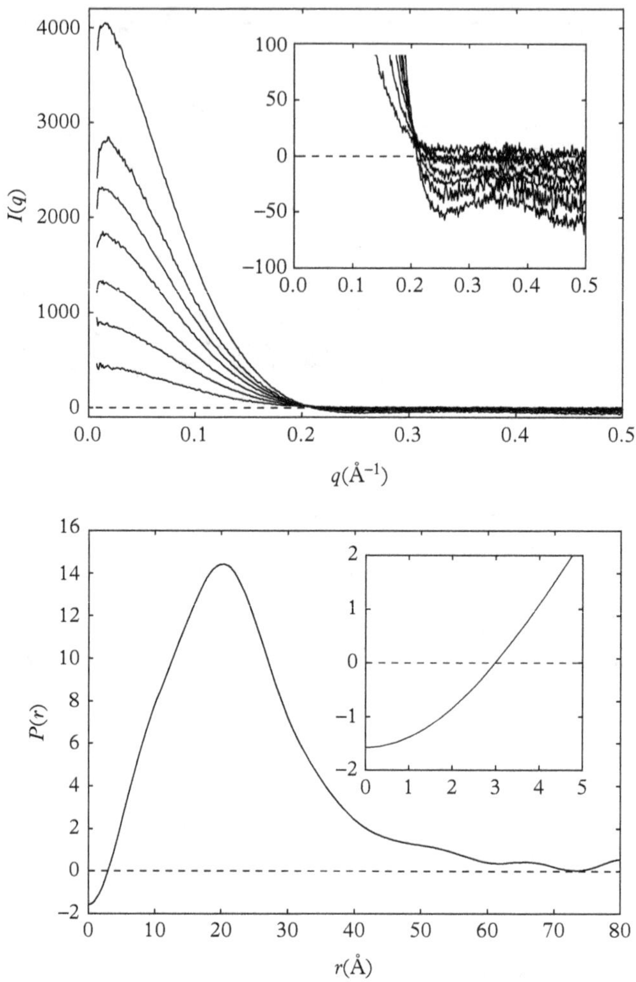

Figure 6.10 Buffer mismatch with lysozyme. The top panel shows buffer-subtracted scattering profiles of a dilution series of seven concentrations from 1–10 mg/mL of lysozyme. Inset shows the same data zoomed in on the y-axis to amplify high-q scattering. Note that as the protein concentration increases (higher overall intensities), the scattering at high q becomes more negative, indicative of greater degrees of buffer subtraction error. Additionally, the sharp decrease in intensity at very low q is also indicative of buffer mismatch. The bottom panel shows the unrestrained $P(r)$ function for the highest concentration of lysozyme and shows that $P(r)$ does not naturally fall to zero at $r = 0$, suggesting buffer mismatch (inset zoomed in near $r = 0$).

suggest the buffer is mismatched. However, this only applies for well-folded proteins. If the protein is not expected to be well folded, the high-q differences between protein and buffer scattering will be greater than expected for folded proteins. Using a dilution series can help identify if any gap between protein solution scattering and buffer scattering at high q is due to buffer mismatch or real particle properties such as protein flexibility. Real particle flexibility will exhibit scattering that scales linearly with concentration throughout the entire q range, whereas mismatched buffer will scale differently depending on the q region. This will be particularly evident at high q, where the small differences between protein scattering and buffer scattering will be amplified.

While perhaps counterintuitive, since the lowest protein concentration contains the greatest percentage of mismatched buffer it will exhibit the *least* amount of buffer subtraction error, since the mismatched buffer blank is what has been subtracted. A concentration series for lysozyme shows that at greater protein concentrations, the high-q scattering becomes progressively more negative, demonstrating that the buffer blank is mismatched (Figure 6.10, top panel, inset). Notice that the highest concentration profile crosses over the other protein concentrations at around $q \approx 0.2$ Å^{-1}, demonstrating that the effect of the buffer subtraction error is different for different q regions. Severe buffer subtraction errors may also manifest at very low q. If the scattering at low q drops steeply just before the beamstop (Figure 6.10, top panel), this may suggest buffer subtraction errors. However, decreases in low-q intensity are more often due to concentration-dependent interparticle repulsion. The $P(r)$ distribution may also be useful for assessing buffer mismatch. Typically, $P(r)$ is calculated while restricting the values of $P(r)$ to be zero at $r = 0$ and $r = D_{max}$. This helps to ensure realistic $P(r)$ distributions for real particles. However, by removing those restraints, one may examine potential errors in the data. If the $P(r)$ function does not naturally fall to zero at $r = 0$ without restraining the function, this suggests potential buffer mismatch (Figure 6.10, bottom panel).

6.16 Summary

The set of examples shown in this chapter illustrate the range of potential avenues for data quality characterization. Some a priori knowledge about your particle shape and size from other experiments is often critical to ensuring the instrumentation setup is correct for your experiment. Accurate buffer subtraction is crucial for correct modeling and analysis of small angle scattering data, and mechanisms for identifying buffer subtraction errors were presented. Examining small angle scattering parameters such as R_g, $I(0)$, and D_{max} as a function of concentration or exposure is very useful for identifying concentration-dependent artifacts or radiation damage that, if unnoticed, can be very detrimental to further analysis, including misinterpreting the results and drawing erroneous conclusions. Small angle scattering is often used for analyzing flexible molecules in solution that may be difficult to study with other structural techniques. Qualitative and quantitative assessments of flexibility were described using Kratky and Porod–Debye plots, yet even more informative analyses can be performed through modeling dynamics (discussed in Chapter 4). Following these protocols for the initial assessment of your data at the beamline can help to ensure successful use of the time and proper characterization of your particle size, shape, and flexibility with solution scattering.

7

Examples of Data Collection and Processing

In the previous chapters, we discussed theoretical and practical aspects of small angle scattering data collection. In this chapter, we present a worked example showing how to process high-quality data and describing features that distinguish high-quality data from data that may be problematic. Several assumptions are made; first, we assume that the beamline or in-house instrument is correctly set up for data collection and is well calibrated. We also assume that dark images are available (images recorded with no X-ray beam to subtract detector noise) and that a concentration series has been collected on a known sample, both to ensure the instrumentation is operational and to serve as a molecular weight control.

7.1 Sample

The worked example detailed in this section describes a SAXS experiment performed on the enzyme aerobactin synthetase IucA from a hypervirulent pathotype of *Klebsiella pneumoniae* (Bailey et al., 2016). The experiment was carried out at the SSRL biological SAXS beamline 4-2 and provided data of very high quality. Proper SAXS experiments should be performed using multiple concentrations of the macromolecule. The stock sample was produced at a concentration of 5.30 mg/mL. The final series consisted of seven concentrations by mixing 40, 30, 20, 10, 5, 3, and 1 μl of stock sample with the complementary volume of buffer to yield 40 μl of sample for each concentration, resulting in 5.30, 3.98, 2.65, 1.33, 0.66, 0.40, and 0.13 mg/mL of protein, respectively. The small volumes for the lowest concentrations have the potential risk of pipetting errors being introduced and, as described below, resulted in errors in concentration estimates. This demonstrates one of the benefits of collecting several concentrations, as inaccurate concentration estimates may preclude the ability to assess important parameters such as molecular weight accurately. To ensure accurate buffer subtraction, elution buffer from SEC, used as a final purification step for preparing the sample, was used as the buffer blank, and to make dilutions. The buffer consisted of 25 mM HEPES, 75 mM NaCl, and 0.1 mM TCEP.

Biological Small Angle Scattering: Theory and Practice. Eaton E. Lattman, Thomas D. Grant, and Edward H. Snell.

DOI: 10.1093/oso/9780199670871.001.0001

7.2 Data

Data collection involves recording scattering data from a buffer blank and then different concentrations of the sample under study. A recommended strategy is to record the buffer blank before and after the data series (ensuring that cleaning of the sample holder occurs between exposures) and to use an increasing concentration step with samples made via a serial dilution of the most concentrated condition. A minimum of three concentrations is required, but more are better. There are multiple strategies for data collection itself; we recommend a series of short exposures such that the series can be summed but that the presence of radiation damage can be detected and those images discarded. Other strategies use, for example, a short, long, then a short exposure, again allowing the detection of radiation damage and a strong signal where no damage is present, but having to discard the long exposure if any damage is detected.

Figure 7.1 shows examples of the raw data that is collected and points out features in the images that are enlarged on the right-hand side. In the top image, the scattering from a buffer blank is shown. This is typically featureless, with a small amount of scattering around the direct beam associated with the sample cell. Enlarging the center of the image reveals some parasitic scatter below the beamstop and evidence of a zinger. Parasitic scatter should be eliminated as much as possible, but, depending on the source, can usually be masked out on a well-conditioned instrument. The zinger is noise caused by high-energy particles interacting with the fiber optic in the detector. Because these are transient events, they can be eliminated by comparing two otherwise identical images and removing any differences recorded, or by removing integration outliers from a series of multiple images.

At low concentration, middle set in Figure 7.1, a faint halo is seen around the beamstop. Enlarging this shows the weak scattering from the low concentration sample that can be compared with that from the buffer blank. As the concentration of the sample increases, the detail in the scattering becomes apparent. The bottom set of images in Figure 7.1 shows concentric rings, clues to the eventual structure. A quick comparison of the images from the concentration series together with a buffer blank recorded before and after the sample, will provide a first indication of any experimental problems, and if both the concentration range and the distance chosen are adequate for the experiment.

The process of image integration and azimuthal averaging is briefly described in Chapter 6, section 6.5. This process may be performed for the user with local beamline software or is often done using the FIT2D software (Hammersley, 2016). The resulting one-dimensional scattering profile will usually be corrected for instrument geometry, detector noise, X-ray beam fluctuations, and other effects. Typically, these files are simple ASCII text files with three columns: q, intensity, and standard error, which can be plotted with most plotting software. However, specialized graphics software such as Primus (Konarev et al., 2003) and ScÅtter (Rambo and Tainer, 2013) can be used to quickly generate a variety of useful plots and calculate many small angle scattering parameters.

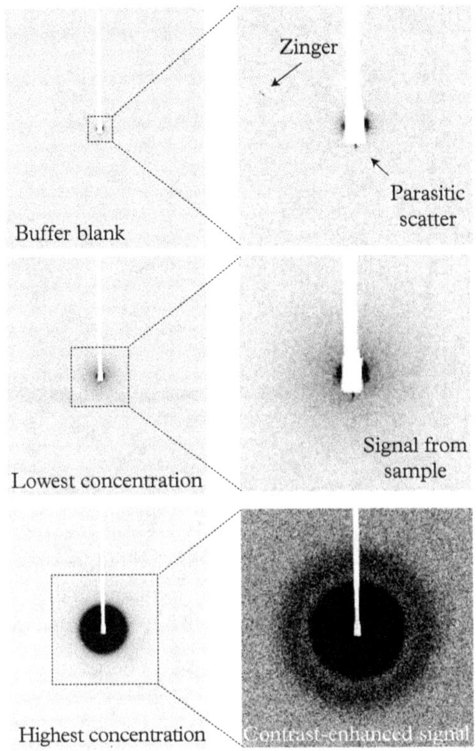

Figure 7.1 Examples of SAXS data recorded from the buffer blank and lowest and highest sample concentrations. Features in each image are enlarged, including the zinger and the parasitic scatter present in the buffer blank. The lowest concentration sample shows a diffuse scatter around the beamstop, and the parasitic scatter can be seen. In the highest concentration image, the brightness and contrast has been adjusted to show concentric rings visible in the data corresponding to sample shape.

7.3 Checking Buffer Blanks

Figure 7.2 shows the integrated scattering profile of the buffer prior to the sample collection and also the buffer recorded after the sample was collected and the sample cell had been cleaned. The profiles are superimposable, suggesting that the buffer data is free from contamination of the sample cell. A stronger signal in the final buffer would be an indication of more material scattering; this could be due to sample adhered to the sample cell surface or other contamination. On a synchrotron, differences can also be caused by variation in beam intensity (e.g., due to a refill or natural beam decay over time). It is important to note the beam characteristics and any changes during the experiment so that these cases do not provide false evidence that the initial and final buffer shots differ.

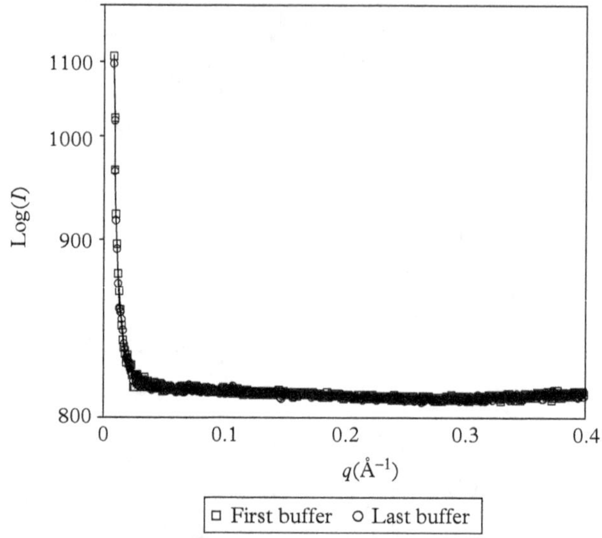

Figure 7.2 The integrated scattering signal from the first buffer and last buffer, illustrating that the buffer signal remains constant and demonstrating with the exact overlay of curves that the sample chamber has been adequately cleaned for subsequent experiments.

Figure 7.3 shows the averaged profile from all exposures of the highest concentration sample, the averaged profile from the buffer blank, and the final buffer subtracted profile to be used in subsequent analysis.

7.4 Avoiding Radiation Damage

In our recommended procedure, multiple exposures for each concentration are recorded. For the current example, 20 one-second exposures were collected and averaged to increase signal-to-noise. Extended exposure to X-rays can damage the protein sample, most often resulting in aggregation of the protein. To mitigate the potential for radiation damage, beamlines often oscillate the sample in the sample cell. One can also reduce damage by including radical scavengers in the protein solution, such as 5% glycerol. To ensure that detectable radiation damage has not occurred after data collection, one can compare each of the exposures to ensure no exposure-dependent changes have taken place. Figure 7.4 shows three of these images displayed with the buffer blank subtracted. The three individual exposures for the same concentration overlap well. An enlargement of the low-q region shows no difference in $I(0)$ between the concentrations, which indicates minimal if any radiation damage. Radiation damage causes aggregation to occur and one would expect a gradual increase in the $I(0)$ as a function of image number. Image 10 in

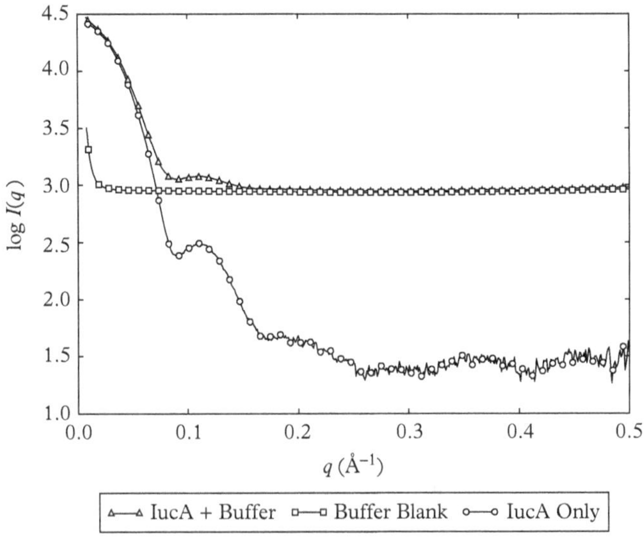

Figure 7.3 SAXS profiles of sample and buffer blank, and the resulting buffer-subtracted profile for the highest concentration of IucA. The sample and buffer blank profiles are each the average of twenty 1 s exposures.

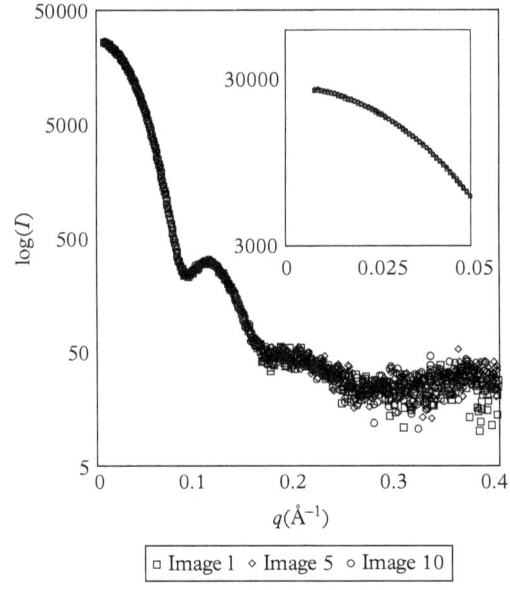

Figure 7.4 Highest sample concentration scattering data with buffer subtracted from a set of three images. The integrated scattering cross section shows the presence of the concentric rings seen in the image in Figure 7.1. The insert enlarges the low-q region to show that there is no visible evidence of the scattering deviating from the first image. For the experiment described in the real data set, 20 images for each concentration are recorded and integrated into a single image for each of those concentrations.

this case, would cross the axis at an increased value, which is not the case. If it were, radiation scavengers may have to be used in the solution, a shorter exposure time employed, or fewer images taken. The set of superimposable scattering profiles is integrated, averaged, and used as a single profile for that concentration.

The integration of the images and reduction to a single scattering curve for each image is a standard procedure for any small angle scattering instrument, laboratory X-ray source, synchrotron beamline, or neutron beamline etc. Dedicated software will typically use calibration images to ensure an accurate beam center, convert from pixel space to q space, and correct for detector characteristics and be able to provide the scattering curve.

7.5 Qualitative Comparison of Concentrations

After ensuring no radiation damage has taken place, the next step is to check whether concentration-dependent changes are occurring. The first and simplest way to do this is by directly comparing the averaged buffer subtracted profiles from each concentration. The buffer subtracted scattering profiles for the seven concentrations in our example are shown in Figure 7.4. For a more direct comparison, one can multiply each profile by a scale factor to display all the concentrations overlaid on top of each other. For the most accurate scaling, a range of data is selected that has the highest signal-to-noise but also avoids the lowest q values, where interparticle interactions would be most noticeable. In this case, the chosen range for scaling was $0.05\ \text{Å}^{-1} < q < 0.1\ \text{Å}^{-1}$. Various software packages, such as Primus (Konarev et al., 2003), can be used to perform this scaling easily. Evaluation of the scale factors can alert one to errors in the estimate of absolute concentration. The scale factors for each concentration were determined to be 1.0, 0.73, 0.50, 0.27, 0.04, 0.13, and 0.05, respectively. Assuming the highest concentration is correct and no concentration-dependent effects are present, these scale factors correspond to true concentrations of 5.30, 3.88, 2.65, 1.41, 0.22, 0.68, and 0.24 mg/mL, respectively. In this case, it is clear that the lowest three concentrations likely suffered from some pipetting error due to such low volumes of protein sample being diluted with buffer, as the scale factors from the data do not match the expected scale factor based on the estimated concentration from the dilution series. Due to this error, and the fact that these lowest concentrations also suffer from significant noise, the lowest three concentrations were not used for subsequent analysis. While these concentrations can likely still be used to gain useful size and shape information, the higher concentrations are likely to yield information that is more accurate. Additionally, inaccurate estimation of concentration precludes the ability to use standards such as water or protein to accurately estimate molecular weight, although other concentration-independent methods of molecular weight estimation such as using the Porod volume or volume of correlation will be unaffected by this.

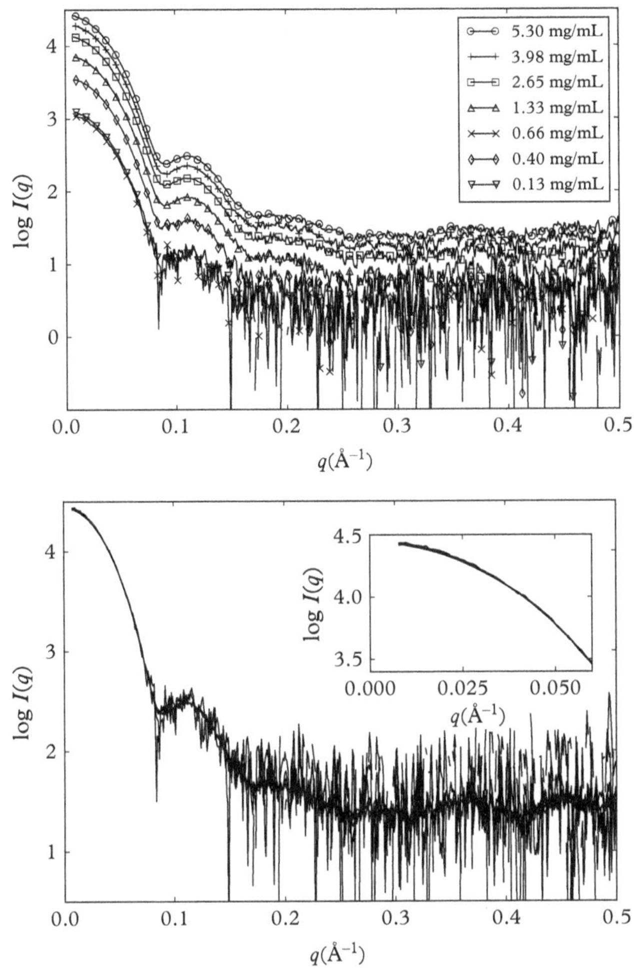

Figure 7.5 SAXS data for each concentration of IucA. On the top, the unscaled data is shown, and on the bottom, the same data, scaled to the highest concentration, using a scaling region of 0.05 $Å^{-1} < q < 0.10$ $Å^{-1}$. The overlay of scattering profiles indicates that no concentration effects are present.

The scaled profiles are displayed below the unscaled profiles in Figure 7.5. A quick qualitative assessment of the overlap of the profiles suggests there are little or no differences in the profiles and therefore no apparent concentration-specific effects. Closer inspection of the overlap in the low-q region, where interparticle interactions are most prominent, also suggests good consistency (Figure 7.5, bottom panel inset).

7.6 Using Guinier Plots To Detect Interparticle Interactions and Calculate R_g and $I(0)$

Another qualitative sample quality check required is to assess whether the sample is likely to contain signal from aggregated protein molecules. As described in Chapter 6, aggregated samples will exhibit upward curvature in intensity at the lowest q values. An easy way of visualizing this is to use a Guinier plot. The Guinier plot is valid only for a limited q range, namely, $q < 1.3/R_g$. If the R_g is unknown, the Guinier region can be estimated iteratively, by first selecting a set of low-q points, for example the first 50 data points. Using a linear fit to this region of the Guinier plot, the R_g can be estimated from the slope of the line, and then the Guinier region can be updated. Performing these steps repeatedly should cause the results to converge to a stable solution. From this final Guinier region, the R_g and forward scattering $I(0)$ can be calculated from the slope and intercept of the line. This is made user friendly in various software packages, such as Primus and ScÅtter, which often include algorithms automating the process. In Figure 7.6, the Guinier plot is shown for the four accepted concentrations of IucA. Primus was used for each profile to select the appropriate Guinier region and calculate R_g and $I(0)$. The

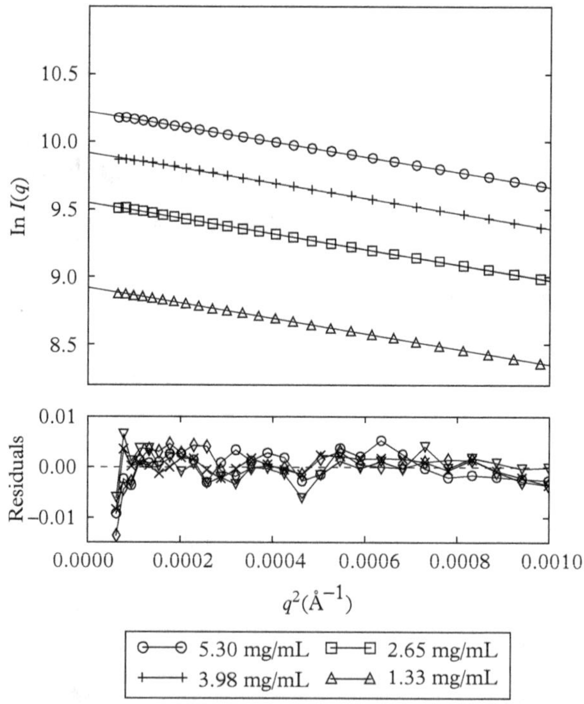

Figure 7.6 Guinier plots for four different concentrations of IucA. The full q range is used up to $q < 1.3/R_g$ (note the x-axis is q^2). The bottom plot shows the residuals of the data to the linear fit.

R_g for each of the concentrations was calculated to be approximately 42 Å, yielding a maximum q value of 0.03 Å^{-1}. Figure 7.6 shows that the Guinier plot for each concentration is linear and lacks upward or downward curvature. The residuals between the data and the calculated fit can be used to assist visualization and assessment of linearity. These results suggest the IucA samples are free from aggregation or severe repulsion.

Section 7.5 described a qualitative assessment of concentration dependence by visually inspecting the overlap of the scaled profiles, in particular the low-q region of the data. For a more precise, quantitative assessment of concentration dependence, we can now plot the R_g and $I(0)$ values as a function of concentration to reveal if any nonlinearities are present in the dilution series. These values, taken from the linear fits to the Guinier plot, are plotted in Figure 7.7. Since $I(0)$ itself is a function of concentration, $I(0)$ must first be divided by the concentration before plotting. Due to the previously mentioned pipetting errors, here the concentrations have been estimated from the scale factors derived from the intensity profiles rather than from the calculations based on pipetted volumes.

Figure 7.7 shows that while the values of R_g and $I(0)$ are mostly linear with concentration, there appears to be a slight decrease in each value for higher concentrations, suggesting minor levels of interparticle repulsion. To mitigate such effects, the data from multiple concentrations can be extrapolated to a theoretically infinite dilution, described in Chapter 6, section 6.13. For further analysis, including shape reconstructions and modeling, this zero extrapolated curve for IucA will be used. In many cases, the curves for low and high concentrations can simply be merged.

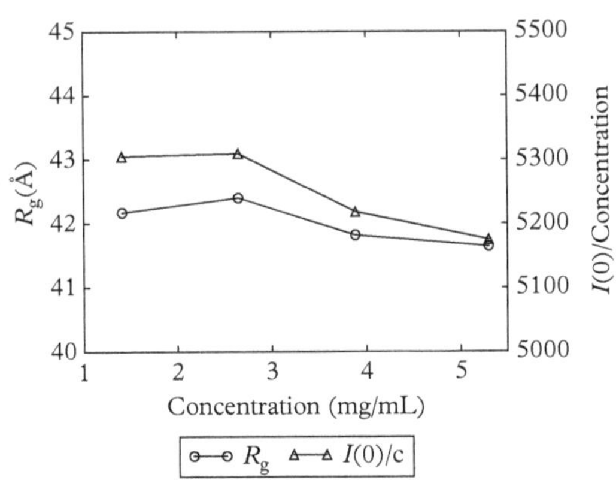

Figure 7.7 Plot of R_g and $I(0)/c$ values obtained from the Guinier plot of each of four concentrations of IucA. The concentration values were estimated from the scale factors derived from the intensity data. From this data, there appears to be a minor amount of repulsion occurring, as evidenced by the slightly decreasing values of both R_g and $I(0)/c$ as a function of increasing concentration.

7.7 Calculating $P(r)$ and D_{max}

One of the first ways to characterize the overall shape of a particle in real space from scattering data is by calculating the $P(r)$ curve (a selection of shapes and their corresponding $P(r)$ curves are shown in Chapter 2, Figure 2.9). Multiple software programs can be used to calculate $P(r)$ curves using indirect Fourier transform

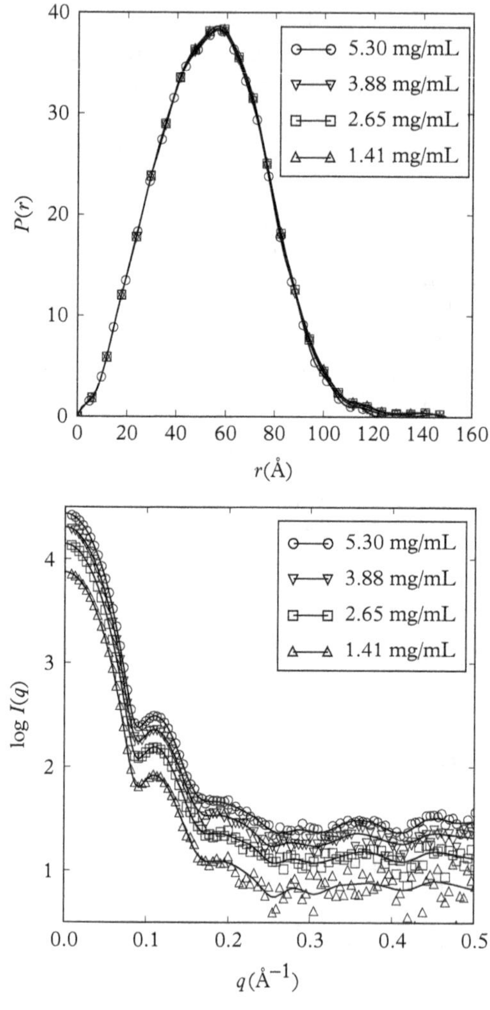

Figure 7.8 Pair distribution functions and associated reciprocal space fits for IucA. The $P(r)$ curves have been scaled by concentration to demonstrate their similarity. For clarity, the $I(q)$ curves have not been scaled. The solid black lines in the $I(q)$ plots are the fits to the data calculated by the Fourier transform of the $P(r)$ plots on the left. The smooth $P(r)$ plots and the good fits of $I(q)$ to the data demonstrate the high quality of the data and are necessary for reliable interpretation of the experimental profiles.

algorithms. To calculate the $P(r)$ curve, D_{max} must be given; however, D_{max} can also be estimated from the $P(r)$ curve. Thus, determining $P(r)$ and D_{max} is an iterative procedure. Some software programs are capable of automating this process, however visual inspection is still strongly recommended to avoid over smoothing or other artifacts in the $P(r)$ curve while ensuring the calculated $I(q)$ profile matches the experimentally determined $I(q)$ data. To calculate $P(r)$ curves for each of the concentrations of IucA, GNOM was used in the Primus software to interactively select D_{max} and view the resulting $P(r)$ and $I(q)$ curves (Figure 7.8).

The $P(r)$ curves from each of four concentrations overlap well, showing no concentration dependence, and have a well-defined D_{max} of approximately 125 Å; D_{max} is often difficult to assess precisely by visually inspecting the curve. Here, the D_{max} given to the indirect Fourier transform algorithm (e.g., GNOM [Svergun, 1992]) for each curve may range from 120 Å to 140 Å, with little to no observable difference in the resulting $P(r)$ curve. This is in fact a sign of high-quality data, that is, that input of a larger than true D_{max} value results in a very similar P(r) curve; D_{max} should be chosen as the smallest value that yields a smooth P(r) curve while still fitting the intensity profile well.

7.8 Assessing Flexibility

The Kratky plot is an important qualitative check for globularity or flexibility in the sample. For a well-folded globular protein it has a characteristic peak at low q and converges to the q axis at high q. It will not converge if the sample exhibits flexibility. Refer to Chapter 3, Figure 3.8, and Chapter 6, Figure 6.7, for examples of a variety of Kratky plots from particles of different shapes. Figure 7.9 shows the Kratky plot for the four concentrations in our example data set. There are two pronounced peaks and a third that is seen in the highest concentration. The plot returns to the q axis and remains relatively flat for the remainder of the

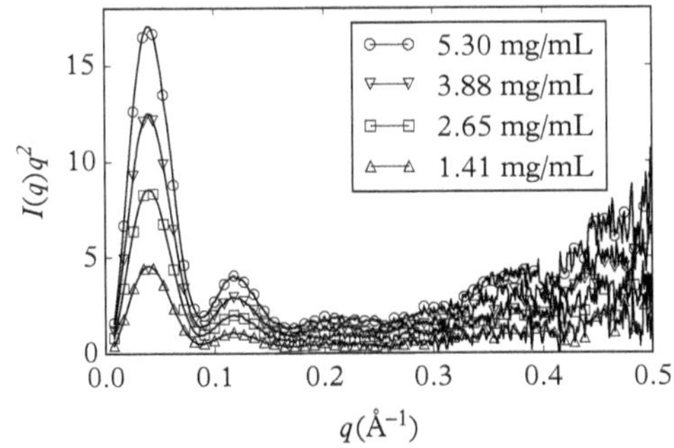

Figure 7.9 Kratky plot for the data from four concentrations, showing that the sample is well folded and globular (see Chapter 3, Figure 3.7, for other cases).

curve, demonstrating the globularity of the sample. The curve begins to rise again slightly beyond a q of ~0.3 Å^{-1}. This small increase at high q is most likely caused by shape effects and is not a result of flexibility in the sample. The plot indicates a globular protein with the additional peaks indicating multiple-domains within the sample.

To obtain a quantitative estimate of flexibility we turn to the Porod–Debye plot. From the Porod–Debye plot, we can extract an estimate of the Porod exponent. For a globular, well-folded protein the Porod–Debye plot will plateau, and the Porod exponent will approach 4 (expected for IucA), whereas for a random chain, the plot will rise without plateauing, and the Porod exponent will approach 2. Like the Guinier plot, the Porod–Debye plot is only valid over a restricted range of data. Unlike the Guinier plot, however, the Porod–Debye plot is not a well-defined quantity and must be selected via visual inspection. The region where the Porod–Debye plot is most effective at assessing flexibility is the first linear region at low q values. The Porod–Debye plots for the four concentrations of IucA are shown in Figure 7.10 for the first 200 data points. The selected region of the Porod–Debye plot should be linear and not have a negative slope. As can be seen from the nonlinearity in the top panel of Figure 7.10, the region containing the first 200 points is too large; however, it does suggest the range of data that should be included (i.e., the region near the first large peak). Restricting the range of points to the first 53 data points (q < 0.056 Å^{-1}), the Porod–Debye plot for IucA clearly shows a plateau, suggesting a compact, well-folded particle with a Porod exponent of 4. From this, the Porod volume can be estimated.

7.9 Checking for Accurate Buffer Subtraction

Now that the profiles have been scaled and concentration dependence and flexibility have been assessed, this is a good time to assess whether the buffer subtraction is accurate. One of the primary methods for assessing buffer subtraction accuracy is comparison of multiple concentrations from a dilution series. As described in Chapter 6, intensity profiles from a dilution series should scale linearly with concentration and any nonlinearity is due to either interparticle interactions or inaccurate buffer subtraction. As described in section 7.6, there may be a slight amount of repulsion found in the low-q region of the profiles; however. there is no steep drop in the very-low-q region (Figure 7.6), suggesting severe buffer subtraction errors are not present. To assess whether more subtle buffer subtraction errors may be present, the entire profile will be examined, in particular looking for nonlinearities at high q values. A direct way of assessing such nonlinearities is to divide each scattering profile by the scattering profile of the highest concentration. Since the concentration of the profiles should be the only difference, the ratio of the profiles should be a flat line whose value is the relative scale factor of the concentrations being compared. Figure 7.11 shows the ratio of intensities of

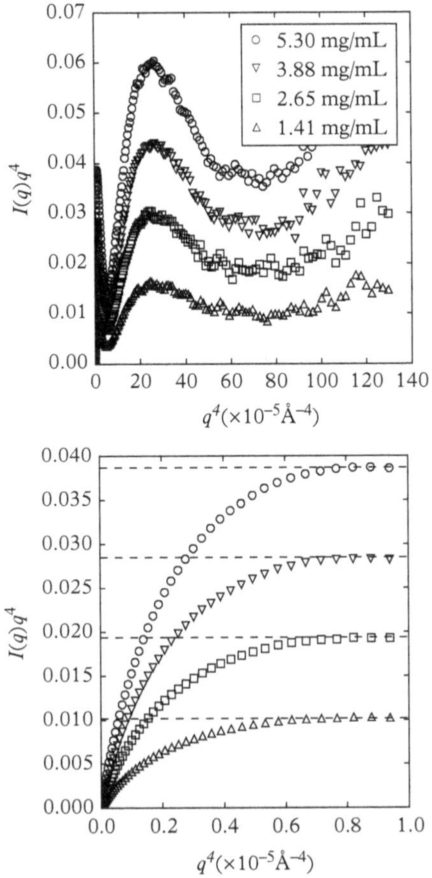

Figure 7.10 Porod–Debye plots for four concentrations of IucA are shown. The top panel shows the first 200 data points. Selecting only the region from the first peak (i.e., the first 53 data points [bottom panel]) demonstrates a plateau in the plots (dashed lines), characteristic of globular, well-folded proteins.

the highest concentration to the next three highest concentrations. For each of the three concentrations, the ratio of the intensities shows a flat line with no nonlinearities, particularly at high q values where buffer subtraction errors are most prominent.

An additional way to inspect the high q region for potential buffer subtraction errors is by plotting the Kratky plot of the scaled profiles. The Kratky plot emphasizes high q intensities, so any nonlinearities present in the data are likely to be identified at high q. Figure 7.12 shows the Kratky plots of each of the four concentration of IucA after scaling. The profiles overlap well even at high q values, demonstrating accurate buffer subtraction.

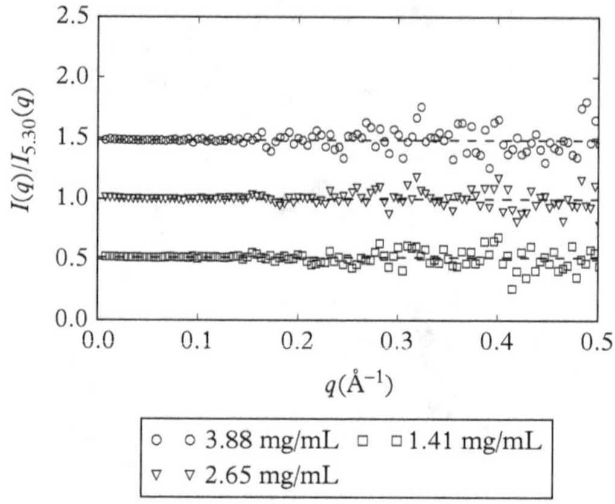

Figure 7.11 Ratio of the intensities of the second three highest concentrations to the highest concentration of IucA. The data has been shifted vertically for clarity. Note that the ratios for each concentration are flat and linear across the entire q range, demonstrating no apparent buffer subtraction errors.

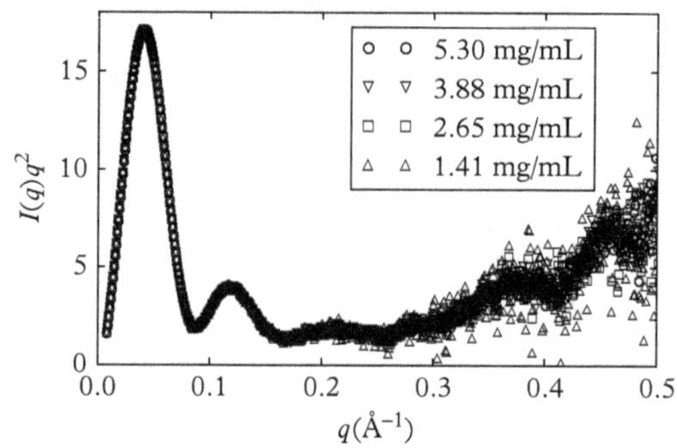

Figure 7.12 Kratky plots of four concentrations of IucA after scaling. Consistent overlap of each profile, in particular at high q, shows the absence of buffer subtraction errors.

7.10 Zero Extrapolation

After ensuring there are no buffer subtraction errors, the data can now be used for subsequent analysis. However, as described in section 7.6, there appear to be slight repulsive interactions at the highest concentrations. One possible route to mitigating such effects in the analysis is to merge data from the highest concentration at high q values with the low-q intensities from the lowest concentration. This ensures high signal-to-noise at the high q values while also avoiding much of the inter-particle interactions prominent at low q values. A second widely used approach is to take advantage of the concentration series and perform zero extrapolation of the data, that is, estimating the scattering profile as though it were performed at infinite

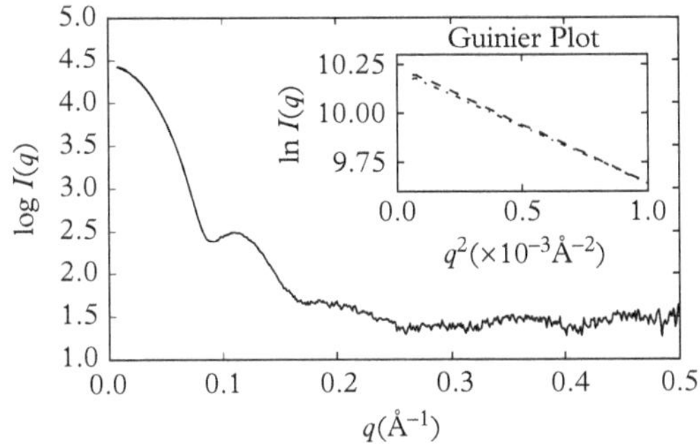

Figure 7.13 The zero extrapolated curve of IucA (dashed line) is shown overlaid onto the highest concentration of IucA (solid line). The inset shows the Guinier plot at low q values, to highlight the slight difference between the zero extrapolated curve (dashed line) and the highest concentration (dot-dashed line). The zero extrapolated curve has a slightly steeper slope (corresponding to a slightly larger R_g) and slightly higher y-intercept (i.e., $I(0)$, corresponding to a slightly larger molecular weight). The slightly larger particle size from the zero extrapolated curve is indicative of the repulsive effects present in the highest IucA concentrations, which have been corrected for by zero extrapolation.

dilution (i.e., from a single molecule). Several algorithms for this process exist, including the ALMERGE program in Primus (Konarev et al., 2003). The zero extrapolated curve generated from the four concentrations of IucA is shown in Figure 7.13. This curve is used for all subsequent analysis.

7.11 Extracting Structural Information

Now that the data has been examined and determined to be of suitable quality, further structural analysis can now be performed. There are many uses for small angle scattering data, but some of the more common are to compare a model to experimental data to test a hypothesis, to look for evidence of structural changes in ligand binding, and/or to reconstruct a molecular envelope to visualize how a sample or complex behaves. In the current example of IucA, one of the main aims for collecting SAXS data was to determine the oligomeric state and assembly of the particle in solution. Computational analysis of the crystal structure of IucA suggested the monomer was the most stable form of the molecule, while homologous proteins were biological dimers with various dimer interfaces, and biochemical analyses of IucA were most consistent with a tetrameric assembly. The first step to determine the oligomeric state of IucA was to estimate the molecular weight. Using the zero extrapolated scattering profile of IucA, the molecular weight was estimated using three methods: by comparing $I(0)$ with that of a known protein standard (lysozyme), by estimating the Porod volume, and by volume of correlation. From these three methods, the molecular weight was estimated to be

240 kDa, 244 kDa, and 242 kDa, respectively. The theoretical molecular weights of the monomer, dimer, and tetramer are 65 kDa, 130 kDa, and 260 kDa, respectively, suggesting that the most likely oligomeric state is a tetramer.

In addition to the oligomeric state of IucA, the oligomeric assembly may also be estimated by SAXS (i.e., the spatial arrangement of monomers within the tetrameric assembly). Since the crystal structure of IucA has been determined, plausible oligomeric assemblies for the monomer, dimer, and tetramer were predicted based on crystallographic packing. To test which of these oligomeric assemblies was most likely to be correct, the scattering profile was calculated from the predicted structures and compared to the experimental data, using the software program CRYSOL. The fits of the calculated profiles to the zero extrapolated scattering profile are shown in Figure 7.14.

Finally, to validate the tetrameric assembly independent of known structure, ab initio envelope reconstruction was performed. Ab initio bead models were reconstructed using the program DAMMIF (Franke & Svergun, 2009). Since the tetramer is of P222 symmetry, ten reconstructions were performed while imposing symmetry to increase real space restraints and improve the ability of the algorithm

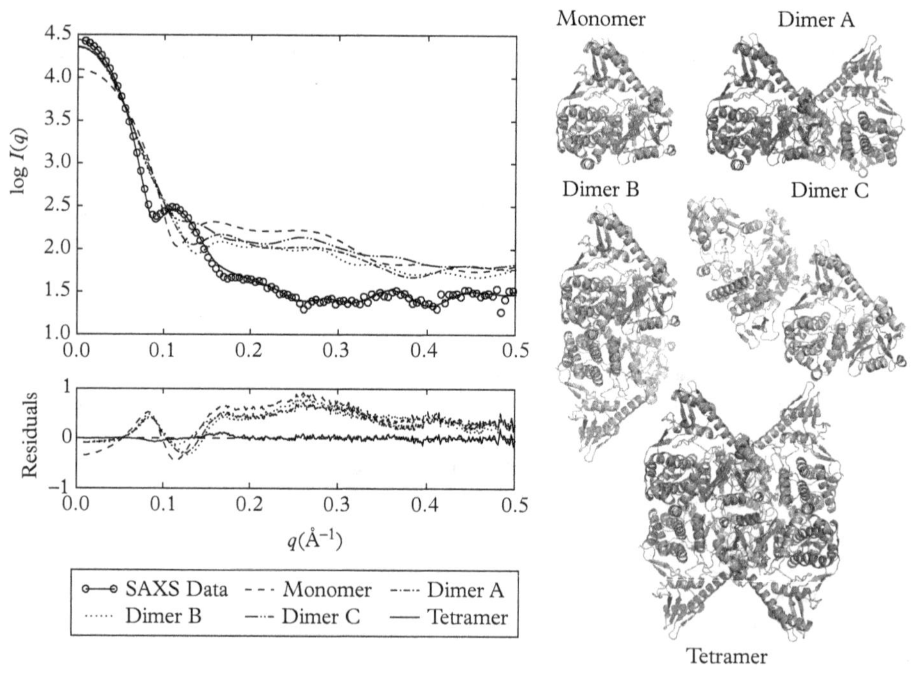

Figure 7.14 Calculated scattering profiles of various oligomeric assemblies of IucA, compared to experimental SAXS data. The tetramer shows the best fit to the scattering profile from among the set of plausible oligomeric assemblies, as shown by the fit and residuals. The molecular structures on the right show each oligomer in cartoon representation.

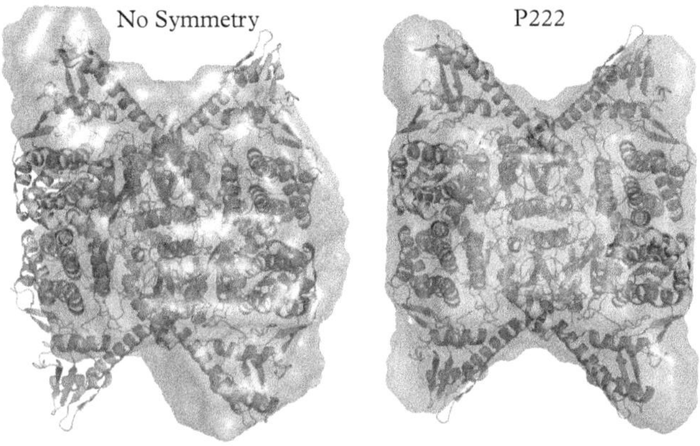

Figure 7.15 Ab initio envelope reconstructions of IucA. Two sets of ten independent reconstructions were performed and averaged. The left panel shows the consensus bead model assuming no symmetry (P1, gray surface), while the right panel shows the consensus bead model while imposing P222 symmetry. The tetrameric assembly derived from crystal packing contacts is shown superimposed to the ab initio reconstructions (cartoon representation).

to reconstruct the envelope. However, to mitigate any potential bias arising from imposing symmetry, ten separate envelopes were reconstructed assuming no symmetry, to be compared with the symmetric reconstructions. For both cases, the ten envelopes were aligned and averaged with DAMAVER (Volkov & Svergun, 2003) to produce a consensus bead model representing the particle. The coordinates for the tetramer were aligned to the bead models by using the program SUPCOMB from ATSAS (Franke et al., 2017). As can be seen in Figure 7.15, the ab initio envelope reconstruction agrees well with the proposed tetrameric assembly of IucA. When assuming no symmetry, the overall size is recovered, but the shape appears only partially correct; however, when imposing symmetry, both the size and the "X"-like shape of the object are recovered well. These results, along with the superior fit of the calculated tetramer scattering to the data, are conclusive that the tetrameric assembly found in the crystallographic packing of IucA represents the solution state of IucA.

7.12 Summary

The main purpose for collecting SAXS data for IucA was the unambiguous determination of oligomeric state and assembly in solution. After collecting the SAXS data, several data analysis procedures were performed to ensure the data was of sufficient quality. These checks included comparing buffer images before and after sample data collection, to ensure no contamination of the sample cell occurred during exposure; checking for radiation damage and concentration dependence; checking for aggregation and buffer subtraction errors; and computationally correcting for minor observed interparticle repulsion. Many of these steps were performed by comparing multiple concentrations from a dilution series, demonstrating the importance of this step in the data collection procedure for ensuring high-quality data. Finally, real space modeling and shape reconstructions were performed to determine the oligomeric state as a tetramer and identify the correct oligomeric assembly from crystal packing predictions.

In many cases, more complicated modeling may be necessary to answer the biological question of interest. However, following the data analysis procedures outlined here can help ensure the highest-quality data is used for answering these questions and provide sufficient information for publication and presentation of results, by using a wide variety of advanced modeling methods.

8

Instrumental and Experimental Considerations

For a typical experiment, after the data is recorded, it is usually processed by software that is specific to the instrument, beamline, and makes use of calibrations made in advance. There are many steps which are invisible to the experimenter but consideration of which may help in understanding the process, improving the outcome, or understanding limitations. In this chapter, we cover detector considerations and the process of going from an image to a scattering profile. Many of these topics hide "behind the scenes," so to speak, having been addressed by beamline scientists or within the data processing software. However, for a full understanding of the instrumentation and data processing aspects, they are useful to know.

8.1 Detector Considerations

Real X-ray detectors are imperfect; they can have variations in response across the detector and spatial distortions that have to be corrected. A detailed review detector considerations is available elsewhere (Pauw, 2014); a summary of the discussion is presented below.

External sources other than the scattering to be measured can produce a signal on the detector. Random signals can be produced by cosmic rays, atmospheric radioactive decay, or electronic spikes. These signals are transient and a process termed 'dezingering' is sometimes used. This can be a user option in data collection. If it is carried out, data is collected with multiple images and all statistically different differences masked or replaced with an average value computed from the other images.

The response across the detector face can be different due to manufacturing tolerances and differences in the electronics. Detectors are usually calibrated with a uniform exposure, with the resulting response surface used as an internal correction, provided as a corrective response table, or used by third-party software for corrections.

Detectors have deadtime: this is where a second photon arriving in rapid succession may not be detected or may cause too big of a signal. This can lead to

Biological Small Angle Scattering: Theory and Practice. Eaton E. Lattman, Thomas D. Grant, and Edward H. Snell.

DOI: 10.1093/oso/9780199670871.001.0001

an overload of the detector or rejection of the signal. Experimentally, with a fast detector, the answer is to reduce the flux, but practically this may cause potential errors in the noise statistics. The response of the detector to different energies can also vary. This is not an issue where a single energy is used, but may impact anomalous studies where scattering from different energies are compared.

Electronic noise and background radiation can lead to a signal even when no X-rays are incident upon the detector. This dark current can be corrected by subtracting an image recorded with the same exposure time but no incident X-rays. A dark image file is often provided by the beamline as part of the data collection protocol. When the beam is incident, the detector the electronic noise is present, and the beam itself can introduce noise due to Poisson fluctuations of the X-ray illumination.

The geometric response of the detector often also needs correction to map detector elements correctly onto two-dimensional space. A regularly spaced mask is used and the experimental signal from the mask corrected to reproduce the physical mask spacing. This is effective for smooth variations, and more complicated procedures exist for other distortions. The result, however, is a correction table similar to that used for detector response.

The physical construction of the detector may mean that there are regions where there is no detector response (e.g., in-between modules that may make up the detector, or individual pixels where there is an imperfection or electronic problem). A software mask is usually constructed that defines if a region or individual pixels are used in integrating the resulting signal. The same mask can also be used to remove parasitic scatter effects from the beamstop.

Different detector types may need different degrees of correction. Typically, an instrument is set up with these corrections applied seamlessly to the data that is output, and the user need not worry about them. Many of the considerations have impact in the choice of detector for a new instrument.

8.2 Integration

In this section we follow closely the clear presentation of Nielsen et al. (2009). Small angle scattering for biological macromolecules in solution is usually isotropic, and the resultant images are integrated circularly (sometimes termed radial averaging) is done in equal steps of $\tan(2\theta)$. The binned θ_i on a flat two-dimensional detector can be determined as

$$\theta_i = \frac{1}{2}\tan^{-1}\left(\frac{iD_{\text{bin}}}{D_{\text{SD}}}\right), \tag{8.1}$$

where D_{bin} is the binned pixel size of the detector, D_{SD} is the sample-to-detector distance, and θ_i is the half scattering angle for bin i. The corresponding scattering vector q_i is obtained from the relationship $q_i = 4\pi\sin(\theta_i)/\lambda$. For isotropic samples, the scattering intensity for a given q_i bin is calculated by averaging the pixel values in a circular band of radius tan(2θ) around the beam center. If A is a two-dimensional matrix describing the recorded image, $A_n(x, y)$ is the intensity value of pixel n at position (x, y), and the beam center is located at $(x = 0, y = 0)$, then the circularly averaged intensity over a band of d pixels wide, centered at a distance q_i from the origin is given by:

$$I(q_i) = \frac{\sum_{N=1}^{N} A_n(x, y)}{N}, \left\{ x, y : q_i - \frac{d}{2} \leq (x^2 + y^2)^{\frac{1}{2}} \leq q_i + \frac{d}{2} \right\}, \tag{8.2}$$

where n runs over the total number of pixels N located at positions (x, y), with the angular band width radius q_i and bin size d (Nielsen et al., 2009). The intensity is corrected for solid angle effects by normalizing by a factor of $\cos^3(2\theta)$.

Note that this assumes a detector that is perpendicular to the beam with a well-defined center. Corrections can be made if the detector is slightly skewed and if the beam center is not at the center of the image. It is always wise to measure the beam center accurately. For a detector that is not perpendicular to the beam, either slightly misaligned or deliberately offset, the mathematics is more complicated but well defined (Boesecke, 2007).

8.3 Limitations on Minimum *q* Resolution

When $I(q)$ is recorded, the X-ray or neutron scattering with a range around q also contributes to the scattering. The minimum q resolution refers to that range and is limited by instrumental geometric and spectral parameters. For a pinhole collimated system (e.g., synchrotron-type beamlines), the q-independent component of the q resolution, W_0, can be approximated by

$$W_0 \approx \frac{2\pi}{\lambda} \left[\alpha^2 + \left(\frac{P}{L} \right)^2 \right]^{1/2}, \tag{8.3}$$

where λ is the wavelength, α is the beam divergence, P is the detector point spread function, and L is the sample-to-detector distance (Pedersen et al., 1990; Paris et al., 2007). W_o thus represents the smearing of q.

Some experiments may use a wavelength bandpass; for example, when a gradient index mirror is in place or for time-resolved studies where a bandpass increases the available flux, there is also an energy spread that reduces the resolution. This is given by

$$W_E \approx \frac{2\pi}{\lambda} \left[W_0^2 + \left(q \frac{\Delta\lambda}{\lambda} \right)^2 \right]^{1/2} \tag{8.4}$$

where the wavelength dispersion term is q dependent (Pedersen et al., 1990). A detailed analytical treatment is beyond the scope of this book and Penderson et al. (1990) provide details for X-ray and neutron cases. Non-pinhole systems are discussed in section 8.4.

8.4 Desmearing

Some laboratory instruments make use of a slit rather than pinhole geometry. The observed intensity that results is similar to that from a pinhole system but averaged over an angular range defined by the slit geometry. Mathematical treatments exists to correct for this smearing (e.g., Lake, 1967) but these are typically part of the instrument data analysis software.

8.5 Signal-to-Noise Considerations

Using a simple model of noise from only two sources, beam (Poisson fluctuations) and detector electronic noise, the signal-to-noise ratio *SNR* in a given q increment can be written as

$$SNR = f_\varphi N / (f_\varphi N + P\sigma_{\text{pix}}^2)^{1/2}, \tag{8.5}$$

where f_φ is the fraction of the azimuth covered, N is the number of photons scattered into the angular increment, P is the number of pixels capturing the angular increment, and σ_{pix} is the root-mean-square read noise of the detector pixel (Koerner et al., 2011). As Koerner et al. note, the read noise in a q increment grows with the sample-to-detector distance as the number of pixels that are recorded to cover the increment increases with the square of the sample-to-detector distance. However, the number of photons scattered into the same azimuth remains constant. *SNR* is reduced by an increase in sample-to-detector distance, which can drive the choice of experimental setup. The experiment should be configured to the largest sample-to-detector distance required but no longer than that. It should be noted that developments in pixel area detectors with low read noise are minimizing σ_{pix}.

For time-resolved data, the signal-to-noise can be low, depending on the temporal interval sampled. The signal-to-noise for both regular and time-resolved data can be estimated from

$$SNR = \sum_i fit_i / \sum_i |fit_i - data_i|, \tag{8.6}$$

where *fit* is the fit to each experimental data point i, and the data points are labeled *data* (Lamb et al., 2008). This quantifies the agreement between the computed fit to the scattering curve and the recorded data over the number of data points, i. Noise in the data manifests as minor features in the fitted scattering profile. Lamb et al. (2008) also define a mean discrepancy, *MD*, such that

$$MD = \left(\frac{1}{N}\right)\sum_{i=1}^{N}|f_i - g_i|, \tag{8.7}$$

where f and g are two curves, the fit and calculated curve from the reconstructed model, and N is the number of data points. These equations can be used to evaluate whether noise is impacting the accurate reconstruction from the data. Calculation of the SNR at different q ranges can be used to evaluate regiemes where different noise sources contribute (e.g., detector read noise at high q, or Poisson fluctuations at lower q).

8.6 Zero Extrapolation

Zero extrapolation is mentioned earlier. Given data from multiple concentrations, the scattering can be extrapolated to a zero concentration to eliminate artifacts caused by interparticle attraction or repulsion. This can be done by solving a set of linear equations assuming the experimental data to be a linear combination:

$$I_{\text{exp}}^{(k)} = v_k I_{\text{sample}}(q) + (1 - v_k) I_{\text{buffer}}(q), \tag{8.8}$$

where v_k is the volume fraction of the sample in the buffer, I_{exp} is the experimental intensity, and I_{sample} and I_{buffer} are the intensities associated with the sample and buffer, respectively (Konarev et al., 2003). There are separate sets of equations for each value of q and each dilution, k, at that q. Solution of the system provides $I_{\text{sample}}(q)$ and $I_{\text{buffer}}(q)$; the latter is useful to subtract the background for measurements made at higher angles and sample concentrations. For proteins, the estimate of the volume fraction of the sample is $7.3 \times 10^{-4}\ c_k$, where c_k is the concentration in milligrams per milliliter, and $0.73\ \text{cm}^3\ \text{g}^{-1}$ is the average partial specific volume of the protein (Konarev et al., 2003). Another approach is to take the scattering profiles after background correction, scale them to coincide at higher angles, and assuming linear concentration dependence, merge the lowest concentration data at low angles with higher concentrations at higher angles. The influence of interparticle attraction and repulsion is minimized while the signal-to-noise at higher angles is maximized.

8.7 Summary

After the raw data has been collected, there are a number of steps that have to be carried out before the scattering profile can be used for further calculations. Detector artifacts and limitations have to be corrected for, and the signal recorded has to be integrated to produce a two-dimensional scattering profile. Instrument characteristics have to be considered and if necessary, desmearing needs to take place. Many of these steps are accomplished by the specific data processing software associated with an instrument and are invisible to the user, but it is useful to be aware of them.

9

SAXS Instrumentation

In biological solution scattering, the data to be recorded occurs at angles that are orders of magnitude smaller than those accessible by common single crystal devices. For a theoretically perfect SAXS instrument, the incident beam should be monochromatic, parallel, and as strong as possible. Unfortunately, we do not live in a perfect world. X-ray sources and optics produce a range of wavelengths ($\Delta\lambda$) and angle ($\Delta\theta$); consequently any measurement records the average (smeared) intensity over a range of momentum transfer (Δq) about the central q. Beam conditioning and data processing is required to minimize these effects. The former drives the design of SAXS instrumentation while the latter helps in optimizing the need for intensity. In this chapter, we focus on the development of laboratory systems, synchrotron beamlines, and the potential of free electron laser sources for SAXS X-ray data collection. We briefly cover neutrons separately in Chapter 11. Synchrotron sources come close to achieving ideal requirements for SAXS experiments. Developments in laboratory systems make them realistic for data collection to complement the synchrotron. The free electron laser offers exciting potential but at the time of writing is not a routine instrument for SAXS.

9.1 Laboratory SAXS

9.1.1 Laboratory X-Ray Sources

There are three types of laboratory X-ray source: the simple electron filament and target (anode) sealed tube source, a source where the anode rotates, and a source where the target is a liquid metal. All produce X-rays by photoelectric absorption with the wavelength of the X-ray produced characteristic to the material of the anode.

For stationary and rotating targets, copper is the typical material producing X-rays at 1.54 Å or an energy of 8.04 keV using the relationship, $E = hc/\lambda$, where E is the energy, h is Planck's constant, c is the speed of light, and λ is the wavelength. (Note that the energy terminology is often used at synchrotron beamlines, and wavelength is used on the home source, for historical reasons). The X-ray intensity is limited by material properties (i.e., eventual melting of the target), with the rotating anode source (see section 9.1.2, Figure 9.3) used to present fresh material to the electron beam while the targeted material cools.

Biological Small Angle Scattering: Theory and Practice. Eaton E. Lattman, Thomas D. Grant, and Edward H. Snell.

DOI: 10.1093/oso/9780199670871.001.0001

Liquid jet X-ray systems use metal alloys as targets with a gallium alloy generating X-rays at 1.35 Å and an indium alloy at 0.51 Å and can produce a flux several times that of a rotating anode source.

9.1.2 Laboratory SAXS Instruments

Laboratory X-ray sources have a spectrum of X-ray emission with a peak at the characteristic X-ray wavelength utilized, peaks associated with other characteristic emission, and a continuous spectrum of X-rays from *Bremsstrahlung* or breaking effects. In the case of copper sources, a beryllium window can be used to filter shorter wavelength radiation and/or reflecting optics used to minimize spectral and geometric divergence.

X-rays scatter from all materials and produce scattered radiation. The first X-ray SAXS experiments were carried out in the laboratory with pinhole-collimated instruments that used guard slits to minimize this parasitic scatter. The early instrumentation is well reviewed elsewhere (Glatter & Kratky, 1982). Current instruments make use of pinhole collimation and the later developed Kratky camera (Glatter & Kratky, 1982). We describe both the pinhole and Kratky camera below and provide examples demonstrating that useful information can be obtained from laboratory systems.

Pinhole systems

Pinhole SAXS systems are simply described and make use of circular pinholes or variable-sized slits (common on synchrotron beamlines). The most common type is a three-pinhole system where the first two pinholes condition the beam, and the third pinhole removes parasitic scatter from the second pinhole. This is illustrated in Figure 9.1. Note that the pinholes are in the same position as the adjustable slits if used.

The scattering angle θ_s between the beam center of the sample and the point of detection (where parasitic scatter has been eliminated) is given by the following equation (Chu et al., 2015):

$$\theta_s = \frac{d_2 + d_3}{2l_2} + \frac{d_3}{2l_3} + \frac{d_s}{2l_3}. \tag{9.1}$$

Laboratory systems with pinhole collimation have been used successfully for many biological studies. Dzananovic et al. (2014) studied the solution conformation of adenovirus associated with RNA-I and its interaction with protein kinase. While no high-resolution structural information was available, the SAXS studies provided a structural framework to understand host immune responses. In another, Bessman et al. (2014) studied ligand binding and dimerization in the epidermal growth factor receptor. SAXS data showed that the receptor dimer became more compact on binding the growth factor and supported a model of ligand binding as a step in dimerization. In both cases, laboratory data provided complementary information to understand biological function.

Pinhole systems define the beam and reduce parasitic scattering but do not eliminate it. Data is lost due to the need for a large beamstop and a reduction in

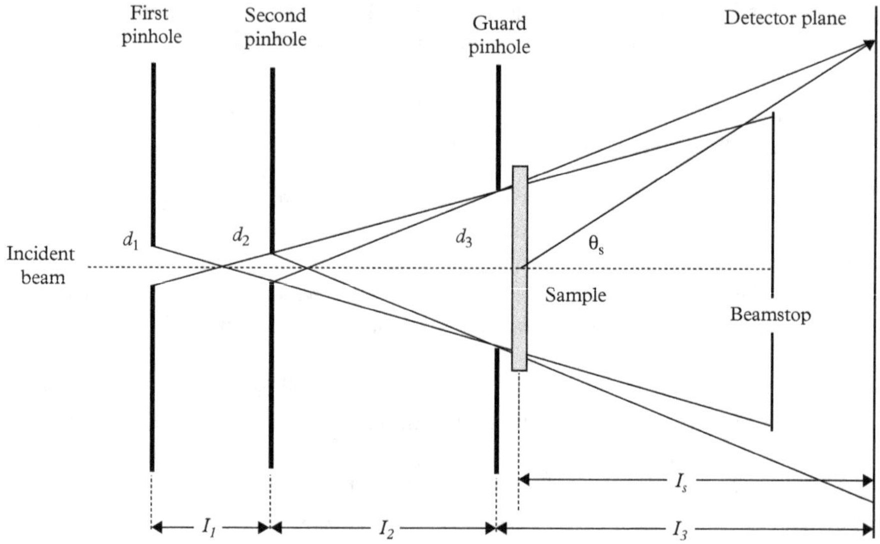

Figure 9.1 Schematic of a three-pinhole SAXS system

signal-to-noise. The pinhole itself has received attention with the development of scatterless slits (Li et al., 2008), potentially improving the q range accessible and reducing noise in the experiment.

The footprint of a pinhole SAXS system can be large, given the distance from sample to detector, to achieve a reasonable q range. Wang et al. (2014) have developed a vertical system, primarily to probe properties that are otherwise affected by gravity, but also enabling the horizontal footprint to be swapped for vertical constraints. Commercial instruments that exploit this for protein SAXS have been produced.

The Kratky camera

The Kratky camera increases the X-ray flux on the sample while keeping the minimal angle low. It achieves this by using a line source with an entrance slit and two blocks: a middle block and a bridge block (see Figure 9.2). The blocks are situated such that they are at the exact coincidence in the horizontal plane and collimate the beam. The operation is illustrated by the cases where the bridge block is positioned too low (Figure 9.2(b)) or too high (Figure 9.2(c)). In the former, there is a minor shift of the beam, and parasitic scattering from the edge of the bridge nearest the incident beam. In the latter, there is parasitic scattering from the far edge of the bridge. Either effect can only be avoided with perfect coincidence of the blocks.

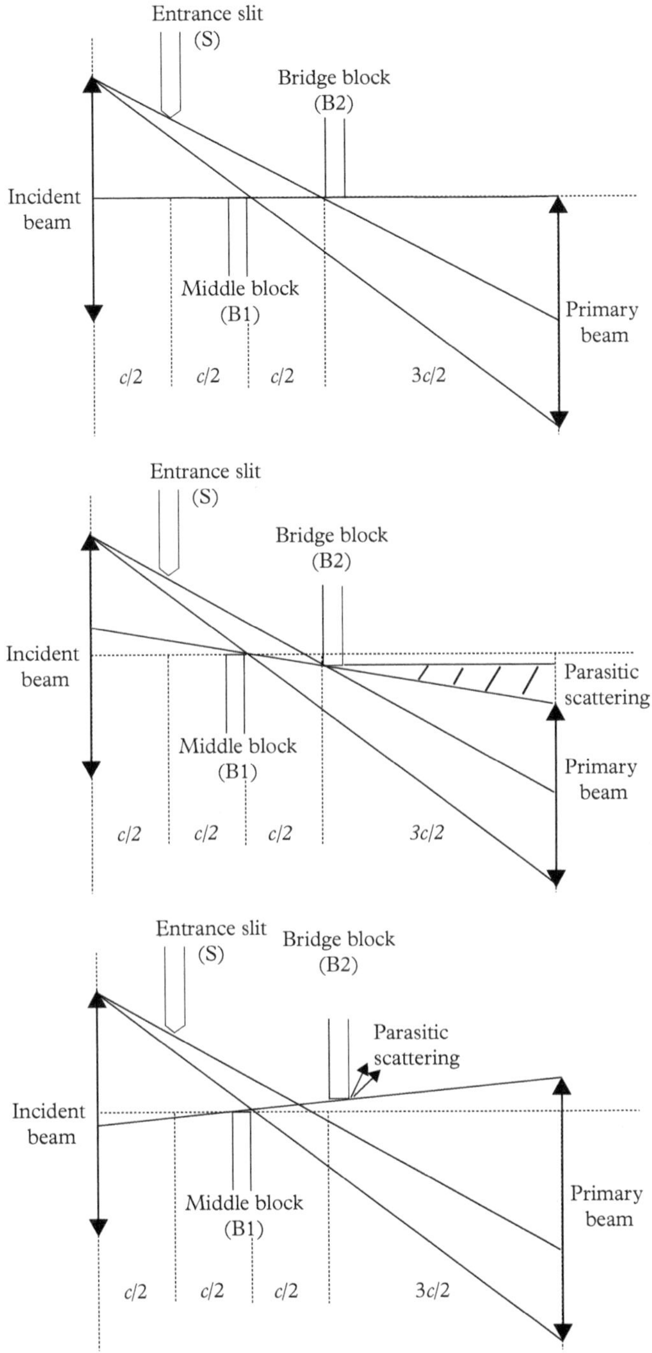

Figure 9.2 Schematic of a Kratky camera showing the correct arrangement: (a) the two blocks used to collimate the beam at exact coincidence in the horizontal axis, (b) the bridge block too low, and (c) the bridge block too high. The vertical dimension is greatly exaggerated. (Kratky and Stabinger (1984) Colloid and Polymer Sciences copyright (1984) with permission of Springer.)

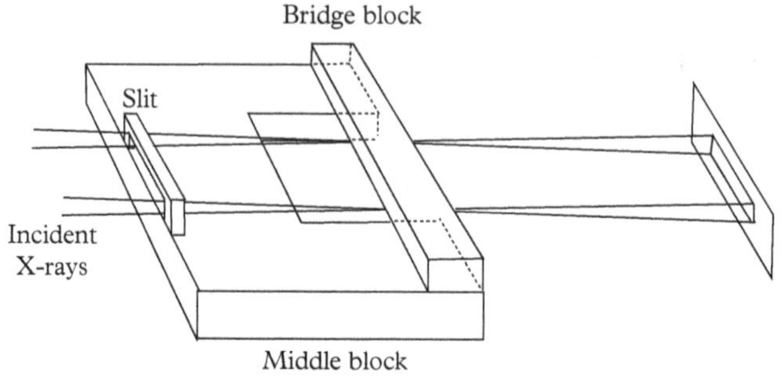

Figure 9.3 Schematic view of the Kratky collimation system.
(Kratky and Stabinger (1984) Colloid and Polymer Sciences copyright (1984) with permission of Springer.)

In practice, only the off-focus component of the middle block is important for the collimation of the beam, so this block can be different from the bridge block, as illustrated in a schematic of a Kratky collimation system (Figure 9.3).

In the traditional Kratky camera, the resulting beam has a line focus and desmearing approaches have to be used or slits incorporated which reduce the overall flux on the sample. Bergmann et al. combined parabolically bent graded multilayer optics with a Kratky camera system and eliminated the smearing (Bergmann et al., 2000), while at the same time obtaining a 40-fold increase in intensity coupled with a rotating anode source without loss of resolution.

Laboratory systems based on the Kratky camera have been successful in providing useful structural results. On a rotating anode source, Edwards et al. used a Kratky based system with a rotating anode generator, gradient index mirror–focused X-rays, and a noiseless detector to help study DNA binding to the nicking enzyme in *Staphylococcus aureus* (Edwards et al., 2013). Three concentrations of sample (10.6, 5.3, and 2.7 mg/mL) were used with six 10 min contiguous exposures for each sample to build up a 60 min scan while ensuring that any radiation damage effects could be identified. The SAXS data confirmed the biological relevance of the crystal structure and supported in vitro and cell-based studies on the importance of the C-terminal domain in the mechanism associated with the N-terminal domain. The system used is shown in Figure 9.4. In another example, heme binding to a disordered region of the transcriptional receptor Bach2 was shown to cause conformational alterations (Watanabe-Matsui et al., 2015). This was a SAXS study with only a single protein concentration (4.7 mg/mL) but looking at samples with and without a heme ligand. Exposure times of 15 min per frame were used with a total exposure time of 150 and 180 min, with and without heme, respectively. These studies and many others demonstrate that laboratory systems can provide data that is structurally relevant to study biological processes.

On a sealed tube source, similar useful data can be collected. Caldwell and Berghuis used a laboratory SAXS system (Caldwell & Berghuis, 2012) to study an antibiotic resistance enzyme. Data was collected in 30 min exposures for the 50 mg/mL and

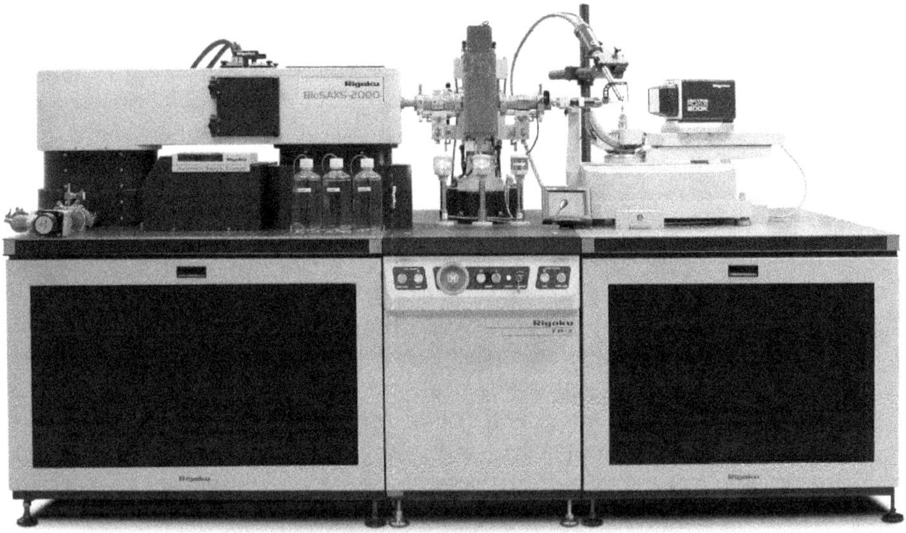

Figure 9.4 A modern Kratky-geometry rotating anode–based SAXS system. In this case, the central module is the rotating anode X-ray generator, with the SAXS system to the left. The Kratky geometry enables the sample-to-detector distance to be relatively short yet still achieve a good q range. Automation is also available in the laboratory, and the system shown has an automated sample changer. Rotating anode sources typically have two ports; in this case, a single crystal X-ray system is shown on the right-hand side. This is a very flexible setup for the home laboratory.

(Figure courtesy of Dr. Angela Criswell, Rigaku Oxford Diffraction.)

25 mg/mL protein samples, 2.5 h for the 12.5 mg/mL sample, 8 h for the 6.3 mg/mL sample, and 10 h for the 3.1 mg/mL sample; 30 min exposures were collected for all 10 mg/mL-substrate dialyzed samples. The results showed a very rigid structure with minimal structural changes on ligand binding.

9.2 Synchrotron SAXS

9.2.1 Synchrotron X-Ray Sources

Synchrotrons are machines where charged particles are accelerated and then stored in a ring. The particles are typically electrons, but positrons are also used. In the ring, the particles lose energy as they change direction, and electromagnetic radiation is given off; some of this in the form of X-rays. There are several components within the synchrotron, and describing them is useful for the understanding of beamline identities. The particles are kept in the ring by bending magnets. As the name suggests, a change in direction is caused by these bending magnets. Between the bending magnets, other devices are inserted. These consist of wigglers, and more recently, undulators—both are termed "insertion devices." A wiggler is an array of a few magnets arranged such that the magnetic field experienced by the particles alternates. An undulator is a similar device, except that many more magnets are present. Both cause changes in direction and act to amplify the X-ray radiation produced at the expense of loss of energy in the particles. An insertion device beamline is typically more brilliant than a bending magnet beamline. Brilliance in the case of a synchrotron is defined as the number of X-ray photons per second per square milliradian per square millimeter per 0.1% relative bandwidth. This also illustrates another property of synchrotron radiation, that is, the bandwidth component coming from the continuous spectrum of the radiation allowing tunability.

Synchrotrons are configured so that at each bending magnet or insertion device, one or more beamlines can be positioned. This gives rise to numbering conventions which may identify the identity of the device type (e.g., B or I, for a bending magnet or an insertion device, respectively), the relative position around the ring with a number or letter, and then, if that beamline is one of several that can come from one device, an appropriate extra identifier.

9.2.2 Synchrotron Beamlines

A typical beamline is illustrated schematically in Figure 9.5. All beamlines have similar components: an aperture to the synchrotron source; optics that condition the beam; a monochromator and, typically, a mirror; slits to define the beam and remove parasitic scatter; shutters; a sample area; an evacuated tube for the scattered X-rays; a beamstop; and a detector.

There are a number of considerations for beamline design, the detailed discussion of which are beyond the scope of this book. Roessle et al. (2007) provide a nice summary of the upgrade path of beamline X33 at the European Molecular Biology Laboratory in Hamburg. Beamline X33 is configured slightly differently from 4-2, the beamline shown in Figure 9.5. The first element that the synchrotron beam hits is a monochromator rather than a mirror. The monochromator will undergo temperature changes due to beam variation and has to be actively temperature controlled and have positional feedback to ensure beam stability. The mirror, which can also be temperature controlled, is used to improve the flux on the sample and, in the case of wiggler and, more importantly, undulator sources, reduce harmonics. Multiple slits are positioned in the beamline. The first slits (or the mirror aperture, as in the beamline shown in Figure 9.5) cut the beam in both the horizontal and the vertical direction. They may be cooled, given the power incident upon them. The slits also control the acceptance angle of both the mirror and the monochromator, provide a beam-forming capability, and cut parasitic scattering. Multiple shutter systems are common as a safety feature. On many beamlines, the sample-to-detector distance can be moved to accommodate differently sized samples. Some beamlines also have a second offset detector for the collection of WAXS data.

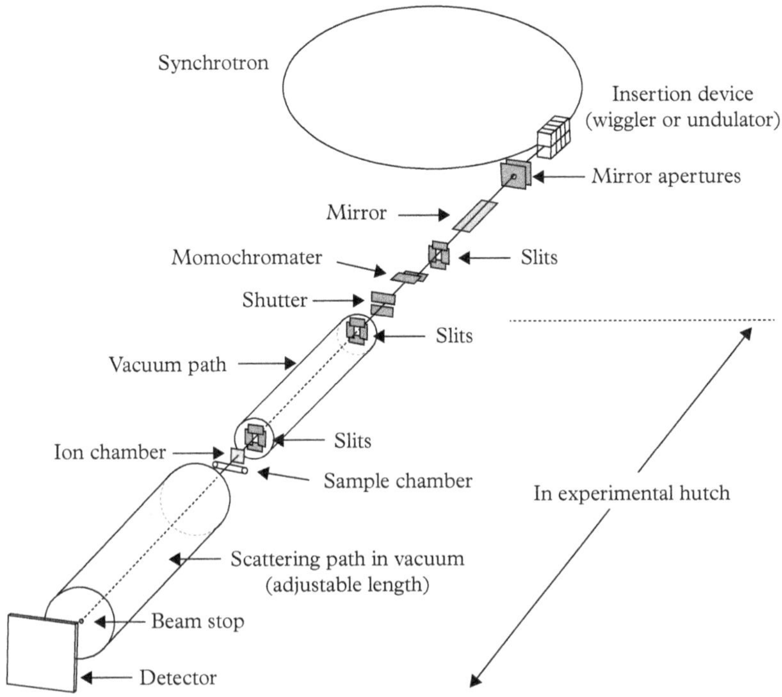

Figure 9.5 Schematic of a typical synchrotron SAXS beamline based on the SSRL beamline 4-2. Several components are shown. Depending on the beamline, some optical elements will be different (e.g., mirrors replaced with multilayers), there are more shutters associated with safety, and there are beam-positioning sensors and feedback devices. In this example, both the sample environment and the detector are shown outside of the vacuum path. Either or both may be in the vacuum path in some facilities.

The sample chamber has multiple embodiments, from a simple, manually filled capillary to sophisticated automated systems that allow high-throughput studies. Some beamlines incorporate a sample chamber positioned in the vacuum path. This can considerably increase the signal-to-noise, with reports of a two- to over tenfold increase, depending on the q range studied (Dubuisson et al., 1997).

From a practical (and potentially historical) perspective, it is useful to identify facilities for SAXS experiments that are currently available. Table 9.1 lists recent and current biological SAXS beamlines together with the X-ray source and references. It is adapted from Criswell and Mooers (2015) and while it will become dated over time, it provides a useful snapshot of current capabilities worldwide, with references on how those capabilities have been developed. Below the table, we have attempted to summarize the beamlines that have been detailed. Where possible, beamline scientists have been contacted to ensure the information is current at the time of writing. Our practical experience is on a limited number of lines, and the state of the beamline may be more advanced than the information we

Table 9.1 *List of biological SAXS beamlines, updated from Criswell and Mooers (2015)*

Source	Beamline/s	Reference/s
Advanced Light Source (ALS) (Berkeley, USA)	SIBYLS 12.3.1	(Hura et al., 2009; Classen et al., 2013)
ALBA (Barcelona, Spain)	BL11–NCD	
Advanced Photon Source (APS) (Argonne, USA)	12-ID-B, 12-ID-C, 12-BMB, 14-ID-B, 18-ID	(Seifert et al., 2000; Beno et al., 2001; Forster et al., 2008; Graber et al., 2011)
Australian Synchrotron (Melbourne, Australia)	SAXS/WAXS SR13ID01	(Kirby et al., 2013)
Brazilian Synchrotron Light Laboratory (LNLS) (Campinas, Brazil)	SAXS 1, SAXS 2	(Kellermann et al., 1997)
Beijing Synchrotron Radiation Facility (BSRF) (Beijing, China)	1W2A	(Dong et al., 1997; Li et al., 2014)
Cornell High Energy Synchrotron Source (CHESS) (Ithaca, USA)	F2, G1	(Englich et al., 2011; Dyer et al., 2014; Acerbo et al., 2015)
DORIS III (Hamburg, Germany)[1]	EMBL X-33	(Roessle et al., 2007; Blanchet et al., 2012)
Dortmund Electron Storage Facility (DELTA) (Dortmund, Germany)	BL9	(Krywka et al., 2007)
Diamond Light Source (Oxfordshire, UK)	B21, I22	(Inoue et al., 2013)
ELLETRA (Trieste, Italy)	5.2L	(Amenitsch et al., 1997; Zanini et al., 1999)
European Synchrotron Radiation Facility (ESRF) (Grenoble, France)	ID02, ID09, BM26B, BM29	(Abernathy et al., 1998; Borsboom et al., 1998; Narayanan et al., 2001; Pernot et al., 2010; Pernot et al., 2013)
MAX-IV (Lund, Sweden)	I911-4	(Labrador et al., 2013)
MAX-lab	I711	(Knaapila et al., 2009)
National Synchrotron Light Source (NSLS) (Brookhaven, USA)[2]	X9	(Yang, 2005; Allaire & Yang, 2011)
National Synchrotron Light Source II (Brookhaven, USA)	LIX	(Fuchs et al., 2014)
Photon Factory (Tsukuba, Japan)	BL-10C	(Igarashi et al., 2011)
PETRA-III (Hamburg, Germany)	P12	(Roessle et al., 2007; Round et al., 2008; Blanchet et al., 2012)
Shanghai Synchrotron Radiation Facility (SSRF) (Shanghai, China)	BL16B1	(Tian et al., 2015)

SOLEIL (Paris, France)	SWING	(David & Perez, 2009)
SPring-8 (Harima, Japan)	BL40B2, BL40XU, BL45XU	(Yamamoto et al., 1998; Fujisawa et al., 2000; Inoue et al., 2001)
Stanford Synchrotron Radiation Light Source (SSRL) (Menlo Park, USA)	4-2	(Smolsky et al., 2007; Martel et al., 2012)
Swiss Light Source (SLS)	cSAXS-X12SA	

[1]The X-33 beamline has been transferred to PETRA-III and is described in the text with the PETRA-III facility.
[2]NSLS has now closed with the commissioning of NSLS-II. Both the DORIS III and the NSLS entries are included, as they describe recent history.

present here. We do not claim to have captured all the available facilities, and any errors are ours alone.

The Advanced Light Source (ALS) has one beamline associated with biological SAXS. The SIBYLS (Structurally Integrated Biology for Life Sciences) 12.3.1 beamline is optimized for both SAXS and macromolecular crystallography (Hura et al., 2009; Classen et al., 2013; Dyer et al., 2014). Unlike many of the other beamlines at the ALS, it uses a walk-in hutch and has a usable wavelength range of 0.73–2.5 Å with the sample-to-detector distance of 1.5 m, allowing q ranges from 0.007 to a maximum of 0.6 $Å^{-1}$. Samples are presented to the instrument in 96-well plates, with ~15 μl required. A Hamilton liquid-handling robot is used to load the temperature-controlled sample cell. The cell itself is constructed with potassium aluminosilicate windows that are 25 μm thick, and it is kept in a helium-purged acrylic box, which, among other things, allows measurements at temperatures below the local dew point. In addition to regular SAXS studies, the SIBYLS beamline is also well suited to SEC–SAXS applications, with a mail-in program. SIBYLS uses an Agilent high-pressure liquid chromatography (HPLC) system designed specifically for SEC–SAXS purpose. The HPLC with an autosampler designed for small volumes allows the beamline to load multiwell plates with unsupervised sample injection. For a relatively homogeneous sample, an injection of 50 μl at a concentration of ~5 mg/mL is required to obtain good scattering data. A post-column splitter creates parallel flow paths after the SEC column and allows the collection of UV and MALS signals simultaneously to the SAXS signal.

The ALBA synchrotron has one SAXS line, BL11. This is predominately used for material science but is capable of studying biological specimens.

The Advanced Photon Source (APS) has a number of SAXS beamlines. Sector 12 has three beamlines that can accomplish biological SAXS: a bending magnet beamline (12-BMB) and two insertion device beamlines (12-IDB and 12-IDC) (Beno et al., 2001). Note, for those unfamiliar with synchrotrons, bending magnets bend the electron beam, and radiation is given off at a tangent to the bend, while insertion devices act to condition the beam, typically amplifying the brilliance on the sample. All of the beamlines have been used to study protein and RNA samples. A noteworthy feature of beamline 12-IDB is the ability to change the resolution range rapidly with the in-vacuum camera system and automated position control (Schmidt et al., 2014). The beamline allows simultaneous SAXS and WAXS measurements. Beamline 14-ID-B in the BioCARS sector of the APS has been used to support time-resolved SAXS and WAXS (Graber et al., 2011). Beamline 18-ID has a sample-to-detector distance of 3.5 m and specializes in time-resolved SAXS and WAXS. Exposures take place on flowing samples with a minimum sample volume of 120 μl recommended, but with data collection possible on a fraction of that. Using an SEC system before the sample capillary is the default experimental setup (Mathew et al., 2004), requiring a little more than the 120 μl recommended.

The Australian Synchrotron has a single SAXS beamline with a maximum sample-to-detector distance of 7 m (Kirby et al., 2013). A wavelength range of ~0.6–2.5 Å is available, with the beamline optics optimized for 1.13–1.52 Å. By using multiple distances, a q range of 0.0015–1.1000 Å^{-1} is achievable for SAXS, with a WAXS detector achieving a q range of 0.5–10.0 Å^{-1}. Multiple detectors are available. Both static and flow measurements in capillaries are possible. The beamline supports multiple types of samples, including biological macromolecules, and is equipped with extensive sample diagnostic equipment.

LNLS has one SAXS beamline, SAXS1 (Kellermann et al., 1997), and has been productive with biological specimens. A new source, Sirius, is currently being built to replace the LNLS. SAXS1 operates at a fixed wavelength of 1.55 Å with a q range of 0.004–0.540 Å^{-1}, with a silicon pixel detector and an automatic sample-loading system suitable for 96-well plates. Minimal sample volume in capillary holders is 80 μl. A stopped-flow system is available for time-resolved studies.

The SAXS station at the Beijing Synchrotron Radiation Facility, beamline 4B9A, used slits to condition the beam and had a sample-to-detector distance of 0.3 –2.5 m in steps of 0.15 m (Dong et al., 1997). It was upgraded to beamline 1W2A, which uses three sets of slits to condition the beam. At the typical sample-to-detector distance of 1.5 m, it offers a q range of 0.0054–0.4300 Å^{-1} (Li et al., 2014). A sample-to-detector distance of 0.5–5.0 m is available, together with a number of different detectors. The temperature-controlled sample cell has a thickness of 1 mm with a volume of ~70 μl and is formed with two parallel windows of mica (Xing et al., 2013). The SAXS beamline is adjacent to a macromolecular crystallography beamline, 1W2B.

MacCHESS, the Macromolecular Diffraction Facility of the Cornell High Energy Synchrotron Source (CHESS), houses SAXS beamline G1 which uses a 49-pole wiggler source (Acerbo et al., 2015). SAXS and WAXS can be collected simultaneously with a combined q range of 0.007–0.700 Å^{-1}. A wavelength range of ~0.95–1.55 Å is available with a beamsize of 250 × 250 μm in automated mode (with a sample robot) or 250 × 500 μm when using an SEC setup.

Beamline BL9 at the Dortmund Electron Storage Facility is a multipurpose beamline that can be used for grazing-incidence geometry diffraction, reflectivity, and powder measurements in addition to SAXS and WAXS. The measurable

q values range from 0.007 to 5.500 Å^{-1}. Biological samples have been studied with success.

The Diamond Light Source has two dedicated SAXS beamlines: B21, which is a bending magnet source, and I22, which is an insertion device source. B21 is a liquid-state SAXS instrument and operates in a fixed-camera-length configuration with an accessible q range of 0.0031– 0.4200 Å^{-1} at 12.4 keV. Samples can be introduced to the beam using a high-throughput sample-handling robot, SEC, or precision syringe pumps. The sample-handling robot can process one 96-well plate and up to eight PCR strips in a single job. Each sample is 25 µl and held at a controlled, specified temperature. The SEC instrument utilizes a dual pumping HPLC that enables two columns to be under buffer flow simultaneously, to reduce column-switching times. A major upgrade is planned that will place the detector directly in the vacuum. Beamline I22 is aimed toward probing extreme sample environment and time-resolved studies. It has a variable sample-to-detector range of 1.9–9.9 m.

The 5.2L beamline at ELLETRA is a simultaneous SAXS and WAXS instrument (Amenitsch et al., 1995, 1997, 1998). Multiple detectors are available, providing an accessible q range of 0.007–0.100 Å^{-1}. A simultaneous WAXS capability extends the q range to ~0.8 Å^{-1}.

The European Synchrotron Radiation Facility (ESRF) has a number of SAXS beamlines. ID02 is a combined ultra-small angle and WAXS instrument with a detector vacuum chamber that is 34 m long and 2 m in diameter. This allows a minimum q resolution of 10^{-4} Å^{-1}. Beamline ID09 is a time-resolved beamline where solution studies can be carried out in the microsecond-to-picosecond time regime. Beamline BM26B is a SAXS/WAXS beamline, but its focus is not on biological studies. Beamline BM29 is dedicated to biological SAXS, with a q range of 0.0012–3.800 Å^{-1} and implements a high-throughput automated strategy allowing a data collection and cycle time of about 1 min (Pernot et al., 2013). As in other beamlines, BL29 can be operated with a gel-permeation/SEC system but with sample cycle times on the order of 30 min.

MAX-lab in Sweden has had a number of X-ray synchrotron sources. The latest, MAX-IV, became operational in 2016. The I911 beamline of MAX-IV has an evacuated chamber (10^{-2} mbar) for SAXS experiments (Labrador et al., 2013), with the ability to vary the sample-to-detector distance from a few hundred millimeters to more than 2 m. This allows an accessible q range of 0.006–2.000 Å^{-1}, with a typical setup of 0.01–0.30 Å^{-1}. Two pairs of X-ray slits act as collimator and guard slits, and a scintillator measures I_0. A fast millisecond shutter and a motorized beamstop with an integrated pin diode complete the system. In-vacuum detectors are used with an offset ability to alter the q range. For diagnostic purposes, an X-ray camera can be positioned to look at the beam, and an aperture and a pin diode can measure the beam profile. The beamline is shared for different sample environments, but for biological samples, the sample chamber is an in-vacuum flow-through capillary.

The National Synchrotron Light Source Brookhaven (NSLS), New York, USA, had a SAXS/WAXS-capable instrument on beamline X9, for which a small component of available time was allocated to biological macromolecules (Allaire & Yang, 2011). A vacuum chamber that was 5 m long was available, with the detectors (both SAXS and WAXS) operating within the vacuum. For biological experiments, the SAXS detector was located 3.4 m from the sample, giving a combined q range of 0.005–2.000 Å^{-1} with an overlap between ~0.12 and 0.20 Å^{-1}. The sample was contained within a capillary in the vacuum environment, requiring a volume of 20–100 μl. NSLS has now closed with the commissioning of NSLS-II. The SAXS beamline at NSLS-II, the LIX beamline (Fuchs et al., 2014; DiFabio et al., 2016), has a 2.0–5.0 m sample-to-detector distance for the SAXS camera (in air) with two WAXS detectors in vacuum. A wavelength range of ~0.7–5.9 Å is available or ~0.77–1.54 Å, with improved beam stability with beam sizes from 1 μm to 0.5 mm.

The Photon Factory operates two SAXS beamlines (BL-10C and BL-15A) and partly uses another beamline, BL-9C, for SAXS (Igarashi et al., 2011). BL-10C was designed for the study of biological macromolecules, and BL-15A for the time-resolved study of muscle fibers. BL-15A has a sample-to-detector distance of up to 3.5 m, with a variable-length vacuum path (Igarashi et al., 2013). During a storage ring upgrade, the BL-15A line instrumentation was moved to BL-6A (Shimizu et al., 2013).

The EMBL X33 beamline located on the DORIS III storage ring was one of the first SAXS beamlines devoted to the analysis of biological macromolecules (Blanchet, Spilotros, et al., 2015). With the construction of the PETRA-III source, hardware was transferred to the P12 beamline on PETRA-III. The P12 beamline has multiple sample-to-detector distances up to 3.1 m, using an adjustable detector tub allowing a q range of 0.001–2.800 Å^{-1}. The fluid capillary is in vacuum, and an automated sample changer is available (Round et al., 2015). More than 200 samples (10–40 ul of solution) can be set up and operated automatically. A full cycle of sample loading, data acquisition, cleaning, and drying takes about 50 s, including 20 50 ms exposures (Blanchet, Spilotros, et al., 2015).

Beamline BL16B1 at the Shanghai Synchrotron Radiation Facility is designed for SAXS/WAXS, anomalous SAXS, and grazing-incidence SAXS (Tian et al., 2015; Hua et al., 2017). It is a microfocus beamline with a focal spot of less than 20 × 20 μm.

The SWING beamline at SOLEIL has a sample-to-detector distance of 0.5–8.0 m. SAXS and WAXS can be carried out simultaneously. Both high-throughput and high-performance liquid chromatography operations are available (David & Perez, 2009). In the high-throughput mode, 14 samples per hour can be measured, with a typical sample volume of 6 μl.

RIKEN Spring-8 beamline BL45XU-SAXS, Hyogo, Japan, is also next to a macromolecular crystallography beamline. The SAXS beamline is optimized to 1.0 Å but can be tuned to energies from 6.5 keV to 14 keV (Fujisawa et al., 2000).

Discrete sample-to-detector distances of 0.45, 1.5, 2.0, 2.5, and 3.5 m are available. Four sets of slits condition the beam.

Beamline 4-2 at SSRL has both high-throughput and SEC operational modes. Time-resolved studies are also possible. Multiple distance configurations are available, with a q range of 0.004–1.300 Å^{-1} accessible in the regular configuration, and an order of magnitude smaller with a Bonse–Hart geometry setup (Smolsky et al., 2007). A fully automated high-throughput capability is available (Martel et al., 2012), together with inline chromatography and stopped-flow systems. The examples described in Chapter 6 were collected at this facility.

The coherent SAXS beamline cSAXS-X12SA at the Swiss Light Source has two sample-to-detector configurations: one flight tube that is 2.1 m long, and another that is 7 m long. These allow combined SAXS and WAXS experiments, with a variety of configurations for biological and general experiments.

9.3 Free Electron Laser X-Ray Sources

XFELs provide high brilliance with incredibly short pulse times measured on the order of femtoseconds. In summary, short pulses of electrons are accelerated into the gigavolt energy range. After being accelerated, these bunches of electrons pass through a long array of opposed alternating magnets, similar to the undulators used in synchrotrons. These alternating magnetic fields exert a Lorenz force on the electron bunch, causing it to follow the sinusoidal path. Accelerating electrons radiate, and so each electron in the bunch emits electromagnetic radiation in the X-ray regime. Initially, the waves from each electron have random phase with those from the others so that the radiation emitted is incoherent. As the bunch proceeds through the magnetic array, it interacts with the radiation field that it itself is creating, so that the electrons are pushed into micro-bunches that are separated by one wavelength of the emitted radiation. Because the micro-bunches are very thin and are separated by one wavelength, all the radiation emerging from the bunch is now coherent, and all the peaks add up in phase. This is illustrated in Figure 9.6, which is taken from the work by McNeil and Thompson (2010). The electrons, shown as a dot on the undulating line passing through the center of the magnets, gradually produce bunches of coherent radiation. The net result of this is that the intensity of the emitted radiation goes up as N^2, the square of the number of electrons, rather than as N. The effect of this is summarized in Figure 9.7; the XFEL (in this case, the Linear Coherent Light Source [LCLS] at Stanford) has a peak brilliance that is many orders of magnitude greater than that of modern third-generation synchrotron sources.

The practical characteristics of the beam are, as described above, short and brilliant pulses. This is exemplified by the characteristics of the LCLS, the first hard X-ray XFEL user facility in Stanford, California. The LCLS produces 120 X-ray pulses s^{-1}. The pulses are as short as 10 fs (10^{-14} s), shorter than the

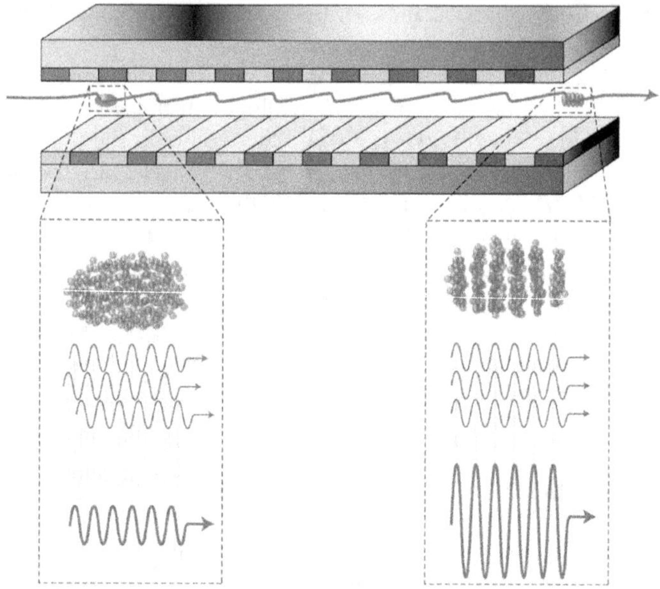

Figure 9.6 When electrons enter the undulator, their initially random phases ensure that mostly incoherent radiation is emitted at the resonant radiation wavelength (left). Because the electrons interact collectively with the radiation they emit, small coherent fluctuations in the radiation field grow and simultaneously begin to bunch the electrons at the resonant wavelength. This collective process continues until the electrons are strongly bunched toward the end of the undulator (right), where the process saturates and the electrons begin to de-bunch. (Reprinted by permission from Macmillan Publishers Ltd: Nature Photonics, (McNeil & Thompson, 2010), copyright (2010).)

timescale of bond vibrations. Each pulse contains about 10^{12} photons, sufficient to produce observable diffraction from a protein crystal smaller than 1 μm^3, or from a comparable amount of material in solution.

9.4 Sample Handling

9.4.1 Capillary Sample Holders

Many SAXS instruments use either a glass capillary to contain the sample or a volume between two flat X-ray transparent plates. In the capillary case, the capillary is typically fixed in position, and the solutions studied are loaded or injected into it. The same capillary is used for both the buffer blank and the sample

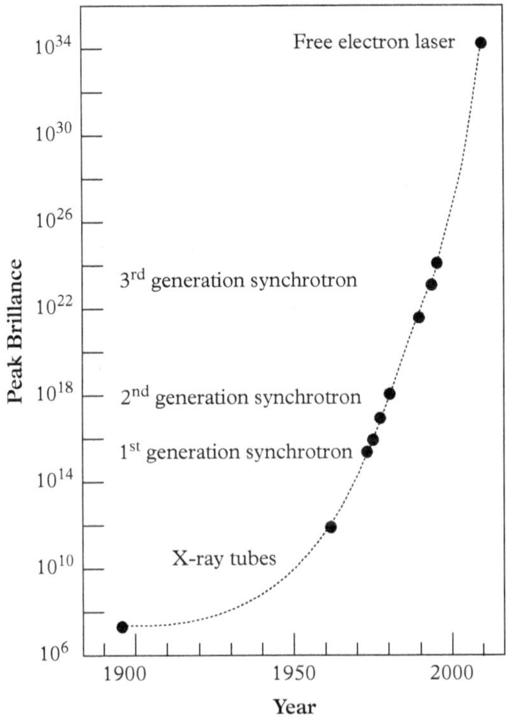

Figure 9.7 Peak brilliance of X-ray sources over the years.
(Used with permission from Chayen et al. (2010).)

images, and therefore should be extensively cleaned between each set of exposures. Flat cell systems vary between fixed and rapidly replaceable. Both fulfill many key requirements for a SAXS experiment. They provide low-background scattering, they make it possible for samples to be loaded and unloaded easily, they can be easily handled (and even be used in vacuum), and they can be cleaned. An optimal path length can be defined by the incident energy and, for Cu, K_α is about 1 mm (Glatter & Kratky, 1982). For higher energies, the path length can be increased (at the expense of sample volume).

On many beamlines, it is possible to visually observe the sample cell. This allows the detection of air bubbles or flow problems in the experiment. Blanchet et al. (2012) describe a system where cell filling is monitored by an inline camera that is rotated 90° out of the beampath during data collection. The camera is integrated with an image-recognition algorithm that checks for bright areas (exploiting the high brightness of the solution–air interface) and automatically flags partially filled cells. A similar algorithm is used on the SIBYLS beamline at the APS (Classen et al., 2010).

9.4.2 Automated Sample Handling

Many SAXS instruments, both laboratory and synchrotron, make use of capillary sample holders. When coupled with injection systems, these can be used in a high-throughput automated process. Examples of this are the capillary setup used at beamline 4-2 of SSRL and the flat window cells used by the SIBYLS beamline at the ALS. The capillary system at SSRL consists of a capillary that is 1.5 mm in diameter and is held in a water-incubated thermal jacket. The sample is in an air gap and there is an opening for sample injection and retrieval, with automated valves connecting to fluid lines for cleaning between data collection. A three-axis - motion stage allows access to a 96-well temperature-controlled sample holder. Samples (10–30 μl) are placed in PCR tubes and sealed with a foil cover. The delivery needle is guided to the sample chamber and pierces the foil to load the sample into the needle. The needle then transitions to the SAXS sample cell and injects the sample. After exposure, the sample can be retrieved (e.g., for radiation footprinting). A liquid delivery and automated valve system allow for a customizable cleaning procedure appropriate to the sample (Martel et al., 2012).

The cells on the SIBYLS beamline are made of aluminum with mica windows that are 25 μm thick. They are arranged in a row of eight and contained within a helium-purged box that prevents condensation below the dew point and allows for anaerobic studies. The cells, similar to the capillary, are cleaned after each use. The SIBYLS beamline uses a standard Hamilton liquid-handling robot to load the cells (Classen et al., 2013). The EMBL beamline X33 at DESY (Hamburg, Germany) first introduced a system capable of handling up to 192 samples in the 80–100 μl volume range. Using a needle, the robotic system extracted sample volumes from microcentrifuge tubes with pierceable lids and delivered them to the flat-walled sample cell via a length of polytetrafluoroethylene tubing (Round et al., 2008). Beamline ID14-3 at ESRF (Grenoble, France), in collaboration with EMBL, has recently commissioned a new sample changer based on this design but which reduces the sample-loading and cleaning cycle down to 30 s (Pernot et al., 2010). The new design minimizes the flow path by moving the sample plates rather than the needle. Beamline P12 at PETRA-III uses a quartz glass capillary that is held in a stainless steel pod and placed within the vacuum chamber. This minimizes air scattering, and the stainless steel fixture is designed to be easily exchangeable in the case of a leak or if the cleaning process fails (Blanchet, Spilotros, et al., 2015). The SWING beamline at SOLEIL uses an HPLC system with its own pipetting robot to load samples into a continuously flowing sample cell (David & Perez, 2009).

9.4.3 Specialized Sample Cells

Specialized sample cells have also been developed. For example, Lipfert et al. (2006) designed a holder and cell allowing rapid interchange for static and flow type experiments. The holder was designed with a cross section larger than the incident beam so that drifts in beam position could be mitigated, and the windows were made from potassium aluminosilicate (mica) sheets that were 25 μm thick, to minimize background scattering and absorption. The sample thickness was 2 mm and, in the static case, required a sample volume of 16 μl with the flow cell holding ~1 mL of solution. The cell was designed to be actively temperature controlled. For static experiments, Hong and Hao used a stainless steel body sandwiched between two mica sheets placed in a water-incubated cell holder, and another cell with a polymer body, again with two mica sheets (Hong & Hao, 2009). An application of SAXS to cryogenic samples (see Chapter 13) involved the development of a microfabricated silicon sample holder (Hopkins et al., 2015). In addition to the regular requirements, the sample holder had to have a small thermal mass and be thermally rigid so that cooling was rapid and that there was no change in exposed sample volume on cooling. This was achieved by etching a chamber from a silicon wafer, with the final design having an X-ray window that was 30 μm thick and opened onto a sample chamber requiring about 800 nl of sample. The design worked well at cryogenic temperatures (100 K) but, due to an open top and small sample volume, was not suitable in the initial embodiment for ambient temperature measurements.

9.4.4 Microfluidics

Microfluidics has been used for many aspects of structural biology, for sample handling, assay techniques, and crystallization. The field offers distinct advantages for SAXS studies in that the X-ray beam can be easily matched to the size of the channels, thus minimizing material not exposed; the material can be flowed, minimizing radiation damage but also minimizing material due to the small volume of the channels used. Finally, microfluidic devices can be used to mix solutions, which provide opportunities for time-resolved studies where reactions are initiated and followed. The subject area is well reviewed by Watkin et al. (2017).

Toft et al. (2008) implemented a microfluidic sample chamber incorporating sample and buffer inlets, a mixing region, and an analysis chamber where the X-rays interacted with the solution. The system allowed for solution flow and dilution during the experimental series. The analysis chamber contained 200 nl of solution, and data was successfully collected in a series of experiments on BSA. While the experiment illustrated the potential of the technique, it required a larger volume of sample to completely fill the channels used, an exposure time of several minutes, and a comparable cleaning time. The total cycle time for a protein and buffer blank measurement was almost 30 min. The system was further developed and, with much shorter exposure times on a more powerful X-ray source and a faster flow rate, the complete experimental cycle time (including cleaning) was reduced by an order of magnitude (Lafleur et al., 2011).

Skou et al. (2014) built on these developments to include a dialysis capability together with UV absorption measurements. This enabled a gradual concentration of the sample so that initial experiments can be made in dilute regimes while the experiment continues to improve signal-to-noise and detect any change in oligomeric state or protein–protein interactions. The same capability allowed buffer exchange studies to be carried out.

An alternative approach has been the development of a compact disk–sized microfluidic device (Blanchet, Spilotros, et al., 2015; Schwemmer et al., 2016). The disk is divided into independent but identical sextants that have three small reservoirs for depositing solutions of sample, mixing buffer, and screening buffer. The disk is rotated to mix the solutions and then transport them to a measurement chamber. Twenty screening conditions are obtained per sextant; these are then used for SAXS collection, moving sequentially from condition to condition.

Other microfluidic applications are described in section 9.7.3, where they are coupled to time-resolved studies, and in Chapter 10, section 10.3.3, which describes how the technology has been used for SAXS measurements that aid crystallization approaches.

9.5 Detectors

Detectors convert the scattered X-ray photons into a readable format so that the scattering can be quantitatively evaluated. There are several considerations that are general for all detectors used for small angle scattering, both X-ray and neutron detection. In no order of preference, the detector should be efficient, that is, the detector should be as sensitive as possible to the scattered radiation. The detector should have a high dynamic range (remember that examples of scattering are given with intensity plotted on a log scale). Information spans many orders of magnitude. The detector should have low noise so that the signal is high while the background is low. The detector should be precise, that is, accurate and stable, so that a concentration series and buffer blanks can be accurately recorded. The detector should have high spatial resolution and a large area. Resolution is not as important as in a detector for single crystal work, due to the nature of the continuous scattering curve and radial integration, but a higher resolution will result in an improved signal where higher frequency information occurs. For time-resolved experiments, an additional factor becomes important in the detector: readout time. Each of these considerations drives cost, and it could be argued that there are no perfect detectors, but, as we will see, there are some that are getting close.

9.5.1 Film, Multiwire, and Image Plates

Historically, X-ray scattering measurements were performed by using X-ray sensitive film; later on, position-sensitive electronic detectors were used. Film has very high spatial resolution and can be made to cover a large area; however, it has poor dynamic range and can have variability, due to the development process. For biological samples in solution, the scattering data is isotropic, and area detectors are used. The first major electronic area detectors were based on thin wires in an absorbing gas atmosphere: X-rays produce electrons on interactions with the gas molecules, which are then accelerated toward the wires, inducing an electrical pulse. With a two-dimensional array of wires, the position of the X-ray pulse can be recorded. While the time resolution is good, the spatial resolution of these detectors is poor, and they are rarely used. Thus, image plates are more commonly used as detectors with laboratory SAXS instruments and SANS beamlines. Image plates contain a photostimulatable phosphor based on barium halide doped with europium (BaFBr: Eu^{2+}) (Dauter & Wilson, 1994). X-rays interact with the material and deposit energy stored in the form of metastable excited Eu states called "color centers." The plate is read by illuminating the material with a laser, which causes fluorescence at a different frequency. This fluorescence is recorded and used to determine the number of photons received in any one pixel. Image plates have a high dynamic range and can have a large area, but their pixel size is also large and the quantum dynamic efficiency is low. They are integrating detectors, and, in their first implementation, had to be removed from the scattering source and read out offline. The properties of image plates for SAXS experiments are discussed by Né et al. (1993). For SANS data collection, a scintillator is used to convert neutrons to electromagnetic radiation; the first example of this was in work by Wilkinson et al. (1993).

9.5.2 CCD Detectors

CCD (for charge-coupled device) detectors use a phosphor to convert X-ray photons into light, with the phosphor usually linked by a fiber-optic taper to the CCD. All the experimental scattering profiles shown in previous chapters have used this detector. With the appropriate phosphor choice, the CCD is sensitive, and it has a high dynamic range. Charged particles can cause artifacts on CCD images (and sometimes images plates). Because of their transient nature, these are easily removed by recording double images and discarding differences—a process termed "dezingering." CCDs have a high count rate, with readouts on the order of a second or less.

9.5.3 Mixed Mode Pixel Area Detectors

Mixed mode pixel area detectors (MMPADs) consist of a silicon detector where X-rays are converted to a charge followed by a CMOS chip used to readout that charge. Koerner et al. (2011) have evaluated SAXS data collection using a MMPAD detector and find that readout noise is significantly reduced compared to a CCD-based detector, with over an order of magnitude improvement in the lowest q range and an overall three-fold reduction in noise. This form of detector has been commercialized as the Pixel Apparatus for the Swiss Light Source, or PILATUS, detector (Broennimann et al., 2006) and the subsequent Eiger detector (Dinapoli et al., 2011).

A common setup is for the detector to sit outside the vacuum path. In order to minimize air absorption and reduce scattering, thus increasing the signal-to-noise, both CCD and MMPAD detectors have been configured in vacuum (Yang, 2013; Wernecke et al., 2014).

9.5.4 The Beamstop

Measuring the direct beam after it has passed through the sample is important to accurately calibrate the collected data. Several approaches can be used to measure the direct beam at the beamstop, including attenuation and use of the primary detector, a scintillator to generate visible light for photodiode measurement, use of the photoelectric effect, and direct detection on a photodiode (Blanchet, Hermes, et al., 2015). The size of the beamstop is important, as it masks out the detector and restricts the lowest scattering angles. Miniature pin diodes have been successfully used for single crystal applications (Ellis et al., 2003), as well as for SAXS beamstop purposes, but are damaged by the radiation available from high-brilliance sources. At the P12 beamline of EMBL Hamburg at PETRA-III, indirect detection of the beam has been implemented. The direct beam is incident on a tungsten alloy beamstop that has a hole drilled down the center. The X-ray beam enters this hole in the beamstop and is partially scattered by an angular end barrier onto a photodiode positioned perpendicular to the incoming beam (Blanchet, Hermes, et al., 2015). This reduces the need for attenuation.

9.6 Data Acquisition and Analysis Software

An important but sometimes overlooked aspect of instrumentation is the control software used to operate the instrument and provide the initial data analysis. In the X-ray crystallographic field, software such as BluICE has been developed for beamline control and data acquisition and extended to perform automated analysis (McPhillips et al., 2002; Classen et al., 2010). This software has been adapted to control SAXS instrumentation on the SSRL beamline 4-2 and at the ALS SIBYLS beamline (Classen et al., 2010). This software acts as a single interface to multiple systems, allowing full control of the data collection and significant automation of steps. For both beamlines, instrumentation setup is restricted to the beamline scientist and is not a user-available feature. In the case of beamline 4-2, the data acquisition software feeds into an automated analysis script that provides basic diagnostic capabilities and an initial analysis of the data.

Inline SEC–SAXS requires a more in-depth analysis than is provided by a simple concentration series. Data acquisition and automation steps have also progressed in this case. Graewert et al. (2015) have developed a data collection and analysis pipeline that significantly enhances both the speed and the success of SEC–SAXS experiments. SEC–SAXS is described further in Chapter 10.

9.7 Instrumentation for Time-Resolved Studies

SAXS can access a wide range of timescales, from days to femtoseconds. Time-resolved studies are possible with the appropriate sample handling, detectors, and data-processing algorithms. Sample handling involves the delivery of the sample to the beam after a reaction has been initiated (e.g., for photo-excited studies) or the mixing of components before delivery to the beam. Photo-excited processes are typically rapid and more the area of developments in XFELs. Instrumentation and time-resolved experiments in this case are described in Chapter 13. The temporal resolution on synchrotron sources is determined by the type of mixer and the measurement time. There are typically two types of mixers used: stopped-flow mixers and continuous-flow mixers.

9.7.1 Stopped-Flow Mixers

Stopped-flow mixers, as the name suggests, mix the solution components and then stop the flow for data collection. The time resolution is limited to a few milliseconds, and the sample required is less than that required for a continuous-flow system. The mixing is turbulent. Stopped-flow mixers have been used extensively for studies on protein folding and apparatus for SAXS is commercially available (Pollack & Doniach, 2009).

9.7.2 Continuous-Flow Mixers

Continuous-flow mixers provide the means to study processes shorter than 10 ms (Pollack & Doniach, 2009). Because of the continuous sample flow, these mixers also provide a means to mitigate radiation damage, because the sample exposed to the X-ray beam is continuously replenished. Continuous-flow mixers bring together the sample with solutions that prompt structural changes such that the combined solution can be presented to the X-ray bean. The initial time point that the instrument samples is dependent on the complete mixing of the solutions, the flow rate, and the distance the fluid has to travel before it interacts with the beam. The time resolution of the instrument, assuming successful mixing, depends upon flow effects (e.g., slower flow at the edge of the capillary and faster in the center), the beam size, and the flow speed. For example, if the stream flows at 10 cm s^{-1} and the X-ray beam profile width is 10 μm, neglecting flow effects, the time resolution of the experiment equals 100 μs. Mixing times can be reduced to microseconds with microfluidic approaches, as described in sections 9.7.3 and 9.7.4.

9.7.3 Microfluidic Implementations

Microfluidic devices allow concentrating, diluting, mixing, and filtering, in line with the SAXS measurement, as previously mentioned. For time-resolved studies, microfluidic techniques can extend the accessible time resolution of SAXS to the microsecond time range. Microfluidics is characterized by low Reynolds numbers. The Reynolds number, Re, is the ratio of inertial forces to viscous forces and is given by

$$\mathrm{Re} = \frac{\rho v w}{\mu}, \tag{9.2}$$

where ρ is the fluid density, v is the flow velocity, w is the characteristic dimension of the flow geometry (inside diameter for a circular channel), and μ is the fluid viscosity. At low Reynolds numbers, the flow is laminar, and mixing is constrained by diffusive time. This makes microfluidic implementations challenging. Mixing can be introduced through turbulent flow generating a high Reynolds number. This was pioneered by Regenfuss et al. (1985), who passed two liquids through an inner and an outer capillary, focusing the tip so that the fluids combined and flowed around a spherical ball at the tip, thus achieving mixing times on the order of tens of microseconds. A detailed description of these devices used in SAXS is given by Watkin et al. (2017). A popular mixer has been that based on a "T" geometry, originally developed to study the folding dynamics of cytochrome c (Takahashi et al., 1997). In this mixer, two solutions enter a "T"-shaped nozzle with high velocity and meet at a point directly below the observation area. As the solutions pass from the narrow flow channels into a wide mixing chamber at the center of the "T", turbulent flow is generated. The mixing rate is determined from the flow rate and the distance between the mixing point and observation. A simple modification of the design involves altering the angle of the input channels such that they form an arrow-shaped junction, which forces the two fluids to undergo a greater change in momentum, theoretically giving rise to more efficient mixing and demonstrating a mixing time of ~50 μs (Bilsel et al., 2005). There are potential limitations in these approaches, including the amount of sample consumed, necessitated by the high flow rates required and the sensitivity of the sample to the shear forces introduced.

A laminar flow–based rapid-mixing approach, referred to as "hydrodynamic focusing," was introduced by Knight et al. (1998). This approach relies solely on diffusive mixing without introducing turbulence. Two input fluids approach a central input fluid from either side, squeezing the central fluid into a thin, focused stream. The thin stream means that diffusion across the stream occurs on a relatively short timescale. SAXS investigations have been carried out using this type of device, yielding a minimum deadtime of 240 μs. Later variants utilizing sheaths of fluid to create a barrier between the sample and fluid to be mixed have reduced this time at least fourfold (Park et al., 2006). The limiting factor has become not so much the time but the volume requirement of the sample to generate enough signal-to-noise in the beam. Diffusion time across thin streams increases with stream thickness but so does the signal.

9.7.4 Combining Microfluidics and X-Ray Microbeams

X-ray sources and optics have rapidly advanced, allowing high-brilliance beams in a very small area. Graceffa et al. (2013) describe the Biophysics Collaborative Access Team (BioCAT) undulator beamline, where a cryo-cooled double-crystal Si(111) monochromator and a Kirkpatrick–Baez mirror bender system provide a 20 μm × 5 μm beam footprint at the turbulent flow mixer on the beamline. The turbulent flow mixer itself is described in Chapter 10, section 10.2, and consists of channels that are 30 μm wide and contain protein and buffer solutions that mix in an arrow-shaped junction, with the arrow shape facing away from the observation region (Bilsel et al., 2005). In the observation region, the channel is 100 μm wide so that the beam illuminates the region but not the edges of the region. Sample requirements are still large in the example given, with ~90 time points in total, requiring ~2 mg per time point.

9.8 Summary

SAXS instrumentation is available for the laboratory, at the synchrotron, and at XFELs. Laboratory systems have limitations, due to the available X-ray intensity and potentially due to smearing effects that can limit the accessible q range. Multiple synchrotron facilities have SAXS beamlines, and the ability to perform SAXS studies on free electron laser sources is growing. Sample handling has adopted automation and, in some cases, microfluidics to reduce sample volume requirements. Sensitive detectors, efficient data collection software, and rapid analysis allow real-time decisions to be made during data collection. Specialized apparatus enables time-resolved studies.

10

Distinct Instrumental Approaches to SAXS

In Chapter 1, Figure 1.1 schematically introduced a small angle scattering experiment as being very straightforward: shining a collimated beam on a specimen to generate an $I(q)$ curve. It ignored the art, complexities, and difficulties of actually generating and validating high-quality small angle scattering data. To some extent, we have underplayed the experiment challenges of small angle scattering in later chapters as well. In this chapter, which introduces a number of small angle scattering experiments that go beyond traditional applications, we also take the opportunity to go through experiments in more detail, to illustrate what an experimental program in small angle scattering involves, and how advances in sources, beamlines, and automation make the road less burdensome.

Most SAXS users are focused on very traditional biological studies. However, new X-ray sources, detectors, and developments in laboratory instrumentation and automation allow for the possibility of experiments that fundamentally expand the abilities and type of analysis that can be carried out. Capabilities have developed for SEC (size exclusion chromatography), for time-resolved studies and analysis, and for the use of SAXS in a high-throughput manner. Each of these areas is discussed in this chapter.

10.1 SEC–SAXS

Some samples can pose problems for SAXS. For example, aggregates may arise at re-concentration of the sample or appear over time, resulting in a mixture in solution; or, complexes may dissociate, also resulting in mixed species. SEC–SAXS can overcome these problems by using SEC to clean up a sample moments before data collection.

The method does have limitations. It requires a large volume of sample, dilutes the sample in the process, and can accumulate radiation damaged sample on the walls of the sample cell, requiring exhaustive cleaning of the sample cell after use. The experimental approach can be very dependent on the sample problems that need to be overcome during the data collection. Optimization and characterization of the purification process is strongly recommended before conducting the experiment.

Biological Small Angle Scattering: Theory and Practice. Eaton E. Lattman, Thomas D. Grant, and Edward H. Snell.

DOI: 10.1093/oso/9780199670871.001.0001

10.1.1 Application to Soluble Proteins

SAXS is very sensitive to interparticle interactions (e.g., trace amounts of higher-order oligomers or the presence of reversible oligomerization and aggregation). Without prior structural knowledge, it is difficult to resolve cases where a ligand may be present but not, for example, 100% incorporated in all the molecules. SEC–SAXS was originally implemented at the Advanced Photon Source (APS) BioCAT beamline (Mathew et al., 2004). It enables successful SAXS analysis of samples that show time and concentration-dependent dynamics (e.g., oligomerization, aggregation, or structural changes). In combination with techniques such as SVD (see Chapter 4, section 4.6.1), it can reveal components within dynamically complex systems. SEC separates particles based on size as they flow over a column packed with a porous gel. Small molecules easily penetrate the gel support contained within the column, thus experiencing large partition coefficients and being strongly retained, whereas large molecules, which are too big to penetrate and are excluded, experience small partition coefficients and elute early, with the mobile phase. By continuously sampling the scattering during elution, oligomeric or contaminating species may be resolved.

Jeffries et al. (2016) provide a detailed description for practical SEC–SAXS protocols and a specific discussion of Superdex column choice and resolving power. They note that SEC–SAXS should not be viewed as a direct purification step but as an analytical procedure to be applied when needed. Samples still need to be pure components resolvable with the column used. Determining the column, load volume, flow, and solvent conditions in advance is a prerequisite to successful data collection. Preparation of an excess of buffer is recommended. The column should be equilibrated before and after the experiment, using a filtered and degassed buffer. Temperature should be constant throughout the experiment. The buffer and column need to be at the same temperature during equilibration as during the experiment. From two to eight volumes of running buffer are recommended for equilibration. Flow rates should be selected that do not exceed the capabilities of the column, are compatible with the amount of sample available, and achieve good signal-to-noise but that are not so slow such that radiation damage becomes a problem. The potential for radiation damage should be assessed by collecting SAXS data from a small aliquot of sample before the SEC–SAXS experiment. If damage is present in this case, it will also be there in the SEC–SAXS experiment, and mitigation measures need to be taken (e.g., the use of scavenging compounds, attenuation, or the reduction of exposure time).

As samples and buffers run through columns, small volume fractionation and/or exchange of the buffer components can occur, which can lead to subtle errors in background subtraction. It is recommended to use dialysis to exchange the protein sample into the running buffer before the SEC–SAXS experiment (Jeffries et al., 2016). Understanding and optimization of the column performance is advised before the SEC–SAXS experiment. Multiple techniques are available, as described in Chapter 6.

In this section, we discuss some specific applications of SEC–SAXS, as the experimental design can be very particular to the sample under study. In the work by Mathew et al. (2004), a fast protein liquid chromatography, or FPLC, system was used with a Superose-6 column coupled to the SAXS flow cell. The sample, after passing over the column and before entering the flow cell, was monitored by UV absorption. The samples, concentrated to ~10 mg/mL flowed at 0.25 mL/min through the 1.0 mm capillary of the SAXS flow cell, such that any given portion of the sample took less than 0.04 s to traverse the 200 μm width of the X-ray beam. As the data was collected, Guinier analysis was used to calculate the R_g. Direct beam intensity was monitored, and solvent and background scatter were assessed by averaging several data sets near the start and end of the run, before the sample had eluted, and after the sample had come off the column, verified by the UV measurements. The samples used were cytochrome c, which is a common SAXS standard resistant to aggregation; plasminogen, which has a variable R_g, depending on concentration; and a mixture of cytochrome c, BSA, and blue dextran (a common chromatographic standard). The UV spectrum for the cytochrome c showed an elution peak corresponding to an R_g of 13.2 ± 0.4 Å throughout the peak. Interestingly, spurious elevated R_g signals were seen before the main UV elution peak, indicating the successful separation of high-molecular-weight impurities or aggregates. The $I(0)$ provided a measure of concentration in the elution, with a fairly constant value of R_g being displayed for 15 time points. The plasminogen sample had multiple elution peaks, with the early peaks having larger and variable R_g's but the main peak showing an R_g that was constant at ~31 Å. The R_g, absorption, and $I(0)$ are shown in Figure 10.1, which is taken from the work by Mathew et al. (2004). The Guinier plots for this data displayed nonparallel slopes; that is, there was a variation as a function of concentration. This is evidence for insufficient resolution in the column in separating different species, or the presence of a dynamic equilibrium of species or conformations. In the final case, with the mixture of samples, the SEC–SAXS approach achieved separation of the three samples, with stable R_g plateaus. The calculated R_g values for cytochrome c and BSA were as expected, but the R_g for the blue dextran could not be reliably determined, due to its size and the angular resolution of the setup. The key to the technique was the identification of the stable R_g region associated with the elution profile. Images recorded in that region can be summed to improve the overall signal-to-noise.

SEC–SAXS, while shown to be very useful in allowing the study of a single species in a dynamic system or mixture, is not without its limitations. The buffer and sample scattering can take place at substantially different times over the elution. Dilution of the sample occurs, and potential changes in the buffer composition can occur during the elution. Although inline UV absorbance measures the sample concentration, the accuracy may be poor. The buffer scattering is collected from the pre-void volume and/or post-included volume or separately. Estimation of the scaling constant for the buffer and/or extrapolation to zero concentration are nontrivial—the sample concentration has to be accurately profiled, and individual species, if present, deconvoluted (Malaby et al., 2015).

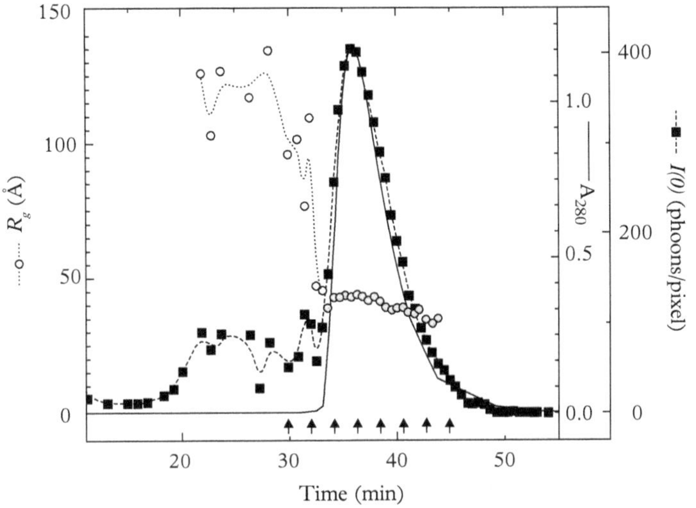

Figure 10.1 A chromatograph as plasminogen eluted from the column, showing both the scattering parameters R_g and $I(0)$ and the UV absorption at 280 nm as a measure of concentration. A major peak at 36 min appears both in A_{280} and in $I(0)$, but substantial high-molecular-mass species were observed in the $I(0)$ trace in the region immediately preceding this, 20–36 min. Since $I(0)$ is proportional to both molecular mass and concentration, whereas A_{280} is proportional to only concentration, the differences in these two traces appear to reflect differences in molecular mass. This is confirmed in the R_g data, which shows plateauing at ~35 Å in the major peak region but much higher and more irregular values in this putative oligomer region. The filled symbols for the R_g trace correspond to the major peak, as defined by the region where the $A_{280}/I(0)$ ratio remains constant, while the open symbols correspond to the putative aggregate region.

(The figure and caption are used with permission from Mathew et al. (2004).)

Because an SEC–SAXS experiment is conducted over a longer time period than a simple concentration series, it can be affected by beam intensity fluctuations and drift. Many synchrotrons minimize this through top-up mode and/or active feedback systems. Even with these considerations, normalization may be needed. Malaby et al. (2015) normalized data using a weighted average of the incident beam, I_0, and the transmitted beam, I_t, where

$$normalized\ I(q) = raw\ I(q) \times C_{norm}/\langle C_{norm}\rangle, \tag{10.1}$$

where

$$C_{norm} = 1(w_0 I_0 + w_t I_t - k). \tag{10.2}$$

Here, $\langle C_{norm}\rangle$ is the average value of the normalization constant, and k is a constant that compensates for the reduced individual contributions of I_0 and I_t in the weighted average. The weightings w_0 and w_t and the constant k were determined by minimizing the beam intensity fluctuations and the drift seen in the data. For online data processing, the authors used a more rapid normalization with $C_{norm} = 1/I_0/I_t$, which was not as effective in reducing beam intensity artifacts but was superior to normalization by I_0 or I_t alone.

Malaby et al. (2015) also used SVD (see Chapter 4, section 4.6.1) to analyze SEC–SAXS data recorded from cytochrome c. The SVD revealed two significant components. The scattering was expressed as a linear combination of the SVD U_0 and U_1 components such that,

$$I(q)_{\text{protein}} = c_0 U_0 + c_1 U_1 \tag{10.3}$$

or

$$I(q)_{\text{protein}}/c_1 = cU_0 + U_1, \tag{10.4}$$

where c_0 and c_1 are coefficients with $c = c_0/c_1$. This enables the total scattering curve to be reconstructed from U_0 and U_1 with an appropriate scaling constant, provided the constant can be determined accurately. With sufficient measurements and multiple concentrations studied, the contribution of multiple components can be resolved within scattering curves.

In the cytochrome c case, the authors systematically analyzed the Guinier region over a range of coefficients and scaling constants using the unweighted merit statistic R^2 from a linear least-squares fit of the data in the Guinier region. The R^2 had a well-defined maximum. This Guinier-optimized linear combination based on SVD analysis was termed "SVD–LC," for SVD linear combination.

The strength of applying SVD techniques to SAXS data is illustrated by a study on a truncated form of the Arf6 GTPase, a well-behaved, structurally defined monodisperse test case (Malaby et al., 2015). Multiple scattering curves were summed to generate a total intensity profile, with each curve normalized according to equation 10.2. SVD analysis indicated two significant components, which accurately reconstructed the scattering. The SVD-aided reconstruction was superior to direct buffer subtraction and better than Guinier-optimized buffer subtraction. The same study looked at an autoinhibited construct of the Arf exchange factor Grp1. This construct is known to be monomeric in the low micromolar range but formed oligomeric species at higher concentrations. It was a challenging sample for both regular SAXS and SEC–SAXS. Direct subtraction of the buffer from the protein signal was unsuccessful. SVD of the data matrix revealed three significant components in the scattering: the scattering expected from the monomer, scattering from a putative dimer, and scattering from the buffer. Protein scattering curves were reconstructed by SVD–LC, Guinier optimization of the scaling constant, and subtraction of the buffer region from the protein. Again, the SVD–LC treatment produced a better fit to the data than the other two methods. The SVD–LC reconstruction suggested that the dimer present was related to the dimer seen in the asymmetric unit of the crystallographic model.

SEC–SAXS can be sensitive to structural and conformation changes, enabling the analysis of dynamical systems. *Arabidopsis thaliana* acyl acid-amido synthetase GH3.12 has both open and closed conformations of the C-terminus, and non-SAXS studies had suggested that adenosine triphosphate (ATP) binds only in the open conformation. SEC–SAXS showed that the C-terminal domain is in the apo-form and assumes the closed conformation on binding (Round et al., 2013).

To perform the experiment on GH3.12, an HPLC system was integrated with the beamline immediately before the SAXS sample chamber. Buffer was degassed online, and a flow rate of 0.5 mL/min was used, with the column being equilibrated with 1.5 times its volume with the buffer. The baseline buffer signal was recorded. Approximately 3,600 images with a sample-loaded concentration of ~10 mg/mL were collected for each sample run. Frames with a constant R_g were merged, and the single averaged frame corresponding to the scattering of an individual SEC purified species calculated. Different combinations of substrate and products were used with the SAXS experiment. The R_g values for open and closed confirmations differed by 3.5 Å, a small but measurable difference.

Scattering calculated from crystal structures with ligands of the open and closed conformations were fitted to the experimental data. The apo-SAXS data showed a poorer fit to the calculated scattering, indicating domain flexibility, but the bound SAXS data was in good agreement. The R_g of the apo form was 30.4 ± 0.4 Å, while the protein in the presence of adenosine monophosphate and ATP gave R_g values of 27.5 ± 0.4 Å. Due to the largely redundant data, these differences fall well above the noise level; they are real.

There were no crystallographic models for the apo form, presumably due to the domain flexibility. SEC–SAXS enabled small conformational changes to be studied, leveraging the complementary biophysical and structural information into a more complete understanding of mechanism. The data used in this case was about one-hundredth of the total data collected, and the authors recommend the use of 10–30 images (dependent on the separation of SEC species) to obtain good signal-to-noise. SEC-based SAXS can separate species but also dilutes the overall sample; therefore, summation of data from an individual species is necessary.

Because of lower flux and brilliance at neutron sources, SANS has historically required a longer exposure time than SAXS. SEC–SANS experiments are not common, due to signal-to-noise requirements and the low time resolution due to long exposure times. With increases in available flux and detector sensitivity allowing relatively short exposure times on small sample volumes, Jordan et al. (2016) have implemented SEC–SANS on the D22 instrument at the Institut Laue–Langevin (ILL). An AKTAprime system was used with a Superdex 75 column 10/300 GL to implement a stopped-flow data collection. Three samples were studied. The first was a gel-filtration standard containing a mixture of six species of different molecular weights: 5 mg of bovine thyroglobulin (670 kDa), 5 mg of bovine γ-globulin (158 kDa), 5 mg of chicken ovalbumin (44 kDa), 2.5 mg of horse myoglobin (17 kDa), and 0.5 mg of vitamin B_{12} (1.35 kDa). A total of 18 mg of standard in 0.5 mL of both water and D_2O-based buffer was used for the experiment. Exposures of 30 s each were used with a flow rate of 0.3 mL/min. The column achieved good separation between 3 and 70 kDa; as expected, the thyroglobulin and γ-globulin eluted in the excluded volume, and vitamin B_{12} was visible in the UV absorbance measurements but did not provide a strong SANS signal. Three exposures were averaged from the UV peak for myoglobin, five for ovalbumin, and ten for the buffer blank. The experimental $I(0)$ and R_g for both ovalbumin and myoglobin were in agreement with literature data, showing that the SEC–SANS yielded signals from each sample within the mixture.

In addition to the chromatography standards, Jordan et al. (2016) looked at two proteins representing SANS experiments that might benefit from an SEC approach. The first, Alba3, was a small aggregation-prone protein. The deuterated protein was studied with both water and D_2O-based elution buffer. The UV absorbance of the sample coming off the SEC column showed three peaks, the first corresponding to aggregates, the second to the Alba3 monomer, and the third presumably to low-molecular-weight contamination. Interestingly, the SANS signal for the H_2O-based buffer showed two peaks, in contrast to the D_2O case, where a third peak of high-intensity resulted. The authors attributed this third peak to the water-based buffer injected with the protein and eluting at the end of the column—suggesting a very efficient buffer exchange. The resultant data in both cases was suitable for Guinier analysis, with the D_2O data being of higher overall quality than the H_2O-based buffer elution. SEC–SANS provided a high-enough signal and sufficient separation to resolve a single oligomer. The second protein, Sir2, shows rapid aggregation even when measured immediately (minutes) after regular SEC purification. In this case, the UV absorbance revealed an asymmetric elution profile, with the Guinier analysis showing a variation in R_g from the SANS data collected across the profile. The protein did not elute as a monodisperse system, and later analysis with other techniques confirmed the presence of both trimers and monomers. Fortunately, the crystal structure was known, and the SANS data could be used with knowledge of oligomeric states of all species, but the SEC approach did not achieve separation of these species in this case. There is plenty of room for optimization of the technique, and SEC–SANS is possible in less sensitive cases or with other separation techniques.

Ion-exchange chromatography SAXS

Hutin et al. (2016) note that SEC–SAXS approaches require at least a 10% difference in molecular mass between different species present, to ensure separation on the column. The high degree of dilution on the column (up to a factor of 10, depending on the column type and sample) requires higher concentrations than a simple concentration series and/or the summation of multiple images to achieve the required signal-to-noise. An alternative to SEC–SAXS is ion-exchange chromatography (IEC), and while the principle is similar, the separation of species is based on net protein surface charge. The resin in the column has a net surface charge, and if the charge is complementary, the protein will bind to the resin. A salt gradient increases the charge in the mobile phase and so causes individual proteins to elute at a specific concentration.

IEC–SAXS has very specific advantages, compared to SEC–SAXS. Proteins of similar molecular weight can have very different surface charges so that they can be readily separated, and different oligomeric states differ in surface area and hence surface charge distribution. From a practical pint of view, a large volume of dilute sample can be loaded on an IEC column and that sample concentrated, since the elution concentration is determined by the capacity of the column and the binding affinity, not the initial concentration. IEC columns can also accommodate high flow rates in the milliliters per minute range. It should be noted that IEC approaches do require optimization of the pH and salt gradient to be most effective (Yigzaw et al., 2009).

Hutin et al. (2016) demonstrated the IEC–SAXS approach, using BSA and the helicase–primase protein fragment $D5_{323-785}$ as examples. They used an HPLC system with an inline degasser, an autosampler, a UV-visible spectrophotometer, and a conductimeter, directly connected to a 1.8 mm–diameter flow through a capillary in the X-ray beam path. Using this, they studied BSA with both a linear and a stepwise salt gradient, and the helicase protein with just a stepwise salt gradient. For the BSA, the linear gradient was run on the IEC with and without the protein sample, to collect sample and buffer blank images, respectively. Several peaks were observed, and SDS–PAGE analysis used to confirm that the first two were BSA while the third was a higher-molecular-mass contaminant. At a flow rate of 1 mL/min, the UV absorption measurements also showed the three peaks. The matching background images were slightly shifted with respect to the buffer run. An alternative approach used differential ultracentrifugation coupled to SAXS (Hynson et al., 2015). For each shift, the ratio of scattering at low q versus mid-to-high q is compared. If the sample scatters in the same way throughout the peak, the shift is valid and can be used for accurate background subtraction. In this manner, the first two peaks (which are overlaid with each other) had a 2.2 mL shift and a constant R_g and mass calculated from $I(0)$—one species elutes in a double peak. No suitable shift was found for the third peak, indicating that the sample was not constant through the peak and that it probably represents more than one species. The frames recorded from the first two peaks were averaged, yielding an R_g of 27 ± 1 Å and a Porod volume yielding a molecular weight of 68 kDa, both corresponding well to the known structure of BSA.

A linear gradient works well for many proteins, but for samples where the elution peak is broad or is not well separated from other peaks, manual steps in the gradient can be used to ensure a purer peak that is rapidly produced. Using the data from the linear salt gradient, Hutin et al. (2016) were able to elute the two subpeaks in a single step, determining values for R_g and molecular weight that were similar to those obtained via the linear gradient method. For the case of the helicase–primase protein, three overlapping peaks were observed in experiments using a linear gradient. Careful design of the stepwise gradient based on the linear gradient results eluted the helicase–primase protein completely at a single condition, concentrating it from the injected concentration of 0.17 mg/mL to 3.4 mg/mL. The resulting R_g was 47 Å, with a Porod volume corresponding to the mass of 338 kDa expected for the hexamer.

The flow rate with IEC–SAXS is typically higher than that with SEC–SAXS, thus reducing the potential for radiation damage and adhesion to the SAXS cell, but background subtraction needs to be considered carefully.

10.1.2 Application to Membrane Protein Systems

Membrane proteins are more difficult to crystallize than soluble proteins, so SAXS and SANS offer a means to obtain at least some structural information. Membrane proteins, because of their hydrophobic regions, are not soluble in typical aqueous buffers and require detergent to avoid aggregation and to maintain correct folding outside of the membrane environment. The detergent molecules interact with the protein, forming a shell around the transmembrane region. The polar heads of the detergent face the aqueous medium. In terms of scattering density, the protein and detergent have different properties.

For the effective stabilization of a membrane protein, the detergent concentration has to be close to the critical micelle concentration (CMC), which is the concentration of detergent above which micelles (an aggregate of detergent molecules) are formed. Such lipid micelles will contribute to the scattering data, making study of protein–detergent complexes challenging.

SEC–SAXS can be used effectively with membrane protein systems and has been pioneered by the Pérez group at the French synchrotron, SOLEIL (Berthaud et al., 2012). In an illustrative example, they studied aquaporin-0 solubilized in *n*-dodecyl-β-D-maltopyranside (DDM) detergent on the SWING beamline of SOLEIL. The sample concentration was 6 mg/mL, and the sample was kept at 10°C with solution flow controlled by an HPLC set up inline with a UV-visible spectrometer and the SAXS experimental cell. Data was collected using a SHODEX KW405-07 column equilibrated for 3 h at a flow rate of 250 μl/min. For data collection, the rate was reduced to 150 μl/min, and 20 μl of the protein solution was used to collect 256 images. To reduce the broadening of the elution profile, a second experiment was conducted with a Superpose 12 10/300 SEC column and the spectrometer removed from the flow path. This column was equilibrated for 40 min with a flow rate of 500 μl/min, and the data collected with 400 μl of sample at a reduced flow rate of 350 μl/min.

The authors derived an equation for the number of detergent molecules as a function of the ratio between $I(0)$ and the protein mass concentration, C_{prot}:

$$N_{\text{det}} = \frac{\sqrt{\frac{I(0)M_{\text{prot}}N_{\text{A}}}{C_{\text{prot}}f^2}} - (n_{\text{prot}}N_{\text{A}} - \rho_s\bar{v}_{\text{prot}}M_{\text{prot}})}{(n_{\text{det}}N_{\text{A}} - \rho_s\bar{v}_{\text{det}}M_{\text{det}})}, \qquad (10.5)$$

where M_{prot}, n_{prot}, and $\bar{v}_{\text{prot}}$ are the molar mass, molecular number of electrons, and specific volume of one molecule of the protein, and M_{det}, n_{det}, and $\bar{v}_{\text{det}}$ are similar for the detergent. This equation comes from the Guinier approximation, specifying $I(0)$ in absolute units.

The refractive index of the solution was measured after the solution left the SAXS chamber. This allowed a separate measurement of the number of detergent molecules using

$$N_{det} = \frac{(1-\varphi)}{\varphi}\frac{M_{prot}}{M_{det}}, \tag{10.6}$$

where φ is the protein mass fraction of the complex, M_{prot} is the molar protein mass, and M_{det} is the detergent molar mass.

UV absorption measurements showed zones corresponding to a small fraction of aggregated protein, the fraction containing the detergent complex and micelles, and finally, the fraction containing micelles alone, from the excess of DDM in the loading buffer. During the protein elution, the $I(0)$ and R_g remained constant, and the refractive index measurements indicated that there were 238 ± 15 DDM molecules per protein over that period. The clear separation of the excess DDM peak allowed the SAXS signal from the protein–detergent complex to be isolated from that of the additional micelles present.

For the analysis to be valid, the authors assumed that the protein–detergent complex mass was constant through the elution (verified by the refractive index measurements of N_{det}) and that there was no significant depletion of the detergent. Data quality checks are especially important to ensure the absence of interparticle effects.

The SIBYLS beamline at the ALS provides a table of the best detergent for SAXS based on the concentration that contributes the least amount of background to the data as a function of the CMC. It divides the detergents into three groups, with the first consisting of those that minimize signal below twice the CMC, the second containing those that can be used at the CMC, and finally those that have to be used well below the CMC. Table 10.1 summarizes the first of these

Table 10.1 *Detergents that minimize the SAXS signal below twice the CMC; data from the ALS SIBYLS beamline*

Group 1 (less than 2× CMC)			
Name	**CMC (%)**	**MW (Da)**	**Type**
1-*s*-Nonyl-β-D-thioglucoside	0.093	322.4	N
n-Decanoylsucrose	0.124	496.6	N
n-Nonyl-β-D-maltoside	0.155	484.6	N
DDAO	0.209	201.4	N
C_8E_5	0.249	350.5	N
FOS-choline®-10	0.420	323.4	Z
FOS-choline®-9	0.588	309.4	Z

Type is N for non-ionic, and Z for zwitterionic detergent.

three groups. Readers are asked to acknowledge the SIBYLS beamline if this data proves useful to their studies. Note that the CMC fluctuates with biochemical and biophysical conditions.

Berthaud et al. (2012) modeled the detergent shell with two types of pseudo-atoms representing hydrophobic and hydrophilic scattering densities, respectively, with the envelope making up a torus positioned at the transmembrane region, with the interior surface conforming to the known protein structure. They initially used a circular section characterized by three geometric parameters. Extending this to an elliptical section with an additional parameter produced a better fit to the experimental data, reducing the χ^2 from 2.29 to 1.37. The parameters used to model hydrophobic and hydrophilic electron density were 0.274 $e^{-}Å^{-3}$ and 0.504 $e^{-}Å^{-3}$, respectively.

It was notable that even at 0.2% (w/v), the SAXS signal for the free micelles was comparable to that for the protein. The number of detergent molecules surrounding the protein was 211 $\pm$ 10 from the SAXS intensity extrapolated to $I(0)$, 238 $\pm$ 15 as noted from the refractive index measurements, and 271 $\pm$ 30 from a coarse-grained reconstruction. The first two values are dependent on the reference values used. The number of detergent molecules surrounding a protein depends on many variables, but this range of values is consistent with those from other studies and different methods of measurement. In a further analysis of the data from this study, the authors explored additional parameters, showing that the pool of parameters providing the best fit between the model and experimental data lies in a global minimum (Pérez & Koutsioubas, 2015).

For the example noted, the protein structure was known, and the information obtained primarily involved the detergent envelope. Nevertheless, this is useful data for modeling detergent systems and distinguishing conformational changes in known structures in the presence of a detergent mixture. Direct electron density determination approaches may be able to extend the information that can be extracted in the future (see Chapter 14).

10.2 Time-Resolved SAXS

Time-resolved SAXS instrumentation was mentioned in Chapter 9. The potential for the technique is illustrated by a study on photoactive yellow protein by Cho et al. (2016). In this study, three separate SAXS experiments were combined to understand the photocycle of photoactive yellow protein, acquiring time-resolved data over ten decades of time from 100 ps to 1 s. By combining the SAXS data with other structural and biophysical studies, a detailed description of the photocycle was provided. In this case, a pump–probe technique was used, with a 390 nm laser pulse followed by a ~117 ps X-ray pulse at variable time points.

SAXS can access a large range of timescales important to biological mechanisms, and various methods of interrogating structures at different time points are available. Using stopped-flow techniques, Oda et al. (2016) studied the early stages of the polymerization of actin. They initiated rapid mixing followed by opening of the X-ray shutter after a variable time delay followed by 10–20 images recorded at 50 ms each, with a gap of 3 ms. Using this protocol with different time delays, they followed the process for almost 2 min from when salt-induced polymerization was initiated. F-actin (the polymerized form) was clearly indicated at 20 s, with an isoscattering point suggesting a two-state transition from the monomer, G-actin. An SVD analysis of the data indicated that there was a third state formed during the polymerization process. Using a time-resolved concentration series, a kinetic analysis of the polymerization could be formed. SAXS data provided evidence for the polymerization process: the addition of neutral salts diminishing the repulsion between G-actin molecules, salt binding causing parallel and antiparallel dimers to be formed, and then trimers and tetramers being sequentially formed from the parallel dimers. Monomers were added to the tetramers, and filaments grew, with the antiparallel dimers being incorporated transiently into the growing filaments.

At the BioCAT beamline 18D of the APS, a turbulent flow mixer is used in combination with a microbeam (Graceffa et al., 2013). The turbulent mixer consists of channels that are 30 μm wide and contain protein and buffer solutions that mix in an arrow-shaped junction, with the tip pointing away from the observation region (Bilsel et al., 2005). After mixing, the solution flows 0.24 mm into a channel, which is 100 μm wide, in the observation region, which is 20 mm long. In this manner, submillisecond mixing times can be achieved with flow rates of 10–20 mL/min. Continuous scanning along the flow direction together with translation of the observation region allows efficient collection of buffer and sample time points. When coupled with a small beam size (~20 μm × 5 μm), high brilliance, and a sensitive detector, short timescales can be accessed with good temporal resolution. This system was tested with cytochrome c denatured with 4.5 M GdnHCl. The sample mixed with buffer with a tenfold dilution provides conditions favoring the native state. The earliest state observed after complete mixing was 100 μs, and data was collected here, at 124 μs, 1.2 ms, and 2.4 s. SVD analysis indicated three basis vectors, consistent with the transition to a native state being a highly cooperative process.

To obtain high time resolution, the exposure time has to be short. This reduces the signal-to-noise and therefore, potentially, the quality of the data. Such noise considerations are important, and equation 8.6 (see Chapter 8) was developed to assess where there was enough signal in the data to for valid structural observations.

SAXS has been used to study protein unfolding and refolding. Chen et al. (1998) used a stopped-flow rapid mixer to collect time-resolved data, diluting by 11-fold initially unfolded lysozyme in a solution of 8 M urea at pH 2.9. Folding studies suggested a burst phase where an intermediate or intermediates are formed comprised of collapsed conformations, followed by folding to the native structure. The low pH considerably prolongs the nonspecifically collapsed ensemble to a lifetime of about 1 s. Stopped-flow data was collected in various timeframes, with a typical data set collected from 25 mixing repetitions. The data from hundreds of mixing events was measured for each condition, to improve the overall signal-to-noise. SVD techniques were used to analyze the resulting data. A major change in shape from the unfolded form to an ensemble of relatively compact globular shapes occurs in the burst phase of refolding, with the R_g reducing from 22 Å to 19–20 Å—a small but significant reduction in an already compact state, due to the four disulfide bonds present. Interestingly, on transitioning from the collapsed ensemble to the native state, with R_g further decreasing to 15 Å, there is a change in $I(0)$. The $I(0)$ depends on the square of the total number of electrons in the scattering molecule and its associated hydration shell, minus that from the same volume of bulk water. The change in $I(0)$ appeared to have been caused by the accessible surface area and therefore hydration shell decreasing on formation of the native fold. The study showed an initial collapsed ensemble that then forms the native state.

Although rare, time-resolved studies have also been carried out using SANS. Ibrahim et al. (2017) studied the unfolding of a tagged green fluorescent protein substrate, GFPssrA, in the presence of unfoldase. Short SANS frames (30 s exposure) were coupled with a fluorescence detector to acquire structural data. This study is notable because of the effective use of contrast matching and the resulting time-resolved picture of the individual components making up the mechanism. However, it also demonstrates the limitation of SANS in this area, with the exposure time required and consequent limitation on time resolution.

10.3 High-Throughput SAXS

High-throughput crystallographic data collection is mature, with sophisticated instrumentation and software developed to control the process. Many of the lessons from the crystallographic field have been applied to SAXS. As in high-throughput single crystal studies, calibration is a most important step. Before investigator data is collected, the incident beam center, energy, sample position, and detector orientations must all be established. Scattering from reference samples such as silver behenate or silver stearate powder allow calibration of the instrument range and determination of the beam center (Nyam-Osor et al., 2012). The calibration data is usually written into a descriptive file or contained in image header details, depending on the beamline. A mask defines regions of the detector where SAXS data is not being collected (e.g., the beamstop shadow). Finally, data from water and stable known molecular weight samples allows placing the scattering on an absolute scale. These calibrations can be used for multiple users, provided the experimental setup is not changed, and enable reliable and reproducible high-throughput use. As for the time-resolved approach in section 10.2, it is useful to discuss successful implementations of high-throughput approaches.

10.3.1 Sample Handling

The SIBYLS line at the APS in Berkeley (described in Chapter 9) has been a leader in developing mail-in processes for SAXS data collection and in providing practical information to enable successful experiments (Dyer et al., 2014). The sample delivery format is through specific, commercially available 96-conical-well plates. The plates are filled at the investigators' laboratory with a defined minimum content and then sealed for shipping. Experience from the beamline suggests that flash freezing is possible but usually unnecessary. The investigator takes the responsibility for ensuring that a suitable concentration series is available and that buffer blanks are matched. Effective high-throughput operations require that expectations for the sample be well communicated to the users. Information handling is an important aspect of the process. The SIBYLS beamline uses formatted spreadsheets filled in by the investigators in advance of the data collection. These describe the nature of the samples and the order of data collection and give practical information on which wells contain buffers and when sample cell washes are required.

Before the sample is studied, the plates are spun in a centrifuge to sediment any large aggregates. If necessary, the original seal is replaced with a pierceable seal to allow the sample-loading needle to access the sample but control evaporation and consequent concentration changes in the samples remaining to be studied.

10.3.2 Data Acquisition and Analysis

Jeffries and Svergun (2015) nicely summarize the data collection and analysis steps for high-throughput automated small angle scattering. Data is first collected from a buffer or background solvent, and then collected from the sample of interest. One-dimensional scattering profiles are produced and the background subtracted from the sample. The resulting scattering profile is then analyzed to determine the Guinier region and $P(r)$ versus r. Molecular weight is assessed based on $I(0)$ and Porod volume, and the R_g is determined. The process is then repeated to form a concentration series.

To carry out data acquisition and analysis successfully requires integration of the sample handling system, beamline and detector control, information management, and data processing capability. Elements of this have been developed by building on developments in the single crystal field (Classen et al., 2010; Blanchet et al., 2012; Blanchet, Spilotros, et al., 2015), and complete integrated pipelines have been developed (Franke et al., 2012). Automation enables high-throughput studies but, more importantly, improves reproducibility and allows real-time analysis of the outcome and optimization of the experimental parameters on the fly.

Many steps happen rapidly during high-throughput data acquisition and analysis. As the images are acquired, they are integrated, and in many cases, the background-subtracted scattering curve is visually displayed. Typically, a series of background and sample images are recorded, and the scattering data analyzed to check that background subtraction has been successful and that no radiation damage is present, identified by differences in the scattering. The R_g can be calculated from each image, and this also provides a signature for potential radiation damage. Data from multiple images that are statistically similar is averaged into a single scattering curve. As mentioned previously, R_g determination is not straightforward, with the Guinier range depending both on molecular weight and on particle shape. Automated procedures and guidelines assist in this process (Petoukhov et al., 2007; Blanchet et al., 2012; Petoukhov et al., 2012; Grant et al., 2015). The results can be used to identify sample issues such as aggregation or radiation damage, both of which can be mitigated to some extent by changing sample conditions or physical parameters such as flow, or by introducing oscillation or reducing temperature or exposure time. A Kratky analysis can reveal other sample issues. The $P(r)$ can be rapidly calculated and compared to the experimental data. This provides the initial structural information on the sample.

10.3.3 Applications of High-Throughput SAXS

Crystallization

As noted earlier, small angle scattering is a useful technique for characterizing sample globularity, flexibility, or unfoldedness. For the purposes of crystallization, the latter two cases can limit or prevent crystal formation. Small angle scattering is sensitive to the interparticle forces of a sample in solution. Light scattering studies have noted a "crystallization slot" defined by the second virial coefficient or B_{22} (Wilson & DeLucas, 2016). The second virial coefficient is a thermodynamic term that provides semiquantitative information regarding the magnitude of weak protein–protein interactions in a given solvent. It is a measure of all of the interactions between two bodies (protein–protein), including contributions from excluded volume, electrostatic factors (attractive and repulsive), and hydrophobic interactions. In crystallization studies, lysozyme and BSA showed a narrow range of B_{22} values that correlated with solution conditions yielding crystals (George & Wilson, 1994). In similar experiments performed for more than 50 different proteins dissolved in a large variety of crystallizing conditions, the solution conditions that produced crystals yielded B_{22} values that ranged between -8.0×10^{-4} and -0.5×10^{-4} mol mL g^{-2} (Wilson, 2003). More than 300 conditions produced values falling outside of this crystallization region (the crystallization slot). Those conditions yielding large negative B_{22} values produced amorphous precipitate, while positive B_{22} values resulted in clear solutions; small angle scattering can identify the "crystallization slot."

Simple crystallization screens are the most economical methods to measure B_{22} but do not easily provide quantitative detail (Dumetz et al., 2007); B_{22} can be directly calculated using small angle scattering data from a concentration series according to

$$\frac{F(q=0)}{I(c,q=0)} = \frac{1}{S(c,\ q=0)} = 1 + (2MB_{22})c, \tag{10.7}$$

where $F(q = 0)$ is the form factor evaluated at $q = 0$, $I(c, q = 0)$ is the forward scattering intensity for each concentration, $S(c, q = 0)$ is the structure factor evaluated at $q = 0$ for each concentration, M is the molecular weight of the particle, and c is the concentration (Bonneté & Vivares, 2002; Pollack, 2011). To estimate $F(q = 0)$ without the knowledge of atomic structure, the lowest concentration is used and assumed to be equivalent to the form factor, since it will experience the least degree of interparticle interactions among the series of concentrations.

Bonneté et al. (2004) used SAXS to study bovine pancreatic trypsin inhibitor (BPTI) and *Aspergillus flavus* urate oxidase (UOX) in various crystallization conditions. BPTI solubility had been well characterized in different salts, and SAXS experiments were performed at pH 4.5 in KSCN from 0 to 350 mM, NaCl from 0.1 to 1.4 M and $(NH_4)_2SO_4$ at 1.25 M. The addition of each of the salts modified the polydispersity of the sample, which prevented B_{22} calculation. However, analysis of the R_g showed an increase from ~13 Å to 24.5 Å, with salt concentration increase corresponding to decamer formation. Crystals have been shown to grow by the addition of decamers, with the nucleation kinetics being strongly related to their presence in solution. In the case of UOX, solutions were prepared in buffers at a range of different pH values and with crystallization cocktails of different molecular weight PEGs and different salts. In 10 mM Na_2CO_3 at pH 10.5, the B_{22} was measured at 0.4×10^{-4} mol mL g^{-2}, a repulsive interaction outside of the crystallization slot. Increasing the salt concentration (sodium acetate and sodium sulfate) did not shift the B_{22} into an attractive regime despite reaching the concentration to mask all surface charges. Varying the pH between 10.5 and 7.5, within the range of protein stability, increased the scattering intensity, but the interactions remained repulsive: 0.86×10^{-4} mol mL g^{-2} to 0.17×10^{-4} mol mL g^{-2}. Varying the temperature from 5°C to 30°C produced a similar trend. The addition of PEG, independent of the molecular weight used (3,350 Da; 8,000 Da; and 20,000 Da) shifted the B_{22} to an attractive region. The nucleation rate in the different crystallization drops followed the variation of the B_{22}, and UOX crystallized in a very narrow range of B_{22} values: -0.80×10^{-4} mol mL g^{-2} to -0.10×10^{-4} mol mL g^{-2}. SAXS data provided an indication of oligomerization needed for crystallization, in the case of BPTI, and the response of the sample to salt concentration, temperature, and PEG, in the case of UOX, enabling a rational design of successful crystallization conditions. A more recent study with lysozyme did not measure the B_{22} but confirmed oligomer formation for successful lysozyme crystallization (Kovalchuk et al., 2016).

Probing a sample with SAXS for crystallization screening requires many experiments to build up a response profile to crystallization cocktail components. High-throughput techniques coupled with intense sources and small sample sizes offer the way forward to rational crystallization approaches. Rodríguez-Ruiz et al. (2017) open up this possibility using a droplet-based lab-on-a-chip system to study the interaction of lysozyme as a function of protein concentration. This study was extended to actual B_{22} measurement, using samples held in a microfluidic chip consisting of a main channel, through which Krytox 100 oil was used as a carrier, and three channels that injected protein, buffer, and salt as a droplet into the main channel (Pham et al., 2017). The channels were of 200 × 200 μm cross section and hydrophobized to ensure stable drop generation. The proteins studied, rasburicase and lysozyme (20 mg/mL and 130 mg/mL, respectively), were formulated with a surfactant to stabilize the droplet interface. The resultant scattering curves were in agreement with the known structures. In the lysozyme case, the protein was studied at a range of NaCl concentrations from 0 to 400 mM, and the B_{22} values were determined. The resultant values were in good agreement with other studies, with a total of only 12 mg of lysozyme used. The B_{22} values showed that, without salt, lysozyme has repulsive interactions but that these change to attractive interactions as the salt concentration is increased. Lysozyme is soluble at high concentrations. As such, it provides a strong signal, even with a limited volume. The extension to more systems will be dependent on the signal-to-noise than can be achieved and the correct estimation of the magnitude of B_{22} coefficients (Scott et al., 2013).

SAXS has also been used to identify lipid mesophases and measure their structural parameters. Membrane proteins have been successfully crystallized in lipid mesophases, and knowledge of the lipid-phase behavior and structural parameters in a variety of conditions is essential for designing new crystallization media with novel lipids or lipid-like soluble additives (Joseph et al., 2011). Conventional SAXS methods have been used successfully for this purpose but are time-consuming and not suited for a large survey of conditions (Cherezov et al., 2001, 2002). High-throughput techniques greatly improve the process, as evidenced by a study from Joseph et al. (2011). In this case, glass sandwich crystallization plates with 96 wells were used, with the glass cover replaced with a cyclic olefin copolymer cover that was 150 μm thick, to minimize absorption and background scattering. Each well was filled with 50 nl of lipidic cubic phase mono-olein and overlaid with 0.8 μl of different crystallization screening solutions, and the plates sealed and imaged visually. SAXS measurements were carried out at 2 and 3 days after setup on beamline 4-2 of SSRL. A temperature-controlled XY stage allowed each well to be imaged in the X-ray beam. Each well was exposed for 1 s, with overall data collection and travel to the next well taking ~28 s and the entire plate taking about 45 min for data collection. The SAXS data was circularly integrated and then analyzed. Unlike the case with regular SAXS data, where particles are randomly tumbling in solution, this study identified crystallization conditions where the lipidic cubic phase was formed through characteristic sharp peaks at low q.

Mapping of macromolecular conformation

Many structural methods capture a single conformation of the macromolecule (e.g., crystallography where the molecular conformation is constrained by the conditions of crystallization). SAXS as a solution technique can access many conformations in a high-throughput mode. Hura et al. studied the ATPase MutSβ, taking SAXS measurements on ten liganded states of the protein with and without DNA substrate (Hura et al., 2013). They used a structural comparison map consisting of a matrix defining a normalized pairwise agreement ratio, R, between each pair of SAXS experiments. A volatility, V_R, is calculated between sequential bins, where

$$V_R = \sum_{i=1}^{n-1} \frac{R(q_i) - R(q_{i+1})}{\left(R(q_i) + R(q_{i+1})\right)/2}, \tag{10.8}$$

where n is the number of experiments. The V_R emphasizes distinctions, similarities, and trends, reducing the influence of specific resolution regions of the data. The result is a graphical representation of solution state conformations.

MutSβ removes extrahelical DNA loops and improperly paired bases (Owen et al., 2009). Human MutSβ undergoes an ordered series of nucleotide-dependent steps to initiate repair. In the SAXS approach, it was revealed that nucleotide binding, not DNA binding, drives MutSβ conformational changes, suggesting that MutSβ sculpts the DNA. The high-throughput analysis of ten states allowed mechanistic information to be derived that was unavailable through other techniques.

A similar study was conducted on apoptosis-inducing factor regulated by allosteric switching driven by an NADPH charge transfer complex (Brosey et al., 2016). A volumetric registration–based map revealed two distinct conformational clusters corresponding to a wild-type monomer and a perturbed dimer state, supporting a proposed mechanism for allosteric dimerization.

Rapid sample characterization

High-throughput small angle scattering techniques show promise for rapid sample characterization. Hura et al. (2009) analyzed 34 proteins that were recombinantly expressed from *Pyroccus furiosus*, and 16 targets from the Joint Center for Structural Genomics (JCSG). Using small volumes (12 μl) at low concentrations (~1 mg/mL) coupled with the automated setup at the SYBYLS beamline and well-developed SAXS analysis methods, they categorized the proteins from the *P. furiosus* set. Of these, three proteins were identified as aggregated (all of which were structurally uncharacterized), six others had sequence homology with known structures, and five out of those six showed similar envelope structure to that of their homologs. In four cases, multimeric forms were indicated by nondenaturing gel electrophoresis, and SAXS could characterize the relevant multimers in the mixture. This type of information is important in understanding the solution state, improving sample preparation, and identifying projects that are not suitable for other structural analysis techniques or where a structural analysis may not provide significant additional information. The proteins from the JCSG set were at a higher concentration, on average 23 mg/mL, and many contained multimerization states or aggregates. Some of these were salvageable with an extra sample preparation step of filtering. Some 41 of the total 50 samples provided usable envelope information.

Grant et al. (2011) carried out a similar study on 28 proteins that had been submitted to a high-throughput crystallization screening center (Luft et al., 2003, 2011) through the North East Structural Genomics group. This set consisted of 15 proteins having a crystallographic model, 2 of them having multiple constructs, 9 with NMR structural models, and 2 with both crystallographic and NMR models available. It was clear that the SAXS bead reconstructions closely matched the known structural data, with the resulting envelopes being extended in regions where residues were not crystallographically resolved. The SAXS results were also consistent with the NMR data. A number of samples had molecular weights associated with oligomers. In these cases, the crystallographic model was used with symmetry operators to see if the oligomer matched the crystallographic oligomer (if appropriate) or presented a different state. For the majority of cases, the crystallographic oligomer matched the solution data but in two cases, a different oligomer was indicated. Where oligomers differ, the biologically correct oligomer cannot be determined with crystallographic and SAXS data alone. The crystallographic data, although constrained by packing artifacts, represents the approximate concentration of the protein in the cell; the SAXS data, unconstrained by packing, represents the protein in a relative dilute solution. Further experimental evidence is required to determine the biological oligomer as neither accurately represent the protein in vivo.

Both studies demonstrated that high-throughput SAXS techniques could provide robust structural information.

The sensitivity of SAXS to interparticle interactions and aggregation has been used in studies beyond those used to inform crystallization. The same sensitivity to aggregation has allowed SAXS to identify distinguishing attributes of mutant superoxide dismutase proteins leading to amyotrophic lateral sclerosis (Pratt et al., 2014). Amyotrophic lateral sclerosis is a progressive neurodegenerative disease, and high-throughput SAXS approaches were used to probe the aggregation sensitivity of specific superoxide dismutase mutations correlating with clinical severity. Substantial differences between the mutants were noted in the low-q region after mild chemical perturbation. The R_g increased with disease severity, with the increase being primarily due to changes in the oligomerization state rather than aberrant folding or unfolding. This application has potential in other diseases associated with protein aggregation.

10.4 Summary

SEC has proven to be useful for studying both soluble and membrane-bound proteins, allowing the study of samples that show time- and concentration-dependent dynamics. It is more time-consuming than a simple concentration series, and there are potential problems associated with accurate buffer subtraction. SAXS and in a few cases, SANS, have both been used for time-resolved studies. With improved X-ray sources, detectors, sample handling, and computing power, the ability to perform SAXS data in high-throughput is available. This enables the use of SAXS to study protein interactions, map macromolecular conformation, and rapidly characterize samples, as well as for other applications.

11 SANS

The majority of this book has been devoted to the discussion of scattering of X-ray photons. Most solution scattering experiments are, indeed, carried out using X-rays. However, the neutron represents another particle used in a variety of scattering experiments in which its unique features allow complementary and often unique experimental results to be achieved. Here, we devote a chapter to neutrons, specifically covering their properties, sources, instrumentation, and use. Table 11.1 compares many of the characteristics of X-ray photons and of neutrons with comparable wavelength.

11.1 X-Rays and Neutrons

X-ray photons can be thought of as a wave phenomenon, so that scattering experiments like crystallography can be interpreted in the classical mode, using the ideas of interference. The initial view of neutrons is much more in the particle mode: they have mass, they are visualized as particles in the atomic nucleus, and their role in nuclear fission, in which neutron collisions split nuclei, is visualized as a ballistic phenomenon. But the laws of quantum mechanics embody any particle with wave characteristics according to the equation first enunciated by Planck, $E = \hbar\omega$. Applying this particle description to the neutron shows that a particle having a wavelength suitable for biological scattering studies, in the range of 1–10 Å, has a very low energy. This energy is on the order of tens of milli-electron volts, perhaps 1 million times lower energy than a photon with similar wavelength. This has important implications for scattering experiments. First, the propensity of such neutrons to cause specimen damage is extremely limited. These energies are much lower than typical energies of the chemical bond, so that the generation of free radicals and other damaging species is basically impossible. Indeed, neutrons interact only with the nuclei of the atoms in a specimen, but negligibly with electron cloud in SANS. Protein crystals used in neutron diffraction experiments seem to last forever.

Second, the energy and frequency of these neutrons are comparable to the energy and frequency of various dynamical modes in the specimen, so that strong signals from dynamical fluctuations in the specimen can be expected.

Biological Small Angle Scattering: Theory and Practice. Eaton E. Lattman, Thomas D. Grant, and Edward H. Snell.

DOI: 10.1093/oso/9780199670871.001.0001

Table 11.1 *Characteristics of photons (X-rays) and neutrons compared*

	Photon	Neutron
Mass	0	1.67×10^{-27} kg
Typical wavelength for scattering experiment	0.1–1.0 nm	0.1–1.0 nm
Typical energy for scattering experiment	1–10 keV	2.5–25 meV, on the order of kT
Sources	Lab sources or synchrotrons	Reactors or spallation sources
Flux normalized to third-generation synchrotron source	1	10^{-7}
Scattered by	Electrons	Nucleus
Dependence on atomic number n	Proportional to n	No systematic dependence on n
Information from isotopic substitution	None	Significant: profound for H versus D
Ability to probe dynamics	Limited	Significant via energy loss spectra
Specimen damage	Serious	~Zero
Penetration	Millimeter	Deep
Energy equation	$E = \hbar\omega = \frac{hc}{\lambda}$	$E = \frac{h^2}{2m\lambda^2}$

Third, the modes of interaction between neutrons and the specimen are quite weak. Most neutrons pass through the specimen without interacting at all. Tiny fractions of them are absorbed by nuclei, creating nuclear transformations such as the conversion of protons to deuterons. Others, those that provide the signal that we will observe, are scattered by a nucleus. Some of these events are elastic, in which energy is conserved and momentum exchanged, and some are inelastic, in which momentum is exchanged but $\Delta E \neq 0$. There is also a scattering process called incoherent scattering, in which the ability of a neutron wave to interfere with the waves from other neutrons is lost, and the waves must be added using intensities without phase interactions. Incoherent scattering tracks the motions of individual scattering centers, without interference from neighbors. The ratio of incoherent to coherent scattering for most isotopes is modest but is exceptionally large for H. The large incoherent scattering by H arises from a transient association between the nuclear proton and the incoming neutron.

Another key difference between the scattering of X-ray photons and neutrons is in the shape of the scattering factor. For photons, the value of the scattering factor as q approaches zero is equal to the atomic number of the atom, and the curve gradually decreases with increasing q, approaching 0 as q approaches 5 Å^{-1}.

For neutrons, the situation is more complex. The curves are often given the name scattering lengths, a term used in quantum mechanics to describe low-energy scattering in which the wavelength of the particle being scattered is much larger than the range of the scattering potential. The scattering length is closely related to the elastic scattering cross-section, the hypothetical area that would produce the same amount of scattering as observed from the particle under consideration.

Scattering length curves show little dependence on q (the scattering center being the nucleus rather than the larger electron cloud), they bear no systematic relationship to atomic number, and they can be negative—a phenomenon impossible for atomic scattering factors. Negative scattering length curves are produced because of the complex, spin-dependent phase relationships in the scattering of neutrons by nuclei.

As Table 11.1 makes clear, there are several other important aspects of neutron scattering that are very different from those of X-rays. X-ray photons are massless and move at the velocity of light. Neutrons have nonzero mass, and the energy relation given shows that the velocity of a neutron in the energy range being used for solutions scattering is quite low—less than a meter per second. A simple implication of this is that neutrons follow curved paths to the detector, and one has to allow for this in the experimental setup and data analysis.

Inelastic scattering of neutrons is very significant. For techniques like crystallography and solution scattering, the inelastically scattered neutrons represent noise, while for studies examining lattice dynamics, they represent signal. Unlike photons, as neutrons lose energy they slow down. Thus, when the specimen is illuminated with a pulse of mono-energetic neutrons, those that are scattered inelastically will arrive at the detector later than those that scatter elastically. By using rotating apertures called choppers, one can allow only inelastically or elastically scattered particles to be detected.

11.2 Neutron Sources

Historically, neutrons for research purposes have been produced at nuclear reactors. Neutrons created in nuclear fission reactions have energies on the order of 10 MeV (million electron volts). To reduce their energy to levels desired for diffraction, the neutrons are equilibrated with a moderator, such as room-temperature D_2O, which slows them through a series of collisions. They are then allowed to emerge through a gap in the reactor wall in the form of a broad beam of very low intensity. The neutron energies in such a beam follow the Boltzmann distribution and are broadly distributed. Figure 11.1 shows the ILL in Grenoble, France, located above the ESRF, also shown in the figure. Clearly, there are only a limited number of such facilities around the world, and access to them is scarce. Since the flux from neutrons is low in comparison to that from X-ray sources, each experiment demands considerable beamtime.

Figure 11.1 An aerial view of the ILL and the ESRF in Grenoble, France. The ESRF is the large ring facility, with the ILL thermal reactor in the domed building above and the instrument hall visible between the reactor and the ESRF.
(Denis Morel)

More recently, pulsed neutron sources, termed spallation sources, have been developed. In these facilities, bunches of protons impinge on a metal targets such as mercury to produce corresponding neutron pulses which appear to emerge from the relatively small source. Spallation sources are exemplified by the fast neutron facility at the Oak Ridge National Laboratory in the US (see Figure 11.2). While still not providing flux close to that available from a synchrotron, pulsed neutron sources have enabled many more experiments than could be previously accomplished at thermal sources.

Figure 11.2 An aerial view of the Spallation Neutron Source (SNS) in Oak Ridge Tennessee. Protons are accelerated from the facility in the distance and travel to a target and experimental hall. The large building in foreground is the office complex.
(Courtesy of ORNL)

11.3 The Technique of Contrast Variation

One of the most powerful and commonly used techniques in neutron studies is termed "contrast variation." Figure 11.3 shows that the average scattering density of various biological substances, such as lipids, proteins, and RNA, are significantly different from each other because their different elemental compositions correspond to different proportions of scattering length curves of different weights. Figure 11.3 reminds us that, at low resolution, we can separate the density of particle into a constant, average density plus a fluctuational component. The observed scattering by the average particle density is proportional to the contrast between that average density and the scattering density of the surrounding solvent, which is, in general, primarily water. By changing the proportions of H_2O and D_2O, one can manipulate the average scattering density of water to match that of each of the major types of biological macromolecules.

The technique can also be illustrated schematically. Figure 11.4 shows the contrast-matching effects of varying the ratio of H_2O to D_2O on a hypothetical particle composed of distinct RNA and protein domains. The observed scattering from a homogeneous particle is proportional to the contrast between that average density and the scattering density of the surrounding solvent, which is, in general, primarily water. By changing the proportions of H_2O and D_2O, one can manipulate the average scattering density of water to match that of each of the major types of biological macromolecules.

We address contrast matching as the first example in Chapter 12, discussing its application not only in neutron studies but also with SAXS examples.

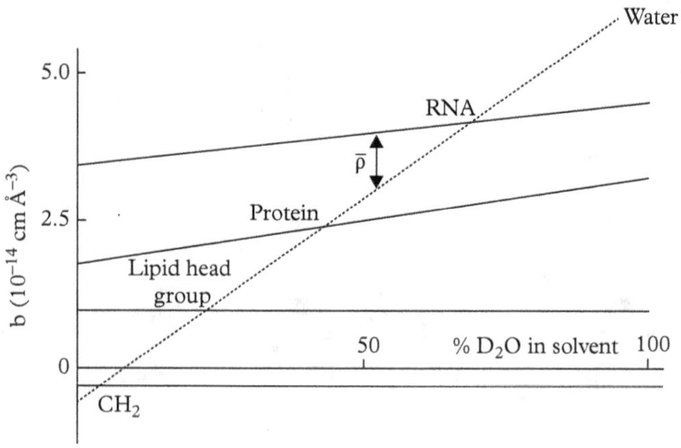

Figure 11.3 The average scattering density for some of the most common molecules found in biological material. It is assumed that all exchangeable protons are exchanged in proportion with the D_2O content of the solvent. The crossing point of the water line (H_2O/D_2O mixture) with that of a molecule is called the matching point, or isopycnic point, where the molecule is masked by the background (density is uniform).

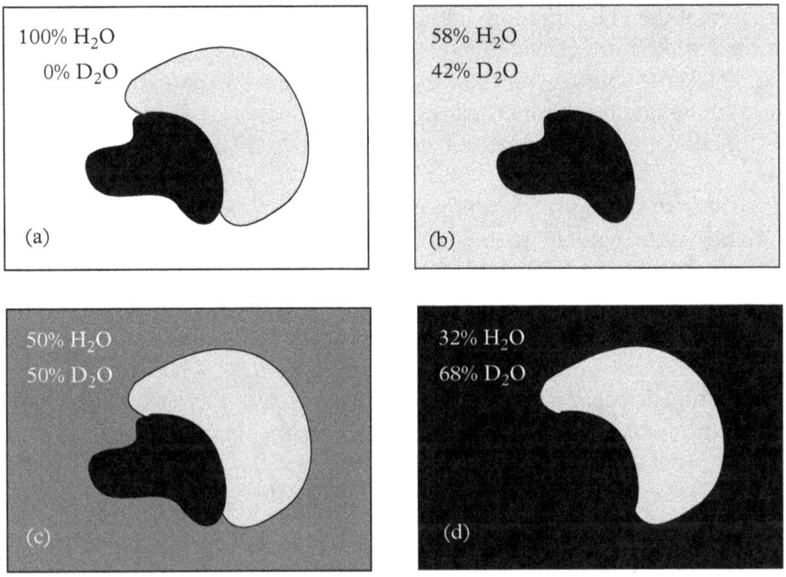

Figure 11.4 A schematic illustration of the effect of contrast matching in the case of SANS. The sample consists of a protein component (light gray) and an RNA component (black). With no D_2O present, the whole sample provides contrast. As the D_2O concentration is increased, the protein component becomes masked (b), an intermediate stage occurs (c), and then the RNA component is masked (d).

11.4 Applications of Neutron Scattering

Engelman and Moore (1972) carried out a landmark SANS experiment in which new information was gained about the structure of the 30S subunit of the bacterial ribosome. The 30S subunit was reconstituted using pairs of proteins that had been deuterated. The purpose of the experiment was to measure the distance between these two subunits using the interference fringes arising because of the scattering between them. Such an experiment requires the preparation of three control solutions in addition to the doubly deuterated one: subunit 1 in the deuterated form, subunit 2 in the deuterated form, and all subunits in the hydrogenated form. The actual experiment involved taking the difference in the scattering profile between two solutions. The first was an equimolar mixture of the doubly deuterated and non-deuterated ribosomes, while the second was an equimolar mixture of the two specimens containing a single deuterated subunit. It can be shown that the difference between the scattering profiles of these two solutions represents only the interference function between the two subunits. If the two subunits are spheres, then the interference fringes take the form of a damped sine wave whose wavelength gives directly the separation between the two spheres. The authors were able to show, however, that even for quite nonspherical particles the position of the first minimum in this fringe pattern was still very good measure of the separation of the particles.

In addition, the authors were able to gain shape information about the subunits involved, by taking the Fourier transform of the scattering curve embodying the fringe function. It was determined that a number of the subunits had a very elongated form.

Altogether, this process determined a total of nine inter-subunit distances. Use of these distances in triangulation algorithms allowed a crude model of the ribosomal 30S subunit to be determined. These measurements also gave rise to an intense effort in the computational community to develop so-called distance geometry to improve our ability to reconstruct objects from such distance constraints.

We conclude this chapter by presenting two contemporary examples of neutron scattering studies that, together, illustrate many of the unique capabilities that this technology provides.

The first of these is a study of an intrinsically disordered protein, the N protein of bacteriophage λ, in solution (Goldenberg & Argyle, 2014). Dunker et al. (2001) recognized the existence of intrinsically disordered proteins, or IDPs, in cells, and pointed out long-standing evidence for disordered regions that had been ignored by the field for many years. Since then IDPs have become an object of serious and intensive study. A key issue has been to decipher what the confirmation of an IDP might be in the very crowded environment within a cell, where macromolecules may occupy up to 30% of the volume.

In the study by Goldenberg and Argyle (2014) the authors used the protein λ N to represent a typical IDP. They mimicked the crowded intracellular environment by using concentrated solutions of either bovine pancreatic trypsin inhibitor (BPTI) or met-myoglobin. These proteins are present in concentrations of 1 to 200 mg/mL. The idea behind the study was to see how the radius of gyration of λ N varied with the concentration of the crowding molecule. A number of theoretical treatments had suggested that radius of generation might decrease as crowding increased. In addition, the authors were able to examine the compactness of λ N by monitoring the *q*-dependent scattering in the intermediate region. In order to develop contrast, λ N was deuterated, while the crowding proteins were not. The proteins were dissolved in mixtures of D_2O and H_2O to match the scattering length density of the crowding protein. Thus, the resultant $I(q)$ curves reflected only scattering from λ N. Solution scattering curves were measured as a function of the concentration of myoglobin or BPTI. Guinier plots were used to monitor R_g, while plots of $\ln I(q)$ versus $\ln(q)$ were used to monitor the power-law fall off at intermediate q. The authors used the following equation to represent the behavior:

$$\begin{aligned} I(q) &= Cq^{-\alpha}, \\ \ln\left(I(q)\right) &= \ln C - \alpha \ln q. \end{aligned} \tag{11.1}$$

In Chapter 2, we discussed the exponent α in terms of particles with well-defined surfaces, such as $\alpha = 2$ for a rod at intermediate q, for example. For coil-like polymers, in which successive segments point randomly in space, it plays a critical role in the analysis of energetics: the famous Flory constant $\nu = 1/\alpha$. In the Flory theory, the radius of gyration of a random-coil polymer is given by

$$R_g = R_0 N^{\nu}, \tag{11.2}$$

where R_0 is a constant that is a function of, among other things, the persistence length of the polymer, and N is the number of monomers in the polymer chain (Kohn et al., 2004).

Figure 11.5 shows both the Guinier plot (left) and the log–log plot (right) as functions of crowding protein concentration. The clear result is that these curves vary very little. The radius of gyration hovers very near 30 Å. In addition, α shows no trend, with a value near 1.8; the corresponding value of ν is 0.56, only slightly

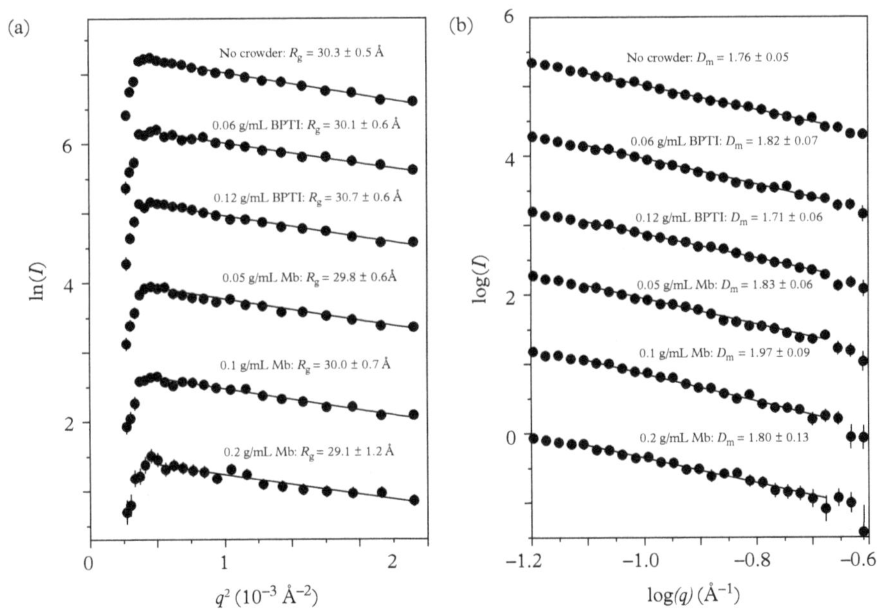

Figure 11.5 SANS from λ N in the presence and absence of globular crowding proteins, presented as Guinier (a) and log–log (b) plots. In (a), R_g was estimated by fitting data for 0.022 Å^{-1} $\leq q \leq$ 0.049 Å^{-1} to the Guinier relationship. In (b), the fractal dimension, D_m, was estimated by fitting the data for 0.078 Å^{-1} $\leq q \leq$ 0.21 Å^{-1} to the power-law relationship of $I(q) = Cq^{-D_m}$. The uncertainties in R_g and D_m are the standard errors derived from the least-squares fits.

(Reprinted from Biophysical Journal, Goldenberg & Argyle (2014) Copyright (2014), with permission from Elsevier)

greater that the 0.5 value that characterized an idealized random-walk polymer. Protein crowding seems to have little to no effect on the λ N IDP ensemble.

Goldenberg and Argyle (2014) pursued the issue of the effects of crowding using computer simulations. Scaled particle theory is a major theoretical tool for describing the thermodynamic effects of steric exclusion on folded globular proteins, which are represented as solutions of hard convex particles. Despite considerable efforts, it is a more of an open question as to how to account for the effects of crowding on unfolded proteins, which are broad ensembles of rapidly interconverting conformations. In an approach whose details are beyond the scope of this book, the authors computationally generated typical members the λ N ensemble, checking their reliability by comparing calculated and experimental *I(q)* curves. The crowding energy of each conformer was evaluated by creating an equivalent sphere and using scaled particle theory to determine the energetic cost of moving that sphere from dilute solution into the crowded one. The ensemble contained conformers with equivalent radii from 17 to 68 Å. Their simulations are sensitive to the value of ΔG_{solv}, the solvation free energy for λ N. Their final simulations show changes in average dimensions of about 10% from crowding, less than others but still below the effective value of zero found experimentally.

The second example comes from the nonbiological world of synthetic polymers, in particular, bottlebrush polymers in which rather short side chains, analogous to the bristles, are grafted into a backbone, which is analogous to the brush body. Zhang et al. (2014) integrated elastic SANS, incoherent scattering, and computer simulations to develop atomic level models for representative ensemble members of bottlebrush polymers with a relatively rigid poly(norbornene) backbone, decorated with poly(lactide) side chains of various lengths.

Figure 11.6 shows models of this bottlebrush with three different side-chain lengths, in idealized conformations used to at the initiation of the authors' molecular dynamics simulations. The authors collected SANS *I(q)* curves for the three polymers at the SNS at the Oak Ridge National Laboratory. The molecules were dissolved in tetrahydrofuran. These curves were compared with those calculated from conformers generated during all-atom molecular dynamics simulations, which were motivated by a desire to move beyond the coarse-grained modeling. The scattering intensity from the trajectories was calculated using

$$I(q) = \frac{1}{V}\left\langle \left| \sum_{m=1}^{N} b_m e^{iq \cdot r_m} \right|^2 \right\rangle, \tag{11.3}$$

where q is the wave vector, r is the position of the atom, b_m is the coherent scattering length of atom m, V is the sample volume, and m is the index of atoms of the polymer, which consists of N total atoms. The angular average over the wave vectors takes into account the random orientations of the polymers in the solution.

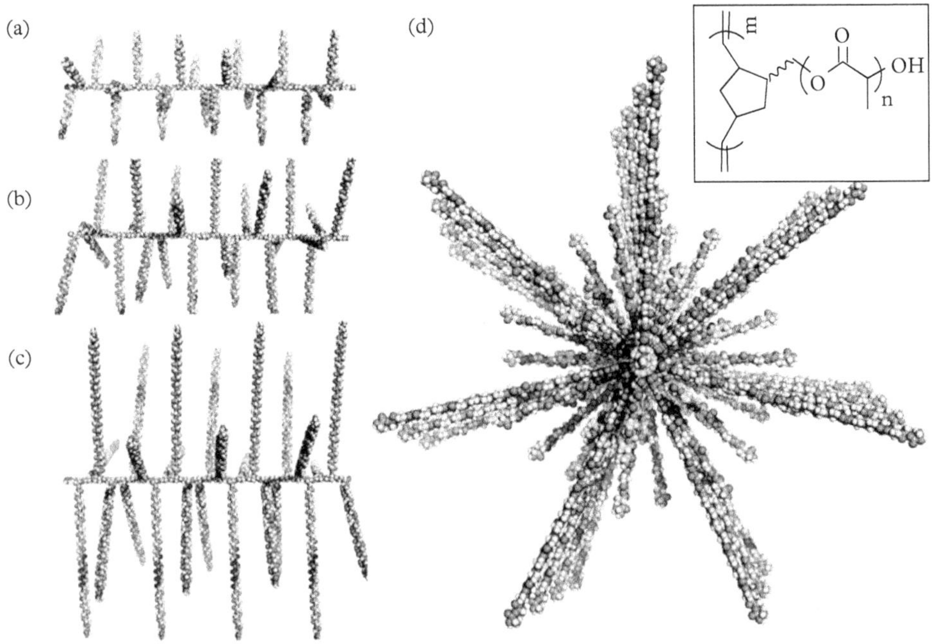

Figure 11.6 Initial atomistic models for (a) PNB25-g-PLA5, (b) PNB25-g-PLA10, and (c) PNB25-g-PLA19. (d) Projected view of model polymers along the backbone. The chemical structure is shown in the inset.

(Reprinted with permission from Zhang et al. (2014). Copyright 2014 American Chemical Society.)

The authors found that early in the simulations, when the conformation had not moved from the initial state, agreement between the calculated and observed curves was quite poor, but that it improved with time, so that at 40 ns, agreement was quite good. It is important to realize that the molecular dynamics trajectory is that of a single molecule, so that at any time point, $I(q)$ is calculated from a single polymer conformation; there is no averaging. A χ^2 analysis revealed that conformations with $R_g \approx 37$ Å gave a best fit to the observed curves. Figure 11.7 illustrates the kind of information derived from the simulation guided by experimental data.

This discussion bypasses many additional aspects of the paper, including the dramatic differences in bottlebrush conformations as a function of side-chain length, and the limitations on conformation sampling of the 60 ns simulations. For our purposes, the illustration of how SANS and computation together can provide both atomic level information and accurate representative conformations is the critical point.

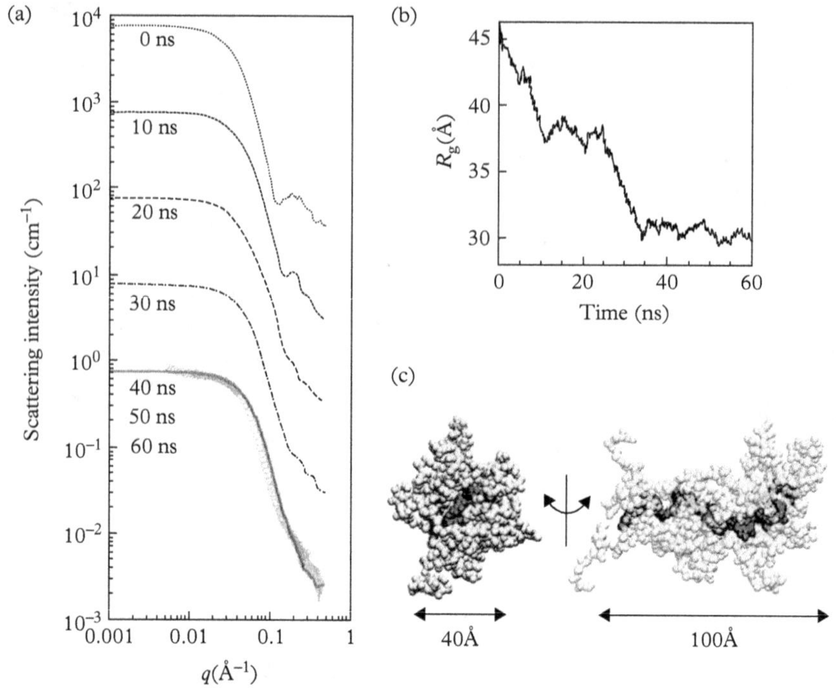

Figure 11.7 (a) Scattering intensities calculated at different simulation times: the values for $I(q)$ at 40, 50, and 60 ns overlap with each other; $I(q)$ from the experimental SANS measurement is shown with blue open circles. (b) Radius of gyration over the simulated time. (c) Snapshot of the simulated polymer at 60 ns (left, axial view; right, side view). Backbones are indicated with dark shading; side chain carbons are light gray.

11.5 SANS Beamlines

The number of neutron facilities and, therefore, beamlines is less than the number of synchrotron SAXS-based facilities. Teixeira et al. (2008) summarize many of these instruments along with details of their capabilities.

The Open Pool Australian Lightwater (OPAL) reactor at the Australian Center for Neutron Scattering has a number of SANS instruments. QUOKKA is used for biological studies, supplemented by an ultra-SANS instrument, KOOKABURRA, and a time of flight instrument, BILBY. At the European Spallation Source in Sweden, two SANS instruments will be available: LoKI is a generally applicable SANS instrument suitable for biological studies, and SKADI enables more diverse sample environments. The NIST Center for Neutron Research in Maryland, USA, has several SANS instruments. The Center for High Resolution Neutron Scattering on neutron guide 6 (NG6) and the SANS instrument on neutron guide

7 (NG7) are general-purpose instruments that have been used for biological studies. Instruments at the ILL include the beamlines D11, D22, and D33; D22 is the main biology instrument. ISIS at the Rutherford Appleton Laboratory in Oxfordshire, UK, is a pulsed neutron and muon source with several SANS instruments. Biological experiments can be conducted on Loq, Sans2d, and Nimrod. In Japan, the Japan Research Reactor-3 reactor has the SANS-J-II instrument and the small and wide angle scattering instrument TAIKAN is available at the Materials and Life Science Experimental Facility, at the Japan Proton Accelerator Complex (J-PARC). At the Oak Ridge National Laboratory in Oak Ridge, Tennessee, the High Flux Isotope Reactor (HFIR) and the SNS both have SANS instruments. HFIR has the Biological Small-Angle Neutron Scattering

Table 11.2 *List of biological SANS beamlines*

Source	Instrument/s	q range (Å^{-1})	Reference/s
OPAL, The Australian Centre for Neutron Scattering, Sydney, Australia	KOOKABURRA QUOKKA BILBY	2.78×10^{-5} 6×10^{-4}–0.7 0.0003–1.3000	(Gilbert et al., 2006; Hamilton et al., 2009; Rehm et al., 2013; Rehm & de Campo, 2016; Sokolova et al., 2016)
European Spallation Source, Sweden (under construction)	LoKI, SKADI		(Jaksch et al., 2014)
NIST Center for Neutron Research, Maryland, USA	CHRNS NG7	0.0015–6.0000 0.0008–0.7000	(Glinka et al., 1998)
ILL, Grenoble, France	D11 D22 D33	3×10^{-4}–1 4×10^{-4}–0.85	(Dewhurst, 2008; Lindner & Schweins, 2010)
ISIS, Rutherford Appleton Laboratory, Oxfordshire, UK	Loq Sans2D	0.006–0.240 (0.150–1.400) 0.002–3.000	(Heenan et al., 1997; Heenan et al., 2011; Duxbury et al., 2014)
J-PARC	TAIKAN	5×10^{-3}–10	
HFIR, Oak Ridge National Laboratory, Tennessee, USA	CG-3	0.0009–1.0000	(Heller et al., 2008; Heller et al., 2014)
Spallation Neutron Source, Oak Ridge National Laboratory, Tennessee, USA	EQ-SANS	0.002–1.400	(Zhao et al., 2010; Liu et al., 2011)
SINQ, Paul Scherrer Institut, Switzerland	SANS-I SANS-II	0.06–54.00 0.2–35.00	(Blau et al., 2009)

instrument (BIO-SANS, for short) and a parallel facility used for general-purpose SANS studies. The SNS has two SANS instruments: the Extended *q*-range Small-Angle Neutron Scattering Diffractometer (EQ-SANS), which is available for biological studies, and the specialized Ultra-Small-Angle Neutron Scattering Instrument (USANS, for short), which is less suited to biological studies. The Swiss Spallation Neutron Source (SINQ) has two SANS instruments, SANS-I and SANS-II, with the SANS-I instrument being used more for biological studies.

Table 11.2 details SANS beamlines worldwide and is not meant to be an exhaustive list but to capture some of the current instruments and provide the reader with references to explore further.

11.6 Summary

SANS is a specialized application of solution scattering limited by source availability and intensity. While not routinely used for structural studies of biological samples in general, it does have unique characteristics that make it attractive to determining the individual positions of components of complexes. This is due to the scattering properties of hydrogen and deuterium, as these allow the technique of contrast matching. SANS is highly complementary to SAXS and provides unique information not available by other techniques.

Part 4

Applications Past, Present, and Future

12

Examples of Biological Small Angle Scattering

12.1 The Use of Contrast Matching

12.1.1 The Neutron Case

In contrast matching, as discussed in Chapter 2, section 2.5, the scattering density of the solvent is adjusted to be equal to that of a compositionally distinct substructure or component, such as lipid. The small angle scattering contribution from such an element effectively vanishes, thus confining the scattering to that from other components, such as protein. It is thus used to select individual components of complexes or different structures within the whole and to mask out the contribution from other components, allowing substructures to be identified. It is more common in neutron scattering due to the ease of application with the opposite signed scattering factors for hydrogen and deuterium (see section 11.3). In the first example, this is used to place subunits of KaiB in the KaiB–KaiC complex involved with circadian rhythms in cyanobacteria (Sugiyama et al., 2016).

Circadian rhythms allow organisms to adapt to daily alterations in the environment. Cyanobacteria are photoautotrophic organisms capable of oxygen-producing photosynthesis. The central oscillator that generates the circadian rhythm in the cyanobacterium *Synechococcus elongatus* comprises three proteins: KaiA, KaiB, and KaiC. The KaiB–KaiC interaction is a key event during oscillation of the cyanobacterial circadian protein system. To characterize the oligomeric state of the KaiB–KaiC complex in solution Sugiyama et al. (2016) employed, native mass spectrometry and sedimentation velocity analytical ultracentrifugation. They indicated that upon titration with KaiB, the KaiC hexamer showed highly cooperative formation of a complex with a molecular mass of 428,600$\pm$410 Da, a 6:6 stoichiometry.

The scattering intensity arising from each domain is proportional to the square of its molecular mass. Scattering from the much larger KaiC (with a molecular mass of 65kDa) dominates that of KaiB (with a molecular mass of 14kDa)—it is difficult to collect structural information on KaiB from the scattering profile of the complex. To overcome this, Sugiyama et al. (2016) assembled the 6:6 KaiB–KaiC complex with deuterated KaiC subunits (KaiC$_{\mathrm{DT}}$) and non-deuterated KaiB

Biological Small Angle Scattering: Theory and Practice. Eaton E. Lattman, Thomas D. Grant, and Edward H. Snell.

DOI: 10.1093/oso/9780199670871.001.0001

subunits. The solvent D_2O ratio at the contrast matching point was determined to eliminate scattering originating from the partially deuterated $KaiC_{DT}$ subunits. SANS was used to study the complex. In 97% D_2O solvent, $KaiC_{DT}$ was virtually invisible in terms of scattering because the scattering length density of $KaiC_{DT}$ was matched to that of the solvent. The neutron scattering of KaiB subunits were clearly observed in the complex. The scattering profile of the $KaiC_{DT}$–KaiB complex approximately followed the q^{-2} power law, indicating that the six KaiB subunits were arranged into a disk-like shape.

With six-fold symmetry of arrangement of the six KaiB subunits along with their KaiC-binding surfaces known from deuterium exchange and mass spectrometric data (Abe et al., 2015), the authors examined three distinct structural models of the KaiB–$KaiC_{DT}$ complex (Figure 12.1). The first model had the subunits forming a hexameric ring on the top of the KaiC C1 ring, the second had the KaiB subunits located on the edge of the KaiC C1 ring, and the third had the KaiB subunits located on the side of the KaiC C1 ring. The scattering profiles were calculated for each model and compared with the experimental data (Figure 12.2). The results clearly showed that the first model illustrated the best structural arrangement of the KaiB subunits in the complex. Furthermore, the experimentally determined R_g, 33.6Å, was consistent with that computed from the first model, 33.2Å, in comparison with those from the second (44.2Å) and third (59.4Å) models. The hexameric KaiB ring forming on the KaiC hexamer was consistent with the C6-symmetrized particle-image-based cryo-electron microscopy structure (Villarreal et al., 2013).

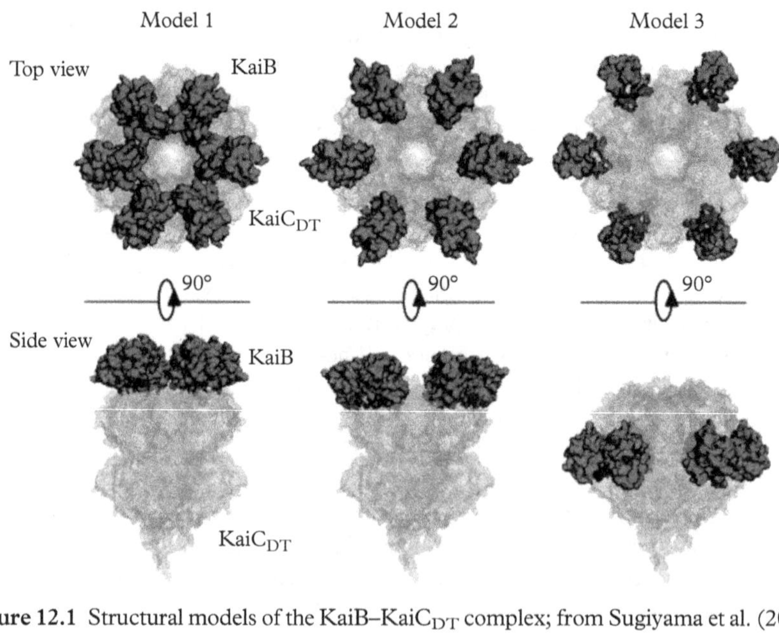

Figure 12.1 Structural models of the KaiB–$KaiC_{DT}$ complex; from Sugiyama et al. (2016).

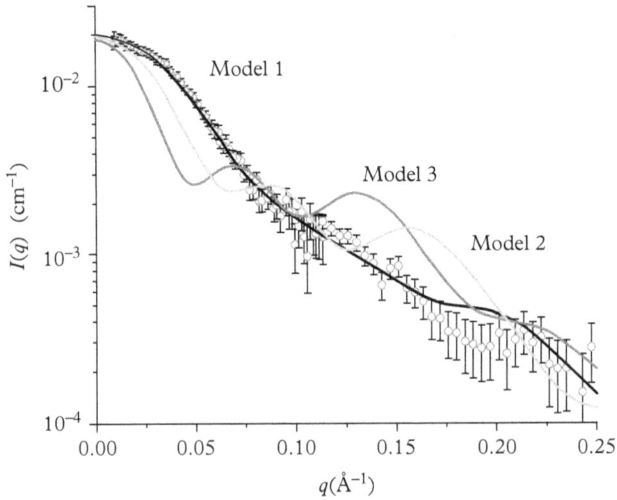

Figure 12.2 SANS profiles calculated from Model 1, Model 2, and Model 3, along with the experimentally obtained profile (open circles); from Sugiyama et al. (2016).

Sugiyama et al. (2016) noted that in this interaction mode, the hexameric KaiB subunits cover the top surface of the KaiC C1 ring, including the interaction site for SasA, a histidine kinase that operates as a clock-output protein in the regulation of circadian transcription (Abe et al., 2015). Therefore, it is conceivable that the cooperative binding of KaiB as a competitive inhibitor against SasA promotes synergistic release of this clock-output protein from the KaiC hexamer. This study illustrated how SANS enabled data from complementary techniques to produce a new understanding of mechanism.

A classic example of the use of contrast matching in SANS is by Jacrot and colleagues (Jacrot, 1976; Jacrot et al., 1977) in determining the radial disposition of protein and RNA in a nearly spherically symmetric particle, the Southern bean mottle virus. Figure 12.3 shows scattering curves from the virus at two different concentrations of D_2O. Shifts in the peak positions are clear. Because the particle is effectively spherically symmetric, the sharp dips in the scattering curve can be identified as places where the underlying spherical-particle transform changes sign. Using this phased $F(q)$ curve, one can reconstruct a radial trace through electron density within this particle as a function of D_2O concentration. This is shown in Figure 12.4. The curve obtained at a D_2O concentration that contrast matches RNA, 69%, shows a spherical shell of density at the periphery of the particle, proving that the protein is on the outside. This study produced critical knowledge on virus structure.

12.1.2 The X-Ray Case

Contrast matching is also possible in the X-ray case, as discussed in Chapter 5, section 5.7. The average electron densities of water, 50% sucrose, proteins, and

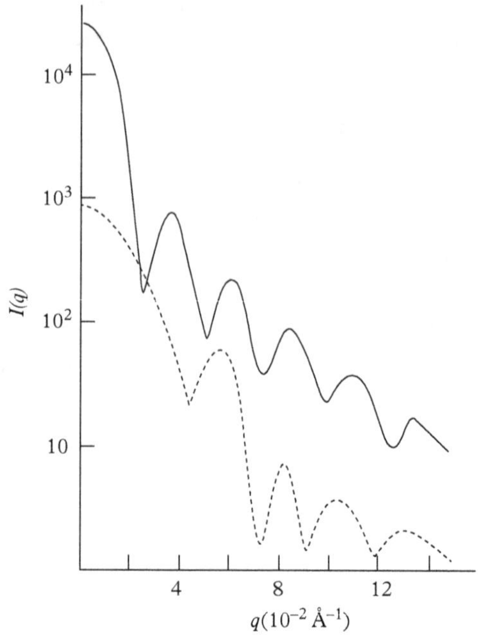

Figure 12.3 Scattering curves from the Southern bean mottle virus in solutions of different D_2O content: the continuous line is with 69% D_2O and scattering mostly due to the protein shell, and the dashed line is with solvent content 42% D_2O and the scattering mostly by the nucleic acid (RNA). The subsidiary maxima are shifted to a larger q; this indicates that the sphere that approximates the volume occupied by the RNA has a smaller diameter than the virus.

(Reprinted by permission from Macmillan Publishers Ltd: Nature (Jacrot et al., 1977), © 1977.)

nucleic acids are 0.33, 0.40, 0.44, and 0.55 e^- Å^{-3}, respectively. Chen et al. (2014) studied the nucleosome core particle, a 147-basepair DNA chain wrapped around a symmetric octamer of histone proteins. The scattering intensity as a function of contrast for a protein–nucleic acid complex can be simplified as

$$I(q) = (\Delta\rho_1)^2 I_1(q) + 2\Delta\rho_1\Delta\rho_2 I_{12}(q) + (\Delta\rho_2)^2 I_2(q), \tag{12.1}$$

where $\Delta\rho_1$ and $I_1(q)$, and $\Delta\rho_2$ and $I_2(q)$, represent the contrast and scattering intensity of the protein and nucleic acid, respectively, and $I_{12}(q)$ is the cross-term (Svergun & Koch, 2003). The authors used this relationship by increasing the solvent electron density to match that of the protein. The first and second terms vanish, and the scattering profile is dominated by the nucleic acid term.

The modulation of DNA accessibility by nucleosomes is a fundamental mechanism of gene regulation in eukaryotes. The dynamics of DNA packaging and unpackaging from the nucleosome core particle is a critical aspect of regulation, but little is known about the structures of the partially unwrapped intermediates

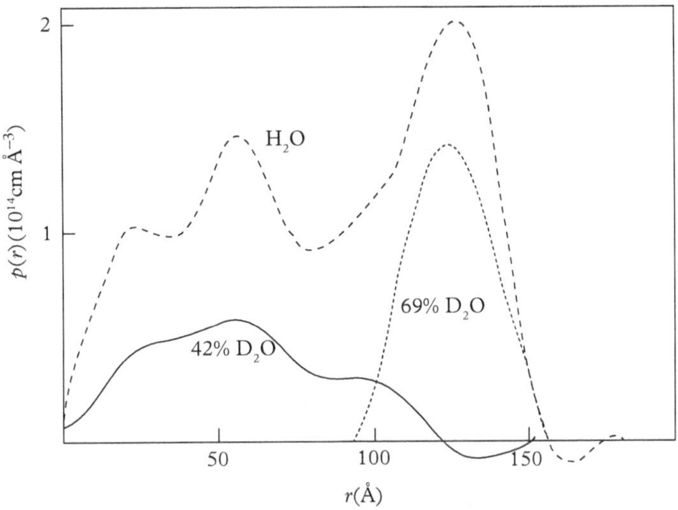

Figure 12.4 The radial density distribution in the Southern bean mottle virus, obtained by Fourier transform (in the spherical approximation) of the curves shown in Figure 12.3. The density is in absolute values. Radial distributions in H_2O, 42% D_2O, and 69% D_2O are shown. The variation in density with contrast allows an unambiguous determination of the localization of protein and DNA.

(Reprinted by permission from Macmillan Publishers Ltd: Nature (Jacrot et al., 1977), © 1977.)

relevant to nucleosome dynamics in processes such as transcription, DNA replication, and repair. SAXS data was collected on the nucleosome core particle assembled from *Xenopus laevis* histones and two well-characterized 149-basepair nucleosome core particle–positioning sequences, Chen et al. (2014). The data was collected at the CHESS G1 station, using oscillation to reduce radiation damage. A concentration series showed negligible concentration dependence. A time-resolved experiment was also conducted by incorporating a stopped-flow mixer mixing the core-particle sample with high-concentration salt buffer.

The initial SAXS data measured the response of the core particles to increasing NaCl concentration, both with and without sucrose used as a contrast agent for the protein. An analysis of the SAXS data revealed dramatic conformational changes from globular to extended structures with the increase in NaCl concentration, as anticipated from the dominance of the electrostatic interaction between histones and DNA. SAXS data with the sucrose present allowed clear and identifiable features to be seen in the NaCl-dependent $P(r)$ curves, revealing specific conformational details on the unwrapping DNA. The time-resolved data gave kinetic data on the process, aided by being able to see the DNA with the protein masked.

Contrast matching is effective in SAXS studies of protein–nucleic acid complexes. This study was extended, combining it with time-resolved FRET to understand the sequence of histone release (Chen et al., 2017).

12.2 Time-Resolved Studies at Synchrotrons and XFELs

SAXS is useful in studying biological dynamics and has been used extensively to study protein folding and denaturation (Millett et al., 2002; Kathuria et al., 2011). The details of this use are somewhat beyond the basic understanding of the technique presented in this book, and it is a specialized area in itself. SAXS and WAXS have been used with good effect in studying light-activated structural transitions. Protein dynamics is often key to function and occurs over a wide range of timescales, from subpicoseconds to seconds, and a similarly wide range of length scales, from the sub-angstrom level to tens of angstroms or more. Small angle scattering is well suited to investigating larger ranges of motions because it is solution based, so these motions are not constrained by crystal packing artifacts. For light-activated processes, the time resolution is largely limited by X-ray source and detector properties.

A number of light-activated systems have been studied at synchrotrons with pump–probe techniques using known structural information and the differences in successive scattering curves. Large-scale conformational changes have been imaged in the signaling of photoactive yellow protein (Kim et al., 2015). The photoreaction of photoactive yellow protein has been studied crystallographically using the Laue technique and, more recently, through pump–probe experiments with XFEL sources (Pande et al., 2016). This bacterial light sensor converts a light stimulus into a biological signal, and its mechanism is well understood. Time-resolved pump–probe SAXS experiments were conducted at the APS on beamline ID14B. An E46Q mutant in solution was irradiated by circularly polarized nanosecond 460 nm laser and probed by 15 keV X-ray pulses incident at well-defined time delays (Kim, Yang, et al., 2016). X-ray scattering patterns were recorded using a CCD detector. The time-resolved difference scattering curves were generated by subtracting a scattering curve recorded at -5 μs (before the laser pulse) from the curves at positive time delays. The data acquisition was implemented at a repetition rate of 2 Hz while translating the quartz capillary containing the sample, allowing for the different time points. At each time delay, over 150 scattering curves were averaged to achieve a high signal-to-noise ratio. Figure 12.5 illustrates the experimental data over time.

To quote the authors, "The difference scattering curves show not only a significant amplitude in the small-angle region ($q < 0.3$ Å^{-1}), which is related to the global structural change, but also oscillatory features in the wide-angle region ($0.3 < q < 1.0$ Å^{-1}), which is attributed to the subtle structural change such as the rearrangement of the secondary structure." SVD was used to identify the principal components of the scattering species in a manner similar to that used in time-resolved crystallographic studies (Schmidt et al., 2003; Rajagopal et al., 2004). Four structurally distinct intermediates were determined with four time constants: 3.1 (±2.2) μs, 143 (±4) μs, 6.7 (±1.0) ms, and ~1 s. Scattering curves directly reflecting the intermediate structures were constructed using PCA (Figure 12.5(b)). The population distribution

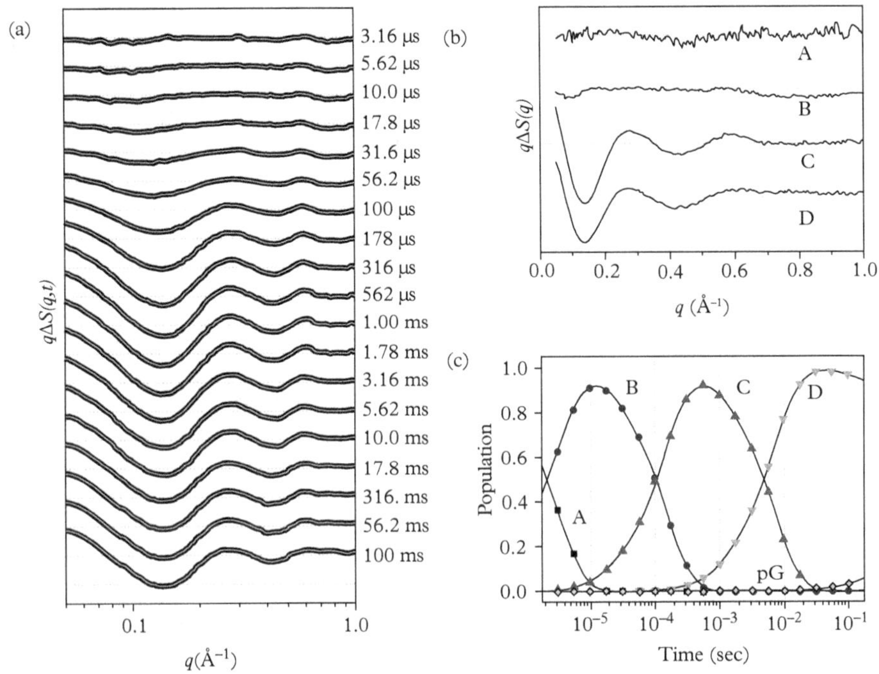

Figure 12.5 Time-resolved X-ray solution scattering data for the E46Q mutant of photoreactive yellow protein, showing photocycle and global kinetic analysis. In (a), the experimental and theoretical difference scattering curves are shown overlaying completely. In (b), species-associated difference scattering curves are extracted using PCA. In (c), the time-dependent population change of the intermediates following the proposed kinetic model determined by PCA is shown.

(Reproduced from Kim et al. (2016) with permission from the PCCP Owner Societies.)

of each intermediate structure was determined (Figure 12.5(c)). The results, when combined with structural knowledge and spectroscopic techniques, allow for a structural understanding of the photocycle. The SAXS and WAXS data provide time points, population information, and structural envelopes that evolve during the photocycle. Distinct structural changes occur in the protein as evidenced by the ranges of R_g and D_{max} observed, 15.08–15.36 Å and 41.02–58.09 Å, respectively. Precautions must be taken to ensure that perceived structural changes are not sample interaction effects.

A subtle structural change occurs in the case of photodissociation of CO from hemoglobin, another system that can be studied with a pump–probe approach. Homodimeric hemoglobin (HbI), which has two subunits, is a good model system for investigating allosteric structural transition as it exhibits cooperativity in ligand binding (Kim, Muniyappan, et al., 2016). The T72V mutant of hemoglobin was prepared in a carbon-monoxide derivative form, T72V HbI$(CO)_2$. Beamline 14IDB was used again, with 532 nm laser pumping and X-ray patterns recorded at time delays ranging from 100 ps to 10 ms. The sealed capillary containing the sample was translated back and forth along its long axis. The movement was synchronized with the pulse trains of laser and X-rays so that a fresh sample was provided for every exposure of X-ray and laser pulses. The individual structural contributions to the evolving scattering curves were extracted using SVD. The SVD analysis indicated that the initial T72V HbI$(CO)_2$ in solution undergoes a photoreaction through three structurally distinct intermediates. The structurally distinct intermediates had time constants of 3.1 ns, 140 ns, 490 ns, 980 ns, 39 μs, 270 μs, and 4.0 ms and were in agreement with a kinetic model of the photo-induced structural transitions. The SAXS data indicated that the first two intermediates were structurally similar but that the third differed significantly. Rigid body modeling was carried out, with the experimental data for this third intermediate using a deoxy-T72V structure, with the helicies and hemes comprising 18 rigid bodies. Multiple refinement process took place, with 600 different initial structures. The SAXS data for the third intermediate gave structures that were closest to the deoxy structure rather than HbI$(CO)_2$ but were structurally distinct with a root-mean-square distance of 0.60 Å. The combination of time-resolved SAXS and complementary structural data allowed the mechanism of action to be understood.

In both the photoreactive yellow protein and the hemoglobin studies, an expanded energy window (bandpass) was used to increase the available X-ray flux to record usable signals at a defined time. The combination of improved X-ray sources, the advent of the XFEL, and novel algorithms which can exploit timing jitter (see Fung et al., 2016) offer the opportunity for even more precision in the exact time points sampled.

12.3 SAXS Providing Information Missing with Other Structural Techniques

SAXS data can be used to study regions that are not resolved by other structural methods (e.g., X-ray crystallography, NMR, or cryo-electron microscopy). Santos et al. (2017) used SAXS to study two DNA binding proteins, Dps1 and Dps2, from *Deinococcus radiodurans*. Each protein assembles in a conserved dodecameric structure with structurally uncharacterized N-terminal extensions. The extensions have been proposed to be involved in DNA interaction and condensation. They are thought to be mechanistically important.

SAXS data was measured at the ESRF BioSAXS beamline BM29 using an inline SEC system. Four samples were studied: full-length Dps1, full-length Dps2, and two truncated constructs without the N-terminal extensions, Dsp1 lacking residues 1–50, and Dps2 lacking residues 1–39. The truncated Dps2 showed the presence of several oligomeric states but the dodecamer was present only in small quantities, making it impossible to study further.

The SAXS data for the truncated Dps1 was consistent with the crystal structure. The full-length Dps1 and Dps2 showed peaks in $P(r)$ at 55 Å with a shoulder at 20 Å, and at 50 Å with a shoulder at 20 Å, respectively. For both, the $P(r)$ also showed extended tails at ~121 Å. This tail was absent from the truncated Dps1, which had a D_{max} of ~100 Å.

The molecular envelopes from the proteins were obtained by ab initio reconstruction with dodecameric symmetry imposed. This proved essential to the analysis. The envelopes are shown in Figure 12.6, with the crystallographic structure placed in the envelope. The residues unresolved in the crystallographic study are clearly seen as tails in the reconstruction. Ensemble techniques were also used and showed more flexibility in the tails than that seen with imposed symmetry. This is consistent with not being able to resolve the N-terminal domain in the crystallographic structure.

Dps2 serves as an iron-storage protein, so additional SAXS data taking advantage of the presence of iron was collected on the ID02 beamline, also at the ESRF. In this case, 50% glycerol was used to increase the contrast of the iron core compared to the protein. Six different iron concentrations were studied: a control with no iron and then 24, 48, 100, 200, and 400 iron atoms per dodecamer. As the protein took up iron, the core radius decreased to 22 Å, compared to 29 Å for the apo protein. The mean radius decreased from 45 Å to 37 Å. There are limitations both in the experiment and in the analysis techniques available, and absent the direct electron density determination described in Chapter 14, the authors were only able to determine molecular envelopes for the apo form, and the dodecamer forms with 24 and 48 iron atoms, respectively.

It is beyond the scope of this book to discuss in detail the mechanistic interpretation of the data, but it is important to note the use of SAXS to provide structural information on regions that were crystallographically unresolved, and the ability to structurally explore different solution conditions unconstrained by a single crystallization state.

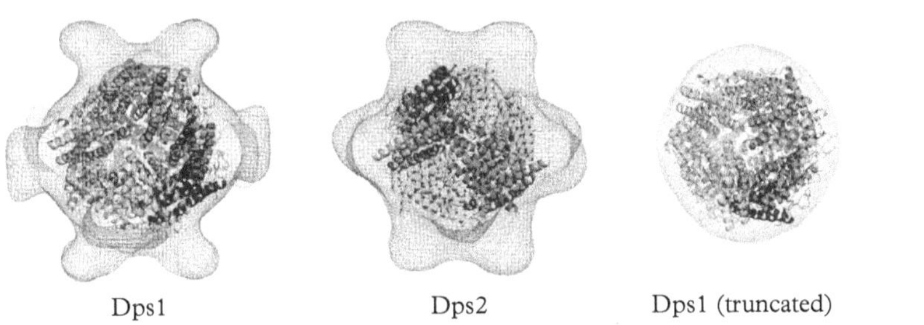

Figure 12.6 Dps ab initio models superimposed on the corresponding crystal structures. Dps crystal structures are represented in cartoon, with each monomer in a different shade. (Reprinted from Journal of Molecular Biology, Santos et al. (2017), Copyright (2017), with permission from Elsevier.)

12.4 Putting Complexes in Context

RNAse III, an enzyme specific for double-stranded RNA (dsRNA) cleavage, is required for RNA maturation and gene regulation. Song et al. (2017) studied Rnt1p, which is a eukaryotic RNAse III, and the Rnt1p–RNA complex using a combination of SAXS, and X-ray crystallography, coupled with existing structural knowledge. SAXS data was recorded at beamline 12-ID-B of the APS on both apo-Rnt1p and the post-cleavage complex. Each sample was recorded at four concentrations: the original 3 mg/mL and then two-, four-, and eightfold dilutions, with the scattering then being extrapolated to an infinite dilution. Thirty 1–2 s images were recorded for each concentration and the buffer blank.

For the complex, the missing terminal residues from the structural model (residues 1–41 and 458–472) and the inter-domain residues (152–196) were treated as random polypeptide chains. For apo-Rnt1p, the crystal structure of the the Rnt1p post-cleavage complex was used as the initial model. The N-terminal domain dimer, the endonuclease domain (RIIID) dimer, and two dsRNA-binding domains (dsRBDs) were defined as rigid bodies for the purposes of rigid body refinement to the SAXS data.

The estimated molecular weight from the SAXS data indicated that apo-Rnt1p forms a dimer in solution. However, the D_{max} was 20 Å larger than the crystallographic model of the post-cleavage structure, indicating an extended conformation in solution. The R_g was also significantly larger than expected.

Figure 12.7 shows the results, with the apo-Rnt1p crystallographic structure overlaid on the ab initio molecular envelope (Figure 12.7(a)). The data is in good agreement with the predicted scattering (Figure 12.7(b)). The rigid body modeling indicates that the apo-Rnt1p adopts an open conformation that is significantly different from that of the complex. There is a 60 Å displacement of the N-terminal domain dimer, and displacements of the dsRDB1 and dsRBD2 by 25 and 20 Å, respectively. In the open form of Rnt1p, the dsRDBs have a sufficient amount of freedom and space for initial substrate selection.

SAXS data showed that the ligand-free protein adopts an open conformation in solution. Crystallographic data and modeling indicated substrate loading with the N-terminal domain moving adjacent to the RNA. Cleavage occurs with the resulting crystal structure matching the solution data, and the process starts again. SAXS data provided the information necessary to understand the first step of the cycle with the open conformation and supported crystallographic data on the last step.

12.5 Functional Changes

Reimer et al. (2016) used SAXS to study nonribosomal peptide synthetases (NRPSs), very large proteins that produce small peptide-like molecules with wide-ranging biological activities. NRPSs contain tailoring domains that help accommodate D-amino acids and other non-canonical structures and are vital for

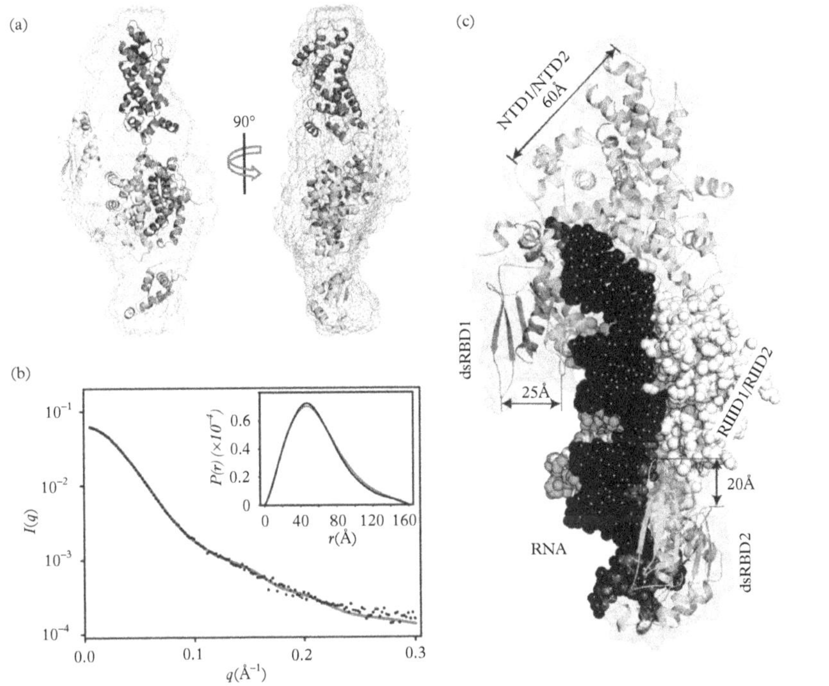

Figure 12.7 (a) Rigid body fits of apo-Rnt1p in the ab initio molecular envelope and (b) overlay of scattering data to the calculated scattering. The insert shows the $P(r)$ for both the experiment and the model. (c) The apo-Rnt1p and the post-cleavage complex are shown. The apo-Rnt1p is in the extended conformation with slightly darker ribbon diagrams. The RIID dimer and the RNA in the complex are shown as atomic sphere models in white and black, respectively.

(Reprinted from Structure, Song et al. (2017), Copyright (2017) with permission from Elsevier.)

the production and bioactivity of the NRPS products. In addition, NRPSs contain a core that consists of an adenylation domain (called the "A domain"), a peptidyl carrier protein domain (called the "PCP domain"), and a condensation domain (called the "C domain"). In order for an amino acid to be incorporated into the growing peptide product, it must be added to this core: the A domain then selects, activates, and transfers the substrate amino acid to the PCP domain, which transports it to the C domain for peptide bond formation.

In linear gramicidin synthetase (comprising the subunits LgrA, LgrB, LgrC, and LgrD), an active formylation domain termed the "F domain" is the first domain of the protein's F–A–PCP initiation module. In an SEC–SAXS experiment performed on the G1 beamline at the Macromolecular Diffraction Facility at CHESS, F–A–PCP was eluted in a single monomeric peak, and 11 peak exposures were averaged. Ab initio models were then generated from the SAXS data. Four independent crystal structures of the initiation module of LgrA were available,

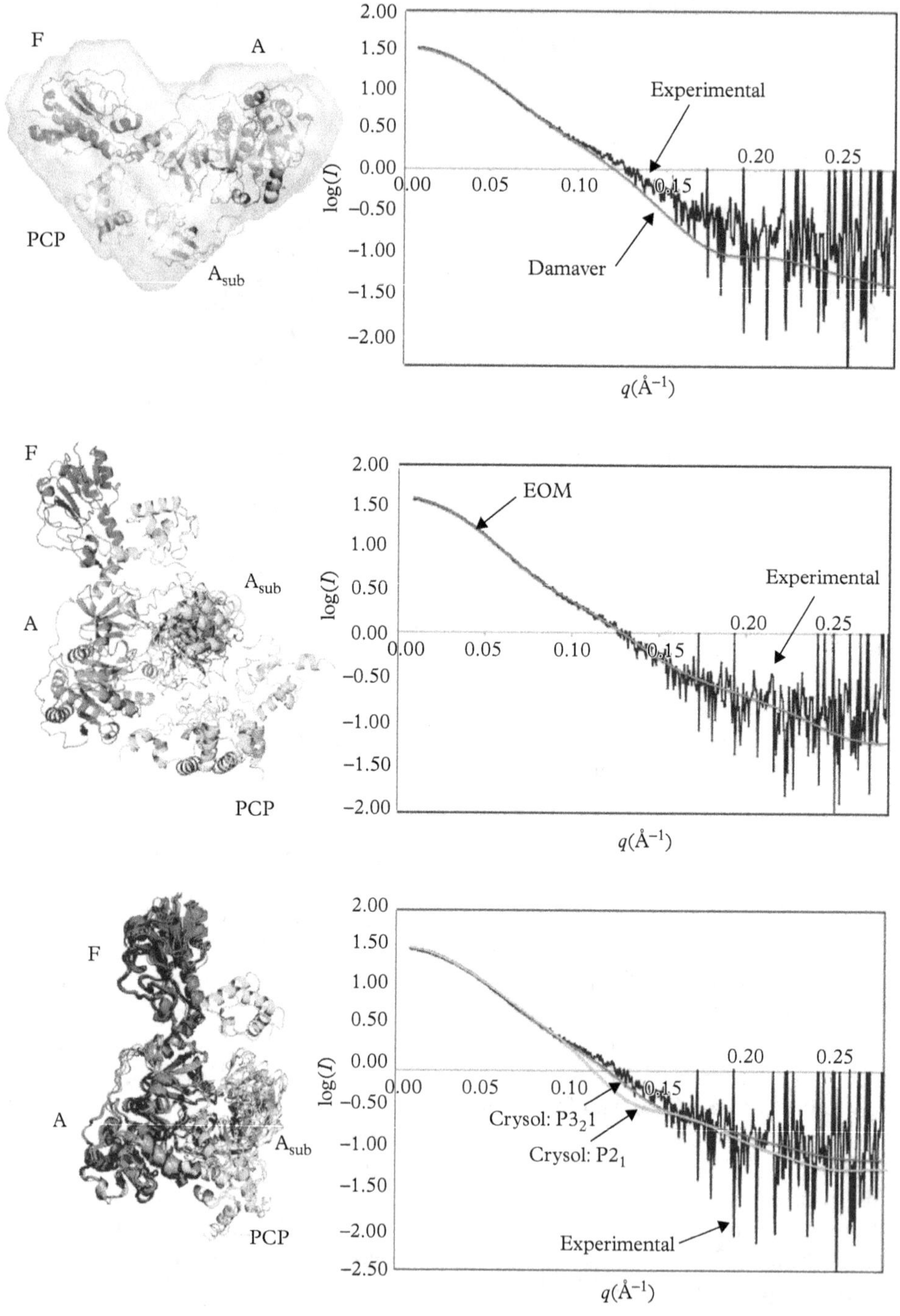
F
A
PCP
A_sub
Experimental
Damaver
log(I)
q(Å⁻¹)
EOM
Experimental
Crysol: P3₂1
Crysol: P2₁
Experimental

showing four different functional conformations: the A domain open (substrate binding), the A domain closed (adenylation), and thiolation and formylation states. Crystal structures of the F, A_{core}, A_{sub}, and PCP domains were used with an EOM (Bernado et al., 2007; Tria et al., 2015) to produce a structural model from the SAXS data. The ensemble generated showed a mixture with the formylation state representing 60% of the models generated, with the remainder resembling the thiolation state. This is summarized in Figure 12.8, which shows the fit of the crystal structure of the formylation state alone to the SAXS data, the ensemble fit, and the formylation and thiolation states overlaid. The SAXS data confirmed flexibility within the system, allowing the crystallographic structures to be put into mechanistic context.

12.6 Protein–Ligand Interactions

Tuukanen and Svergun have reviewed the use of SAXS for studying weak protein–ligand interactions (Tuukkanen & Svergun, 2014). They note that when weak interactions are studied, the components are typically present in the solution both as individual molecules and in complex, with the ratio of individual molecules to complexes depending on the binding constant K_d. Thus, K_d specifies the relative population of the molecular species, with the protein A, the ligand B, and their complex present in an equilibrium; K_d is defined as

$$K_d = \frac{[A][B]}{[AB]}, \tag{12.2}$$

where [A], [B], and [AB] are the molar concentrations of the protein, the ligand, and the complex, respectively. The general scattering form of a mixture can be represented by a linear combination of individual scattering intensity contributions

Figure 12.8 In the top panel, the crystal structure in the formylation state is shown, superposed on the averaged filtered ab initio SAXS model with a normalized spatial discrepancy (NSD) value of 0.819 ± 0.052 (see Kozin & Svergun, 2001). The calculated scattering curve is overlaid with the experimental scattering with $\chi^2 = 3.010$, where I represents scattering intensity, and q is equivalent to $4\pi\sin(\theta)/\lambda$. To understand the flexibility of F–A–PCP better, an EOM was performed, which generated five different ensembles; these are shown in the middle panel. The ensemble resembling the formylation state structure represented over 60% of the optimized models generated, while the remaining <40% resembled the thiolation state structure. The calculated scattering of the EOM model has a $\chi^2 = 1.028$, which demonstrates that F–A–PCP has flexibility. The data is consistent with extreme flexibility for A_{sub} and PCP domains, and limited flexibility in the F–A_{core} domain. In the bottom panel, all independent molecules from the crystal structures were overlaid to further illustrate the flexibility of the system. CRYSOL was used to generate predicted scattering curves for the formylation state and thiolation state crystal structures, with $\chi^2 = 2.12$ and $\chi^2 = 5.54$, respectively.

(Reprinted by permission from Macmillan Publishers Ltd: Nature, Reimer et al. (2016).)

$I_i(q)$ from the different species, if the scattering intensities of the individual components are known from experimental measurements or theoretical calculation:

$$I(q) = \sum_{k=1}^{K} \eta_k \cdot I_k(q), \tag{12.3}$$

where η_k is the mass fraction of species k, and K is the total number of components.

Experimentally, a set of experiments with varying protein or ligand concentration can be used to obtain K_d, looking for ratios of A, B, and AB that minimize the goodness of fit of the calculated profile to the experimental data.

As Tuukanen and Svergun (2014) note, if the number N of different molecular species is unknown, a series of measurements with varying amounts of molecular species together with SVD or PCP (see Chapter 4, section 4.6.1) can be employed.

SAXS has been used to investigate K_d in several studies (Bernado et al., 2009; Kutter et al., 2009; Polizzi et al., 2013; Altieri et al., 2016). Bernado et al. (2009) studied STAT5a, a mammalian signal transducer and activator of transcription (STAT) protein. STATs are thought to convert from an antiparallel inactive dimer into a parallel one following phosphorylation. SAXS data was collected from STAT5a at three different concentrations: 4.6, 2.3, and 1.15 mg/mL. Attempts to fit the available crystallographic dimer structure were unsuccessful at any of the concentrations studied. PCA was then used (see Chapter 4, section 4.6.1) to analyze the data from the three concentrations. The results suggested that the scattering could be explained by two species. Potential dimeric arrangements were explored and the experimental data fit to a mixture of the crystallographic monomer with each one of 5,000 rigid body dimers generated; K_d was the only parameter fitted using χ^2 to assess the merit according to

$$\chi_i^2 = \sum_c \frac{1}{N} \sum_q \frac{(I_c^{\text{expt}}(q) - k \cdot I_c^{\text{theo}}(q))^2}{\sigma(q)^2}, \tag{12.4}$$

where

$$I_c^{\text{theo}}(q) = \omega_c^{\text{mon}} \cdot I^{\text{mon}}(q) + \omega_c^{\text{dim}} \cdot I^{\text{dim}}(q) \tag{12.5}$$

and

$$\omega^{\text{mon}} + 2\omega^{\text{dim}}. = 1 \tag{12.6}$$

Here, c runs over the concentrations studied, N is the number of experimental points in the scattering curve, and $I(q)$ and $\sigma(q)$ are the intensity and experimental uncertainty, respectively; k is a scaling factor, $I^{\text{mon}}(q)$ and $I^{\text{dim}}(q)$ are the simulated curves from monomer and dimer, respectively, and ω^{mon} and ω^{dim} are the relative population of monomers and dimers, respectively.

Bernado et al. (2009) found that the top 50 fits to the experimental data defined a narrow range of dissociation constants with $K_d = 86 \pm 11$ μM. A good fit could not be obtained by simply adjusting the concentration of the species using the correct K_d but also required choosing the correct structure. The 50 dimers that provided the best agreement were clustered in five structurally related families, with the most populated highly similar to the unphosphorylated STAT1 and STAT5a models that were available. The unphosphorylated STAT5a dimeric crystal structure perfectly reproduced the experimental data when the system was modeled as a mixture of dimer and monomer.

Alteri et al. (2016) studied the influence of a small protein called *Thermococcales* inhibitor of PCNA (TIP) on the proliferating cell nuclear antigen (PCNA). PCNA forms a trimeric ring encircling duplex DNA and acts as an anchor for a number of proteins involved in DNA metabolic processes. Most proteins that bind to PCNA do so via a PCNA-interacting motif (PIP). TIP binds to the PCNA but does not have the PIP motif. Crystallographic studies show that it binds to the PIP interaction site but also extends beyond it. TIP alters the relationship between the two structurally similar domains that make up PCNA. SAXS data were collected from different concentrations of a PCNA–TIP peptide complex, using an N-terminally truncated TIP construct that enhances solubility. SAXS data were also collected on a complex of PCNA and flap endonuclease 1 (FEN-1) peptide residues 330–340, and a PCNA mutant engineered to be monomeric. The SAXS data for the PCNA:FEN-1 peptide and the free PCNA fit well to known structures. The trimer in *Thermococcus kodakarensis* PCNA is known to be less stable than other PCNAs, so fits of equilibrium models of monomer–trimer and monomer–dimer were made to the experimental data. Both resulted in poor fits, suggesting that PCNA retained its trimeric structure. The experimental data for the PCNA:TIP peptide complex showed that it was not a trimeric ring, and models of the known structure showed poor agreement with the SAXS data. Fitting the experimental SAXS data to the crystallographically determined 1:1 complex and a 2:2 complex present in the asymmetric unit provided good agreement with 69% of the 1:1 complex and 31% of the 2:2 at a 2 mg/mL concentration. SAXS data at higher concentrations showed an increase in the self-association of PCNA:TIP, with the K_d estimated at ~100 μM. The combination of X-ray crystallographic studies and SAXS data provided a structural basis for the inhibition of PCNA by TIP.

12.7 Dynamical Systems

The majority of the book has focused on compact globular proteins but as we have alluded to in several places, small angle scattering is sensitive to dynamics and flexible or even disordered systems (see sections 12.8 and 12.9). We have not covered these in detail beyond identifying such cases through the Kratky plot and the use of ensemble methods, as they comprise a specialized area of study with good guides on how to approach these cases (Kikhney & Svergun, 2015) and how to extract dynamical information with complementary scattering techniques (Meisburger et al., 2017).

Dynamics, identified from the Kratky plot, can be present in different embodiments. Kikhney and Svergun (2015) recommend a dimensionless Kratky plot where $I(0)$ is normalized to unity, and R_g multiplied by s, ($q = 2\pi s$), to remove information about the size of the protein but maintain the information on the shape. For an unfolded protein (i.e., a random coil), Flory (1953) defined a power-law relationship between the length and R_g, where $R_g = R_0 N^v$ (see equation 11.2; Kohn et al., 2004), where R_0 is a constant (in Å), N is the number of amino acid residues, and v is an exponential scaling factor; R_0 and v were determined as 1.330 ± 0.076 and 0.598 ± 0.028, respectively, for chemically denatured proteins (Kohn et al., 2004), and 2.54 ± 0.01 and 0.522 ± 0.01, respectively, for native IDPs (Bernado & Svergun, 2012). This suggests that IDPs are, in general, more compact than chemically denatured proteins (random coils).

Dynamic and flexible systems, by their nature, are polydisperse. Scattering from them is calculated as the summation of the scattering from each component, with each component's scattering weighted according to the presence of that component. As Kikhney and Svergun (2015) note, there are two main approaches to studying dynamic or disordered proteins. For a globular protein where dynamics is indicated, if the structure is known, the protein can be modeled by breaking it into globular units based on prediction of flexible regions. For a disordered protein, models representing different states can be constructed. This approach has been used with algorithms assembling the protein chain based on Ramachandran constraints and was successful in describing both NMR and SAXS data (Wells et al., 2008; Jensen et al., 2009). Another approach is to use an EOM (Tria et al., 2015, Bernado et al., 2007). Ensemble methods are discussed in Chapter 4, section 4.7. An EOM computes a large number of random models of the sample and then selects a sub-ensemble from those random models, based on the fit to the experimental data. This sub-ensemble is either the complete pool that best fits the experimental data or a minimum set that mimics systems adopting a few defined conformations.

EOM was used in the study of a putative DsbA-like protein from *Proteus mirabilis* (Furlong et al., 2017). This sample was termed "PmScsC," based on the source and the fact that the gene is located within a cluster of four predicted suppressor of copper sensitivity (Scs) genes. DsbA is an *Escherichia coli* monomeric thioredoxin-fold oxidase that introduces disulfide bonds into protein substrates. Furlong et al. determined three crystal structures of the PmScsC: the compact state structure, the transitional state structure, and the extended state structure. None of these individual crystal structures was compatible with the SAXS data that was recorded. Ensemble optimization was performed, and a final ensemble of six structures was obtained that yielded an excellent fit to the data. The ensemble spans a diverse range of conformations, supporting the notion that the PmScsC trimer is highly dynamic in solution. The authors mutated PmScsC to replace a flexible region with a rigid helical peptide and studied this with SAXS. The resulting ensemble contained four structures consistent with the reduction in flexibility and therefore different potential states present in solution.

Through a combination of complementary techniques including EOM, the authors showed a mechanism for how redox proteins with multiple active sites are able to shuffle incorrectly folded substrates.

In another study, Holdbrook et al. (2017) combined molecular dynamics, NMR, and SAXS with known crystallographic structural information to study the trimeric periplasmic holdase chaperone Skp. The outer membrane of gram-negative bacteria protects the cell from potential harmful agents. Outer trans-membrane proteins act as selective and nonselective gateways in the membrane but are synthesized within the cell. Skp binds and stabilizes unfolded outer membrane proteins in its central cavity, thereby preventing premature folding of outer membrane proteins and holding them in a folding-competent state prior to arrival and release at the outer membrane. Skp is known to be a trimer with a "jellyfish"-like architecture with a head domain and three long "tentacles" or "arms" that define the outer boundaries of a central cavity (Korndorfer et al., 2004; Walton & Sousa, 2004). The cavity has a clamp-like binding site typical of those observed in chaperones (Stirling et al., 2006). In the structure of Skp, the arms are are proximal to one another, although this may be a result of interactions between neighboring Skp trimers in the crystallographic unit cell. NMR studies identified a hinge region that allows the tentacles to open up, which is also consistent with missing electron density for parts of the tentacles in X-ray structures.

Molecular dynamics studies by Holdbrook et al. (2017) suggested that the tentacles could separate from each other on the nanoscale timescale. The average R_g of 32.8 Å determined from the molecular dynamics studies was in good agreement with 33 Å obtained from earlier neutron scattering (Zaccai et al., 2016). The SAXS data obtained gave an R_g value of 36 Å, which is in good agreement with the molecular dynamics and neutron values when the ~3 Å hydration shell is taken into account. The individual conformers obtained from the molecular dynamics trajectories had substantial variability in the conformations adopted and, therefore, in the calculated scattering curves. No single conformer fit the experimentally obtained SAXS data within experimental error. The EOM method was used to generate a pool of potential Skp structures, together with an equal number of structures with a symmetry element imposed. The combined pool had R_g values in the range of 27.5–42.5 Å, (Figure 12.9(a)), which fit the experimental data well (Figure 12.9(b)). Representative models, shown in Figure 12.9(i) and (ii), were similar to the extreme open and closed states from molecular dynamics simulations. The SAXS data indicated that Skp adopts a conformation ensemble composed of open and closed structures and intermediate states. The EOM study still showed a discrepancy at the lowest q, so another pool of models was constructed (using the RANCH (for *ran*dom *ch*ain) routine; Bernado et al., 2007; Tria et al., 2015), allowing the position of the tentacles to vary with respect to the head domain. The resulting ensemble better represented the experimental data and showed a bimodal distribution, with the first peak corresponding to an R_g of 32 Å, and the second to a range of R_g's from 38 to 41 Å. This is compatible with

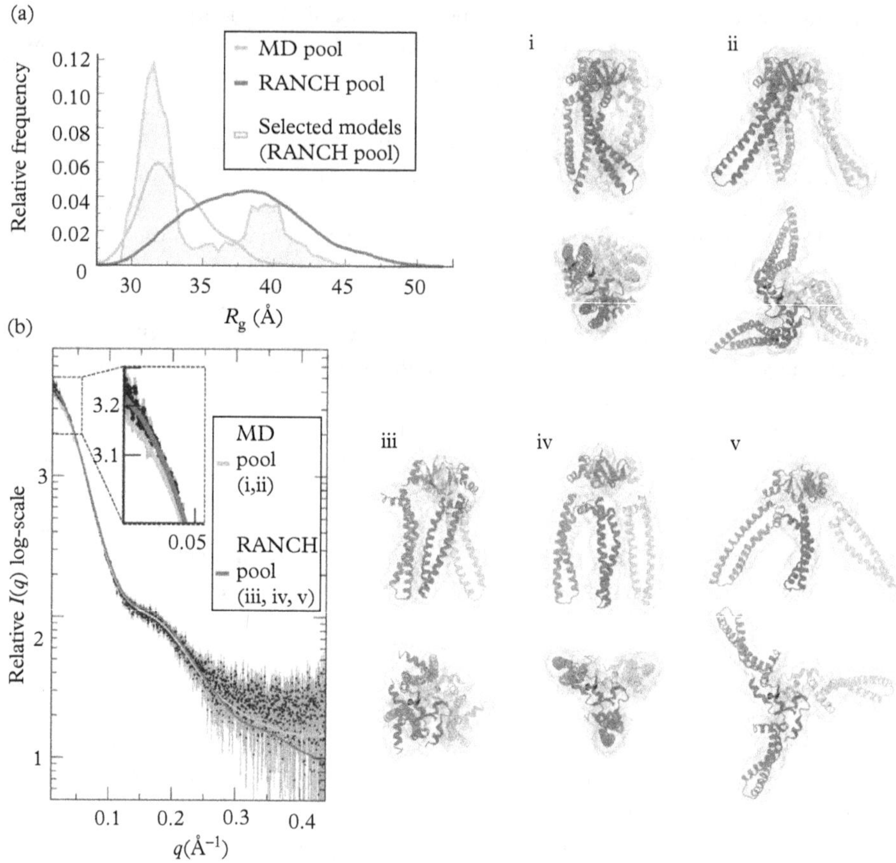

Figure 12.9 Ensemble fitting to the experimental SAXS intensities. (A) Relative frequency of structures with a particular R_g in the molecular dynamics pool (MD) and the RANCH pools. The frequency of models selected from by EOM is shown for the RANCH pool. (B) The optimized ensemble of RANCH-pool and MD-pool structures fitted to the experimental SAXS profile. Representative structures selected by EOM are shown for both the MD-pool (i and ii) and the RANCH-pool (iii, iv, and v) models. The protein structure is shown in cartoon format with transparent molecular surfaces overlaid, with each monomer of trimeric Skp shaded differently.

(Reprinted from Structure, Holdbrook et al. (2017), Copyright (2017) with permission from Elsevier.)

both the open and the closed model. The study suggested that Skp exists as a dynamic ensemble of conformational states with variable access to the central cavity; in addition, it showed how SAXS could be used to access a dynamic process.

SAXS and SANS are low-resolution techniques, and Panjokovich and Svergun have demonstrated that a coarse-grained approach to modeling dynamics can also be applicable (Panjkovich & Svergun, 2016). In this approach, normal mode analysis is used to model harmonic and linear motions, breaking the structure up into pseudodomains that can be refined against experimental SAXS data with rigid body restraints and localized movements. While normal modes do not represent nonlinear and anharmonic motions, which are both common in biological macromolecules, the technique shows considerable promise. A benchmark data set containing structural models of 44 representative proteins that had at least two distinct conformational states used as a test set showed distinct improvements with the approach.

12.8 Systems with Limited or No Structural Information

We have discussed the use of SAXS for characterizing samples, particularly in the application to crystallization screening in Chapter 10, section 10.3.3. For structural studies, the most effective use of SAXS is where complementary structural information is present or, as theoretical scattering curves can be accurately calculated, to test hypotheses by eliminating incorrect models.

Burger et al. (2016) have made use of SAXS for quantifying conformational flexibility in proteins independent of structural information. They formulate the scattering intensity integrated over all conformations in the solution, using a probability term to define the likelihood of each conformation. Protein flexibility or disorder can be calculated by calculating the entropy, which is a function of the probability of any given conformation, $P(\bar{x})$. To simplify the calculation, they make use of the R_g, reducing the dimensionality of the problem. The R_g criterion is convenient because in the low-q region, the intensity is mainly dependent on the dimensions of the protein. The scattering intensity in the context of this R_g-distribution (R_gD) model is given by

$$I_{\mu,\sigma}(q) = \int_0^\infty I_S(q, R_g)P_{\mu,\sigma}(R_g)dR_g, \quad (12.7)$$

where $I_S(q, R_g)$ is the scattering intensity of a sphere radius R_g with homogeneous charge density and $P_{\mu,\sigma}(R_g)$ is the probability distribution over the different R_g's that the protein can adopt in solution. The authors use a log-normal distribution for $P_{\mu,\sigma}(R_g)$:

$$P_{\mu,\sigma}(R_g) = \frac{1}{\sigma R_g\sqrt{2\pi}} e^{\frac{-(ln(R_g)-\mu)^2}{2\sigma^2}}, \quad (12.8)$$

where μ and σ are the mean and standard deviation of the log-normal distribution, respectively. The log-normal distribution is only defined for positive R_g and the $P(0) = 0$. The calculated intensity is fit to the experimental intensity using values of μ and σ that minimize the difference between calculated and experimental data.

The optimized values of μ and σ are denoted as $\hat{\mu}$ and $\hat{\sigma}$, and the entropy S is calculated from

$$S = -\int_0^{\infty} P_{\hat{\mu},\hat{\sigma}}(R_g)\ln\left(P_{\hat{\mu},\hat{\sigma}}(R_g)dR_g\right)$$
$$= \frac{1}{2}\left(1 + \ln(2\pi\hat{\sigma}^2)\right) + \hat{\mu}. \qquad (12.9)$$

The differential entropy expressed by S is a quantitative estimate of the R_g diversity of the sample in solution.

To test the $P_{\mu,\sigma}(R_g)$, three proteins were examined: a globular folded protein, a partially unfolded protein with a globular core and two disordered tails, and an IDP. The R_gD model produced entropies consistent with the disorder or lack of it, using only the scattering information. An analysis of 226 experimental data sets showed agreement between the R_gD model results and the qualitative information obtained in Kratky plots.

To assess the sensitivity of the R_gD model, five systems were studied where experimental data was available on the protein and a protein–ligand complex. In each case, the R_gD results identified a difference between the native and the ligand–bound protein, with ligand binding reducing entropy. Kratky plot analysis showed qualitative agreement in some cases but was not sensitive for all.

The heterotrimeric protein complex containing the integrin-linked kinase (ILK), parvin, and PINCH proteins, termed the IPP complex, is an essential component of focal adhesions, where it interacts with many proteins to mediate signaling from integrin adhesion receptors. Steiger et al. studied the minimal IPP complex and used SAXS to determine that IPP adopts a predominately compact conformation (Stiegler et al., 2013). The minimal IPP complex was expressed and purified migrating as a single species on native gel electrophoresis. The migration was unchanged with addition of excess components of the complex, suggesting that the complex is stable and monodisperse in solution. SAXS data on a concentration series showed that the complex did not have any interparticle interactions over the concentration range sampled and Kratky plots suggested that the protein was globular.

Structural models were available for an ILK–PINCH complex consisting of residues 1–170 and 4–68 of each, respectively, and an ILK–parvin complex consisting of residues 185–452 and 246–372, respectively. These were used with rigid body modeling and produced an envelope similar to that produced by the averaged ab initio reconstruction. A predominately compact structure was seen. EOM modeling with an initial starting ensemble of 10,000 models yielded an optimized ensemble of 20 models as the best fit to the experimental data, with just two fitting the data equally as well as the full 20. The EOM modeling also suggested a compact structure with a small population of a second structure.

A crystallographically unresolved flexible linker spans two domains of ILK. The SAXS results suggest partial structure in this linker, with potential interactions with the N and C terminal domains. This guided further studies of this and provided the structural foundation to understand conformational restraints that govern the IPP complex.

12.9 Intrinsically Disordered Proteins

Intrinsically disordered proteins (IDPs) range from fully disordered proteins similar to random polymers, to molten globules that contain some or all secondary structure, and, finally, to multidomain proteins containing disordered regions (Dyson & Wright, 2005). They are extremely abundant in the cell and fulfill many essential functions. Unstructured proteins tend to have a lower specific volume than globular folded proteins, and this is reflected in a slightly lower $I(0)$ (Harpaz et al., 1994). Bernado notes that features of SAXS that indicate the potential presence of IDPs are the attenuation of fine structure that becomes pronounced in Kratky representations, and $P(r)$ functions that have large D_{max} values and a smooth decrease to zero. Receveur-Brechot and Durand (2012) have reviewed this area, with a discussion on evaluating SAXS and SANS data for the presence of intrinsic disorder, and discuss examples in the field.

As noted, typical scattering curves of IDPs are characterized by their lack of features. (Bernado, 2010). As we have discussed in Chapter 3, section 3.2.3, a Guinier plot can be used up to a qR_g of 1.3 for a well-folded protein. For an IDP, the Guinier region is reduced to $qR_g < 0.8$ or less. This can limit the accuracy of the measurement, depending on the instrument characteristics and the number of data points gathered in the limit of the effective Guinier region.

The R_g follows the power law $R_g = R_0 N^v$. The variable v depends on the structural behavior, with $v = 0.38 \pm 0.05$ for native globular proteins (Millett et al., 2002), and $v = 0.60 \pm 0.03$, with an R_0 of 1.93 for completely denatured proteins. For proteins under strongly denaturing conditions, an R_0 and a v of 2.23 and 0.58, respectively, were observed (Wilkins et al., 1999). For IDPs in an aqueous buffer, an R_0 and a v of 2.54 and 0.52, respectively, were observed (Bernado & Blackledge, 2009), values closer to that expected for a random coil. As Receveur-Brechot and Durand (2012) note, there is a continuum of ordered and disordered proteins, and predictions from R_g represent upper and lower bounds—differences from these reveal global or local structural restraints, implying deviation from a random coil. There is also a relationship between the R_g and the hydrodynamic radius, R_h, which is measured by DLS or pulse-gradient NMR. The R_g/R_h ratio is $(3/5)^{1/2}$ or 0.77 for a globular protein, and approximately 1.4 for a denatured one. A ratio in-between these values suggests the presence of residual structure.

IDPs also have a signature in the $P(r)$, showing an extended tail due to the variety of extended conformations that are present. Unfortunately, this is also a signature of aggregation, and careful sample preparation and characterization are necessary to differentiate the two causes. The $P(r)$ can be used as an independent check on the R_g, according to equation 3.12. This is a model-independent calculation and yields slightly higher values than the Guinier approximation, due to being better suited to the treatment of unfolded chains. As described in Chapter 3, section 3.4, the Kratky plot can characterize a dynamic or disordered sample. A normalized Kratky plot has a maximum value of 1.104 for $qR_g = \sqrt{3}$, whatever the sample size. For a random chain, the maximum increases beyond this.

The flexibility of fully disordered samples can be assessed. The behavior of highly unfolded or disordered chains can be described by the Kratky–Porod model, which assumes a worm-like chain with a persistence length (measuring stiffness) that takes into account the local rigidity. Two parameters describe this chain: the contour length, L, and the statistical length, b, which is twice the persistence length. The persistence length is the defined as the length over which the chain remains straight determined by stiffness or rigidity. The contour length is the length of linear extended chain without stretching the backbone. The scattering intensity can be expressed in terms of these lengths by

$$\frac{I(q)}{I(0)} = \frac{2(x-1+e^{-x})}{x^2} + \frac{b}{L}\left[\frac{4}{15} + \frac{7}{15x} - \left(\frac{11}{15} + \frac{7}{15x}\right)e^{-x}\right], \tag{12.10}$$

with $x = q^2Lb/6$ (Receveur-Brechot & Durand, 2012). The authors note that the formula is valid for $L/b > 10$ and for $q < 3/b$. For a completely disordered coil, b is expected to be 18–20 Å, and L equal to $Nl_0 f$, where N is the number of residues, l_0 is the length between two Cα's (3.78 Å), and f is a geometrical factor equal to 0.95 (taking into account that an unfolded chain is not linear).

The R_g for a random coil can also be inferred from L and b, according to

$$R_g = b\left[\frac{y}{6} - \frac{1}{4} + \frac{1}{4y} - \frac{1}{8y^2}\right]^{1/2}, \tag{12.11}$$

where $y = L/b$.

Receveur-Brechot and Durand (2012) note examples where the signal transduction chemotactic protein CheY, when chemically denatured, has a b of 28 and a lower value of L than that expected for a random coil (Garcia et al., 2001) and when heat-denatured neocarzinostatin exhibits values of b and L close to that of a random coil (Perez et al., 2001).

Interestingly, whereas for a globular protein, ab initio envelope reconstruction is usually robust for multiple calculations, the envelope reconstructions for an IDP or even flexible regions vary dramatically run from run (Receveur-Brechot & Durand, 2012).

12.10 Summary

The use of small angle scattering is on the rise, with many examples of how it can provide biological information that can complement or inform other studies. A comprehensive review of this area is presented by Svergun et al. (2013). In this section, we have described how contrast matching used in the neutron case was able to resolve the structural placement of smaller components of a complex, as well as a historical study showing demonstrating a protein capsid surrounding the RNA in a virus particle. Contrast matching can also be used with SAXS but to a far smaller extent. Time-resolved studies have helped reveal the mechanism of photoreactive yellow protein and hemoglobin using synchrotron sources. The application of X-ray free electron sources is discussed in Chapter 13. SAXS or SANS as a solution technique can access information missing from other studies, in particular, about residues that may be crystallographically undefined. It can also help put known structural information in context (e.g., allowing structural information from components of complexes to produce an atomic level picture of the complete complex). Similarly, small angle scattering can be used to follow large-scale functional changes or show when distinct populations coexist in solution. An interesting potential exists for the study of protein–ligand interactions, especially in the case of allosteric systems. Finally, the technique is sensitive to dynamic processes and in the study of non-globular proteins. The list of examples provided is not exhaustive and should be seen as a glimpse of the potential of biological mechanisms where SAXS or SANS can provide unique information complementary to other methods and help define function and mechanism.

13

Developments on the Horizon

This book is intended to be an accessible introduction to X-ray solution scattering, particularly as applied to biological systems. Thus, there are many advanced topics that are beyond the scope of such an introduction. At the same time, dramatic advances in technology, such as XFELs and algorithmic developments, promise extraordinary new applications for SAXS and solution scattering in general. In this chapter, we provide vignettes of a number of these new approaches to provide a sense of the excitement that is permeating the field now and as an introduction to the huge potential for the technique in the future. This section is less formal and detailed than presentations in the remainder of the book.

13.1 The Impact of XFELs

13.1.1 Time-Resolved Applications

A key application of XFEL properties is to use the pulses to freeze motions, almost akin to a high-speed camera, and to analyze the individual images that result. For biological studies, generically, a specimen is pumped or activated into a functional state that will evolve with time, and is then probed or observed with the X-ray pulse to image the pumped system at sequential times after initiation.

A group led by Neutze has used multiphoton absorption to induce an extremely rapid structural perturbation in the photosynthetic reaction center (Arnlund et al., 2014). The photosynthetic reaction center faces unique problems of energy management. The energy of a single photon of green light is approximately equal to the activation energy of the unfolding of the molecule. In the photosynthetic complex, photons absorbed by the light-harvesting antennae are rapidly funneled to the reaction center via dedicated, specialized channels. It has been hypothesized that excess energy, which may also be deposited in the protein, is dissipated before damage can be done by a process that has been named "a protein quake." The authors measured this process using time-resolved WAXS at the LCLS. The perturbation appears with a time constant of a few picoseconds and decays over a time of 10 ps. Importantly, the protein quake perturbation precedes the propagation of heat through the protein.

Biological Small Angle Scattering: Theory and Practice. Eaton E. Lattman, Thomas D. Grant, and Edward H. Snell.

DOI: 10.1093/oso/9780199670871.001.0001

Experiments were performed at the coherent X-ray imaging station at the LCLS (Boutet & Williams, 2010), using micro-jet injection of solubilized protein samples, instead of the liquid in the beam or flow through a capillary that is typical of synchrotron-based SAXS or WAXS experiments. The sample was pumped using an 800 nm optical laser pulse of 500 fs duration and calibrated so that a heating signal was observed in the difference WAXS spectrum. The XFEL was operated to produce 40 fs pulses at 120 Hz. Illuminated and dark samples were interleaved, each at 60 Hz. The delay time between the laser and the XFEL pulses were calibrated to within 5 ps, and scattering patterns were collected over a series of 41 time delays to a maximum of 100 ps.

The results are shown in Figure 13.1. In this case, computer simulations and modeling played a key role in interpreting the data and developing an understanding

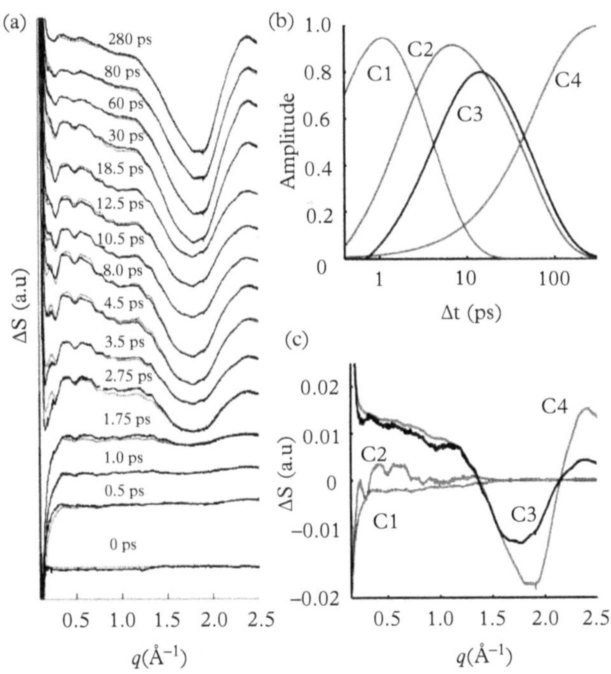

Figure 13.1 (a) Time-resolved WAXS difference data, $\Delta S(q, \Delta t) = S_{\text{light}}(q, \Delta t) - S_{\text{dark}}(q)$, recorded as a function of the time delay between the arrival of the 800 nm pump laser and the XFEL probe for 15 of 41 measured time delays. Linear sums of the four basis spectra shown in (c) are superimposed upon the experimental difference data. (b) Time-dependent amplitudes of the components C1–C4 used to extract the basis spectra. (c) Basis spectra extracted from the experimental data by spectral decomposition: C1, an ultrafast component; C2, a protein component; C3, non-equilibrated heating components; and C4, equilibrated heating components.

(Reprinted by permission from Macmillan Publishers Ltd: Nature Methods, (Arnlund et al., 2014))

of the "quake." A combination of molecular dynamics simulations of heat deposition and flow in a molecule and spectral decomposition of the time-resolved difference scattering curves in Figure 13.1 (a) provided a strong basis for a detailed understanding of the energy propagation in the system. Because the light pulse is tuned to the frequency of the photosystem antenna, cofactors were instantaneously heated to few thousand kelvins, which decayed with a halftime of about 7 ps through heat flow to the remainder of the protein and the surrounding medium. As another example, the basis spectrum C2 in Figure 13.1 (c) contains oscillations in the range $q = 0.2–0.9$. These are signatures of protein structural changes. This study illustrates not only the rapid evolution of the technology and experimental prowess in the field, but also how it is brought to bear on a problem for which biological relevance for extremely rapid dynamics seems to be clear.

13.1.2 Snapshot SAXS

In our introduction, we noted that scattering data is recorded from the spatially averaged scattering of many particles over the exposure time. With the use of XFEL sources, SAXS data can be obtained on timescales that are fast relative to the rotational motion of the particles. In addition, the brilliance of the beam means that scattering patterns can be obtained from volumes containing only a handful of particles. In this manner, averaging over time and the number of particles is greatly reduced, and the scattering patterns from such specimens are two dimensional, containing much more information than the usual $I(q)$ curves. This type of experiment has come to be called fluctuational solution scattering (FSS).

A theoretical treatment of this possibility (Kam, 1977, 1980) and later practical demonstration with frozen solutions suggested promise (Kam et al., 1981), but suitable sources and detectors were not available to fully exploit the potential. More recently, with the advent of XFEL sources, Kirian has produced a detailed tutorial on the method (Kirian, 2012), and this is highly recommended to the reader. For a basic understanding of the method, it is useful to note some of the key elements here.

The method of Kam makes use of a rotational correlation function, which is shown in two dimensions in Figure 13.2. Figure 13.2(a) shows one annulus from a single-particle solution scattering curve that contains two peaks near the top, plus the centrosymmetrically related mates at the bottom. Figure 13.2(b) illustrates the angular correlation function C, which multiplies the value of the scattering curve at two points separated by an angle $\Delta\varphi$ and integrates this pair product over all angles. Figure 13.2(c) shows a more realistic version of Figure 13.2(a); in this panel, the scattering patterns from many molecules in random orientations are overlaid to create $I(q)$. The two panels below show how a set of such $I(q)$ curves can be deconstructed to provide the single-particle correlation function $C(q, q, \Delta\varphi)$. Figure 13.2(d) shows 50 independent patterns like that in Figure 13.2 (c), with the patterns unfurled along the horizontal axis. Figure 13.2(e) shows the

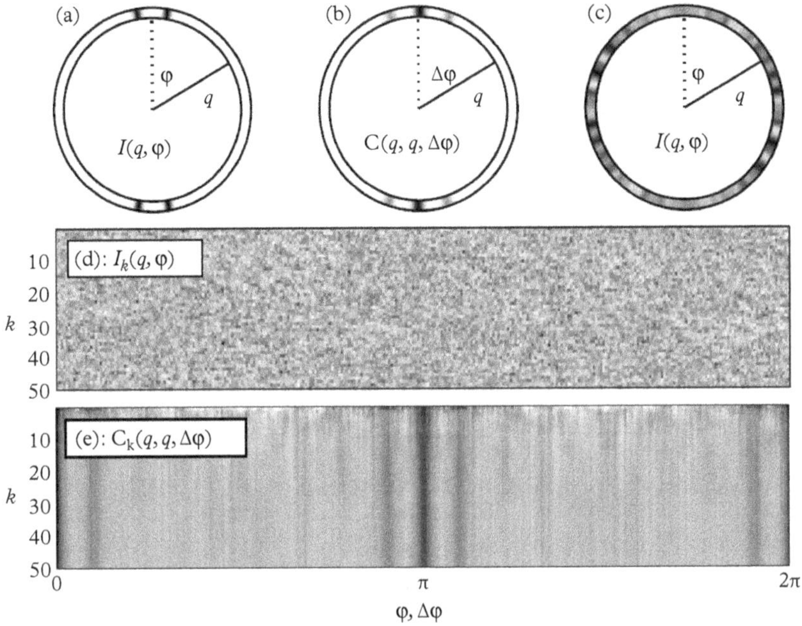

Figure 13.2 Angular autocorrelation from noisy intensity data. (a) The ideal diffracted intensity arising from a very simple test object. (b) The angular autocorrelation of the pattern shown in (a). (c) Diffracted intensity from a superposition of 100 particles illuminated simultaneously; each particle is in random orientation about an axis parallel to the diffracting beam. (d) A series of 50 patterns similar to those shown in (c) but with uniform background and Poisson noise added. (e) The cumulative average autocorrelation of (d). (Figure from Kirian (2012)[1], © IOP Publishing. Reproduced with permission, All rights reserved.)
[1] DOI 10.1088/0953-4075/45/22/223001.

correlation function C calculated for the first k $I(q)$ curves and then added up. Clearly, as k increases, the ensemble correlation function approaches that of a single particle, as seen in Figure 13.2(b). The next steps in this method would be to deconvolute C to arrive at the single-particle scattering curve $I(q)$ and then generalize the method to particles that are randomly oriented in three dimensions rather than only in the plane.

At the time of writing, experiments are being conducted to test the potential of fluctuational scattering on biological systems and to create novel algorithms to advance FSS (Donatelli et al., 2015; Malmerberg et al., 2015). Although we are a long way from routine use, the potential is very promising; a small number of feasibility studies have been carried out on test specimens. One study made use of dumbbell-shaped particles formed from two polystyrene spheres of equal size that were dispersed in micrometer-sized aerosol droplets and was able to create an image of the specimen. The cylindrical symmetry of the object greatly simplified the data analysis and the reconstruction process, but this paper appears to represent the first reconstruction in three dimensions of an object using FSS two-dimensional data. Another test specimen comprised a set of gold rods approximately 90 nm × 25 nm, lying on a plane so that all the particle orientations are related by rotation about a single axis. The authors formed the angular multiparticle correlation functions for each of the observed diffraction patterns and summed these to derive the single-particle correlation function from which the low-resolution structure of the rods was determined. Kirian et al. (2011) have analyzed and simulated FSS signal, noise, and resolution and suggest that it may be possible to achieve a signal-to-noise ratio of better than one at sub-10 Å resolution.

13.2 Working with Cryocooled Samples

While we have talked about solution techniques, single-crystal structural crystallography currently makes use almost exclusively of cryocooled specimens. Crystals are cooled to temperatures sufficiently low that interstitial solvent is transformed into vitreous ice, in which OH radicals formed during the X-ray irradiation are immobile. Cryocooling is enabled by cryo-protective agents, such as ethylene glycol, that promote vitrification and eliminate the formation of crystalline ice, which can dominate the scattering pattern and destroy the lattice periodicity. Small samples are preferred. Cryocooling reduces secondary radiation damage and effectively prolongs the crystal lifetime in the beam by approximately 70-fold, while also providing a practical advantage of greatly simplifying the handling of samples.

For solutions, radiation damage is a more significant concern. At doses where data from crystals is still usable, aggregation is seen in the solution state (Kuwamoto et al., 2004). Successful cryocooling for SAXS could reduce the radiation damage effects, allow longer exposure times (and therefore reduced concentrations) or reduce the sample size. However, in the SAXS case, the protein concentration is lower and the volume is larger than a single crystal. Successful cryocooling (vitrification) without ice formation is therefore more challenging.

Cryocooling with SAXS has been carried out at CHESS. Windowless cylindrical polyimide tubing sample cells were used with the X-ray beam passing through the axis and open ends of the tubing, and the outside cooled by a (crystallography) standard cryogenic gaseous nitrogen stream (Meisburger et al., 2013). The sample dimensions are constrained by the tube but unconstrained in the axis of the beam. Therefore, a normalization procedure had to be used to ensure accurate buffer subtraction from the sample signal. The results, while requiring relatively high levels of cryoprotectant, showed significant promise, and with increases in flux from new and upgraded sources, and developments in novel fluidic handling, the approach may become more common in the field.

13.3 Anomalous or Resonant Scattering in SAXS

Anomalous or resonant scattering of X-rays occurs when the energy of the X-ray photon is closely matched to an inner-shell electronic transition in the atom. Both the amplitude and the phase of the scattered radiation vary steeply with energy over a narrow energy window. Thus, if one subtracts diffraction patterns collected at two energies within this window, the resultant signal comes only from the resonantly scattering atoms—the scattering from the other atoms effectively cancels out. This method has been used extensively in protein crystallography, often aided by the exogenously introduced element selenium, to provide phase information for diffraction patterns.

Resonant scattering can also be used in the area of SAXS. For example, in nucleic acids, the strong negative charges attract a condensed cloud of counter ions, such as Na^+ and K^+, that are critically involved in nucleic acid stability and folding. These ions, and their waters of solvation, are disordered and mobile structures and therefore are not readily detected by X-ray crystallography. Indeed, SAXS is an ideal method for probing the structure of ion clouds around nucleic acids. Scattering profiles for short DNA duplexes measured in the presence of a variety of counter ions can be mathematically dissected into solution scattering curves representing the contributions from water, DNA, and ions (Pollack, 2011). In this case, the rubidium ion, Rb^+, was used as the source of anomalous signal. This signal is critical not only in making the decomposition of the scattering curve into its three components more robust but also in allowing the measurement of both the number and the distribution of ions around the nucleic acid molecule. These measurements are able to distinguish between theoretically proposed models for the ionic distribution. The potential for anomalous or resonant scattering techniques with SAXS is well reviewed by Pollack (2011).

Makowski et al. (2012) point out that, in solution, anomalous diffraction has two components: a set of cross-terms that depend on the anomalous scatters and the rest of the sample, and a set of pure anomalous terms that depend on the distance between the anomalous centers. The latter are of great interest due to their potential to act as "molecular rulers" in analyzing dynamic shifts in the sample. The authors demonstrated that anomalous signals could be recorded from iron in both hemoglobin A and myoglobin, but, unfortunately, the signal from the anomalous center cross-terms was not measurable with current capabilities. The potential for a molecular ruler using this technique is high, but instrumentation capabilities will need to advance before that potential is realized. While it has not been explored to date, XFEL sources may also play a role in this area.

13.4 Summary

Solution scattering has been introduced as a technique in this book, with typical uses described. The field of application is broader than we can cover, and, with new instrumentation, that field is only growing. XFEL sources offer the potential for time-resolved studies exquisitely sensitive to rapid structural biological processes while at the same time potentially extending the information beyond that provided by current solution scattering methods. Cryogenic approaches offer the potential to increase signal and/or improve sample demands, opening up the technique to a wider array of studies. Finally, use of the atomic scattering properties reveals biological information in the case of nucleic acids and has the potential to provide molecular rulers able to resolve dynamic structural changes as they occur. We make no claim to have covered more advanced applications of solution scattering completely and, in fact, have largely ignored cases where anisotropic data also yields structural knowledge. We encourage the reader to explore further and hope we have provided sufficient detail to prompt suitable curiosity to enable this.

14 Pushing the Envelope

14.1 Introduction

Developments in XFELs have led to the possibility of structure determination via single-molecule X-ray diffraction. This has led to a renewed interest in iterative phase retrieval methods for determining the structure factor phases from such experiments. Recently, the application of such methods to solution scattering data has enabled for the first time the calculation of the electron density function directly from the one-dimensional scattering profile (Grant, 2018) albeit at low resolution. This method improves our ability to access multiple distinct electron density regions, such as from proteins, nucleic acids, and/or lipids, by removing the restriction of uniform density from envelope reconstruction. Envelope modeling algorithms to date assume uniform density throughout the envelope (Svergun & Stuhrmann, 1991; Henderson, 1996; Chacon et al., 1998; Svergun, 1999; Walther et al., 2000). At low resolution, below about 10–15 Å, this assumption is approximately correct for particles containing a single molecular species, such as proteins. However, at resolutions higher than about 10–15 Å, the assumption of uniform density begins to break down and become unreliable, as fluctuations in the density contribute significantly to the scattering profile (Tuukkanen et al., 2016). By directly calculating the electron density, we allow the three-dimensional reconstructions to be limited simply by the available information content and not by the implicit limitations of the modeling algorithm, which is dependent on the particle size and shape as well as the experimental geometry.

In Chapter 2, we described the theory behind small angle solution scattering. We explained how the scattering in crystallography is based on the same underlying principles as solution scattering. In particular, both techniques attempt to gain information about the molecular transform, that is, the Fourier transform of the electron density function. This function is a continuous, complex-valued function in three-dimensional reciprocal space, where each point in reciprocal space has an associated amplitude and phase. In crystallography or single-molecule diffraction, the amplitudes of the structure factors in three dimensions are determined experimentally via the measured intensities. The main aim of structure determination and refinement then is the determination of the missing phases in three dimensions, the so-called phase problem. In crystallography, this is often done by either

Biological Small Angle Scattering: Theory and Practice. Eaton E. Lattman, Thomas D. Grant, and Edward H. Snell.

DOI: 10.1093/oso/9780199670871.001.0001

experimentally causing the diffracted intensities to change in a predictable way, such as through anomalous scattering (Bijvoet, 1997), or by guessing the initial phases using a similar structure already solved, in a process known as "molecular replacement" (Rossmann, 1972). In single-molecule diffraction, there is enough information in the oversampled measured intensities and additional real space restraints that phases can be retrieved ab initio without additional experiments (Miao et al., 1998).

In solution scattering, the experimentally measured intensities comprise a one-dimensional function representing the spherically averaged intensities in three-dimensional reciprocal space. Thus, not only are three-dimensional phases of the structure factors unknown, but the three-dimensional amplitudes are also unknown. Instead of having the "phase problem" as in crystallography or single-molecule diffraction, in solution scattering we have the "structure factor problem." Unfortunately, there is not enough information in the one-dimensional scattering profile to *uniquely* determine the three-dimensional structure factors to calculate the electron density function. Various algorithms have been developed to extract three-dimensional structural information from the one-dimensional solution scattering profile by adding in significant additional restraints (see Chapter 4 for examples of conventional modeling of solution scattering). All such algorithms rely exclusively on modeling, that is, by creating millions of atomic or coarse-grained models and comparing the calculated one-dimensional profiles with the experimental data to select the best model. If additional information about the structure is known, for example, from crystallography or NMR, this information can be utilized by imposing these structural restraints when generating the models (Mattinen et al., 2002; Petoukhov & Svergun, 2005). This approach is markedly different to that used in crystallography or single-molecule diffraction, where phases are estimated, electron density is calculated from the estimated phases and the experimentally determined amplitudes, and atomic models are generated to fit the electron density function. Phases from the models are then used to generate new electron density functions, and the process is repeated iteratively. In contrast, when modeling solution scattering data, there is no electron density function to fit a model to, only the comparison of the experimental and the calculated one-dimensional scattering profile.

14.2 Retrieving Three-Dimensional Structure Factors from One-Dimensional Solution Scattering Data

To describe the process of retrieving three-dimensional structure factors from one-dimensional solution scattering data, we first describe the application of iterative phase retrieval to single-molecule imaging. Single-molecule X-ray diffraction experiments result in a set of intensities in three-dimensional reciprocal space.

The structure factor amplitudes are then calculated as the square root of the intensities. This set of three-dimensional amplitudes is used as the restraint in reciprocal space. To start, an electron density function is generated, typically as a random set of densities in real space. The Fourier transform of this electron density function is calculated to yield the initial set of structure factors. The amplitudes are then replaced by the values determined from experiment, and the phases are retained from the initial set of random densities. From this modified set of structure factors, a new electron density map is calculated. This new electron density map is incorrect, since the phases are still random. Real space restraints are then applied using a support. The support is the region of space where the electron density of the particle is present. The region outside of the support is defined as the solvent region. Since the electron density is known to be real and positive, all negative regions of the density are set to zero. Additionally the solvent region is known to be approximately flat (i.e., a constant value), so all density outside of the support is flattened by setting the densities in this region to the average value of this region. Then a new set of structure factors is calculated from these modified densities. The amplitudes are once again set to the values found by experiment, and the cycle is repeated iteratively until the solution converges and the phases have been retrieved.

To apply these basic principles to solution scattering, we must first recognize that we do not have the set of three-dimensional amplitudes from experiment, as is the case for single-molecule imaging. However, we do have a restraint on the intensities. In solution scattering, the reciprocal space restraint is that we know the spherical average of the intensities at a given q value from the one-dimensional scattering profile. Thus, we can use this knowledge to change *all* three-dimensional intensities at a given q value such that their average value agrees with experiment. To do this, we first start with a random set of densities in real space. Then we calculate the structure factors in reciprocal space and, subsequently, the intensities (Figure 14.1). Since we do not know the amplitudes, we cannot simply set the three-dimensional amplitudes of the structure factors to the correct values. Instead, we scale each set of structure factors that are at the same q value by the square root of the ratio of the experimental to the calculated intensities. This effectively scales the amplitudes while retaining the phases. This ensures that each set of intensities at each q value has an average value agreeing with the experimental scattering profile.

To impose the real space restraints, we must first define the support region . In single-molecule imaging, this region is often known ahead of time from other experimental data, such as microscopy, or can be determined ab initio using an algorithm known as "shrinkwrap" (Marchesini et al., 2003). Shrinkwrap attempts to update the support iteratively by convolving the density with a Gaussian to blur out its high-resolution features. A threshold is then applied where the support is defined as all regions of space where the blurred density is greater than some fraction of the maximum density, typically 20% of the maximum. In most cases for solution scattering, the envelope is not initially known, and, in fact, determining this may be the primary goal of the experiment. Thus, shrinkwrap is used to

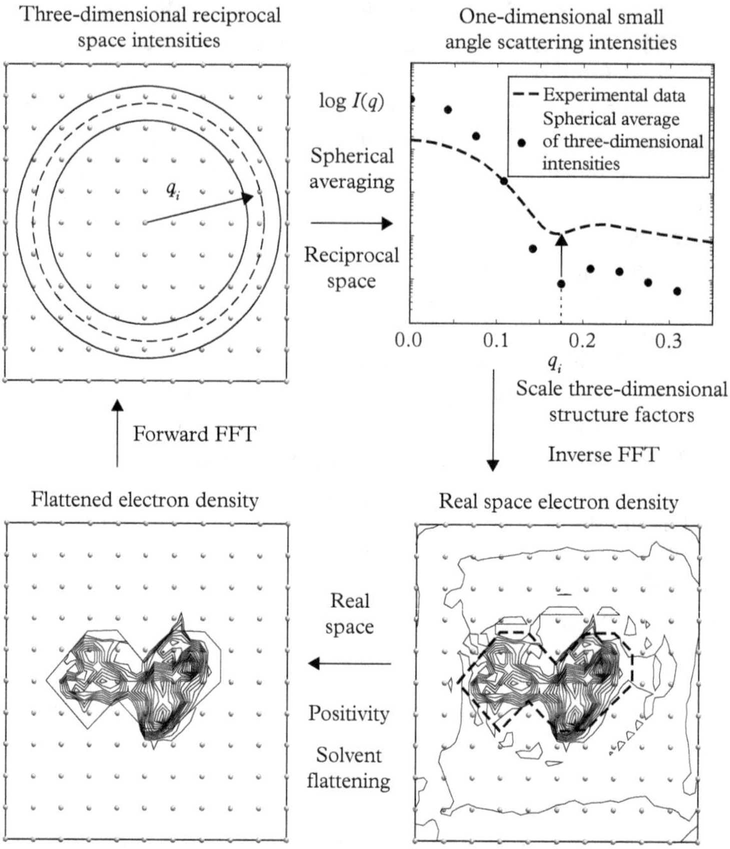

Figure 14.1 Diagram of iterative structure factor retrieval in solution scattering. Initially, a random set of electron densities is assigned on a three-dimensional grid of points. The Fourier transform of these electron densities returns the structure factors in reciprocal space from which the three-dimensional intensities can be calculated (top left panel). The average intensity at each resolution bin (q_i) is calculated and compared with the experimental solution scattering profile (top right panel). The three-dimensional structure factors in each q_i bin are all scaled by the same factor—the square root of the ratio of the experimental intensity to the calculated intensity—to ensure the calculated scattering profile matches the experimental scattering profile. The inverse Fourier transform is calculated to reveal a new electron density map (bottom right panel). Real space restraints are then applied so that inside the support (dashed line), density is positive and, outside the support, density is zero (bottom left panel). A new set of structure factors is calculated, and the process is repeated until convergence; FFT, fast Fourier transform.

(Reprinted by permission from Macmillan Publishers Ltd: Nature (Grant, 2018), © 2018.)

identify the envelope ab initio and update it iteratively as the electron density improves. As in iterative phase retrieval in single-molecule diffraction, real space restraints such as positivity of the density are applied, as well as solvent flattening outside of the support. This process is repeated, typically for thousands of cycles, until the solutions converge. The process is collectively referred to as "iterative structure factor retrieval" (Figure 14.1).

In contrast to single-molecule imaging, the final electron density function obtained from solution scattering is not unique. There are multiple electron density functions that agree with both real and reciprocal space restraints. Thus, running the algorithm multiple times with different random starting densities will result in slightly different answers. This is similar to the case of ab initio bead modeling, where multiple runs with different random seeds leads to slightly different bead models each time. The solution for bead modeling was to determine a consensus envelope, i.e., an envelope containing only the beads that are most commonly found from models reconstructed from many different random seeds (Volkov & Svergun, 2003). This ensures that features that are common to most envelopes, and thus those most likely to reflect the true particle shape, are retained, whereas beads that are artifacts due to the initial random seed are rejected. A similar approach is needed for determining electron density from solution scattering. However, as electron density maps show continuously varying values and do not assume uniformity throughout the envelope, we can perform direct averaging. In this case, some number of electron density maps are reconstructed (typically, 20 or so) and aligned. All the electron density maps are then averaged to result in a final electron density map agreeing with the true particle to low resolution.

14.3 Application to Real Data

While the method is still in its infancy, initial results with real experimental SAXS data show great promise. A set of previously published SAXS data (Grant et al., 2011) was used to show the agreement between electron density maps reconstructed ab initio from the SAXS data alone and the known high-resolution structure determined from crystallography (Figure 14.2). One of the biggest advantages of directly calculating electron density is the ability to model complex objects whose contents are not a single molecular species, as is required for accurate bead modeling, assuming uniform density. An example of such a case is given in Figure 14.3, that of the endophilin A1 BAR protein dimer connecting two micelles made of arachidonyl–coeynzyme A in a dumbbell shape (Petoukhov et al., 2014). In addition to its complex shape, this object is further complicated by the presence of three distinct electron density regions: the protein region connecting the two micelles, the high-density head group forming the outer shell of the micelles, and the low-density lipid tails forming the inner core of the micelles (Figure 14.3).

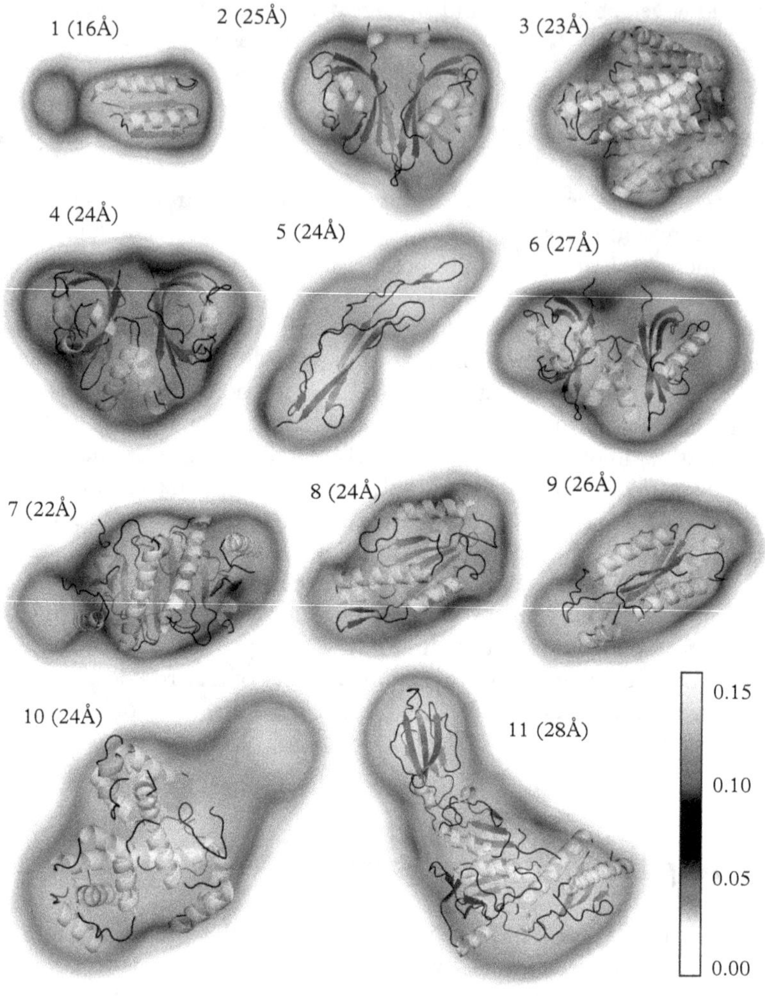

Figure 14.2 Electron density calculated from experimental SAXS data from a set of 11 proteins whose crystal structures are available. Electron densities are displayed as volumes colored according to density (gray shaded bar corresponds to density in $e^-/Å^3$). Crystal structures are shown in cartoon format overlapping the density. Estimated resolutions of electron density maps are shown in parentheses next to the sample ID.

(Reprinted by permission from Macmillan Publishers Ltd: Nature (Grant, 2018), © 2018.)

As the method provides actual electron density maps, estimates of resolution can be obtained in a manner similar to that performed in electron microscopy (Penczek, 2010). The set of ab initio envelopes is split into two sets (even and odd), the average from each set is calculated, and the Fourier shell correlation

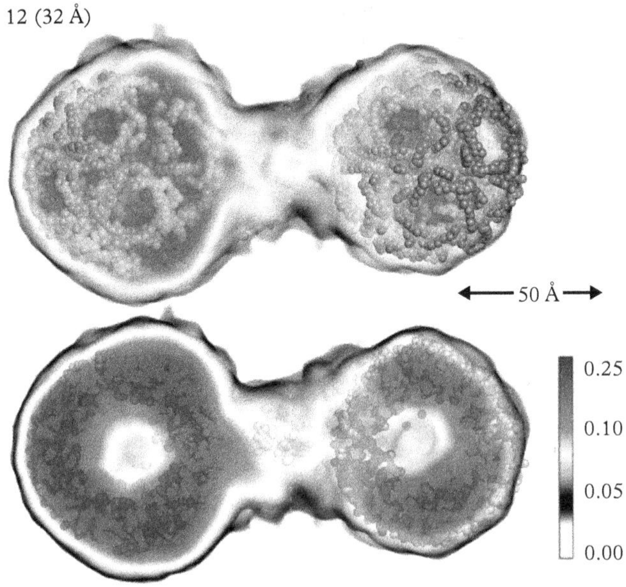

Figure 14.3 Electron density map calculated from experimental SAXS data of a protein–micelle complex in a dumbbell shape. Electron density is displayed as a volume colored according to density (gray shaded bar corresponds to density in $e^{-}/Å^{3}$). The published atomic model is shown in gray spheres overlapping the density. Estimated resolution of electron density map is shown in parentheses next to the sample ID. The scale bar is 50 Å. (Reprinted by permission from Macmillan Publishers Ltd: Nature (Grant, 2018), © 2018.)

(FSC) is calculated between the two averaged maps. The resolution can then be estimated where the correlation curve falls below 0.5.

14.4 Summary

The ability to calculate density directly from solution scattering data opens up the technique to a host of new opportunities. The ability to model density at high resolution and the implicit direct estimation of solvent terms such as the hydration shell may enable high-resolution wide angle scattering data to be used to calculate density when combined with additional structural information. Additionally, other diffraction methods that do not measure three-dimensional intensities, such as fiber diffraction, may also be able to take advantage of iterative structure factor retrieval. In such cases, the additional information provided by the two-dimensional scattering patterns will likely allow higher-resolution density maps to be reconstructed than possible with solution scattering. While the ability to reconstruct electron density ab initio is a major breakthrough in the field of solution scattering, the potential of the technique has yet to be fully uncovered. Additional structural information from techniques such as crystallography, NMR, and electron microscopy and density modification procedures popular in crystallography can now be integrated to perform advanced modeling of the electron density function at high resolution, pushing the boundaries of solution scattering further than ever before.

Epilogue

This book is intended to provide a basic grounding in the theory and practice of small angle scattering, both SAXS and SANS. We have attempted to provide an approach covering theory, practice, and application at a basic level while illustrating the power of the technique to other studies. In our own research, we have used small angle scattering in a complementary approach, and the subject matter extends far beyond the glimpse we have provided.

We have intentionally avoided detailed descriptions of specific software packages, knowing that the field is developing rapidly and that others have done so before (Schneidman-Duhovny, Rossi, et al., 2012; Graewert & Svergun, 2013). For those that are looking for specific software, we highly recommend the ATSAS package (Franke et al., 2017), a workhorse for SAXS and SANS, but other software is also available. Similarly, we aim to complement another highly recommended International Union of Crystallography text in the same area (Svergun et al., 2013), while avoiding unnecessary duplication.

The field is advancing on many levels. New X-ray sources provide access to brilliance and time resolution that were unheard of when the technique was first developed. This opens the door to the study of lower concentrations, shorter exposure times, and time resolution on the scale of many biological mechanisms, especially with the advent of X-ray free electron sources. Detector technology improvements provide complementary advances in reducing noise, extending both dynamic range and area, and enhancing the time resolution. Computational developments have both greatly increased the speed of analysis and enabled techniques to study ensembles of systems in a reasonable time frame. New algorithms, such as the direct determination of electron density described in Chapter 14, and techniques such as FSS present research avenues for the future, with applications that we have only glimpsed so far. Sample-handling automation, developments in areas such as microfluidics, and mechanisms to initiate processes (e.g., rapid mixing) promise new avenues for experiments previously impossible. We see many directions that will rapidly advance the scope, ability, and application of small angle scattering over the next decade.

Biological Small Angle Scattering: Theory and Practice. Eaton E. Lattman, Thomas D. Grant, and Edward H. Snell.

DOI: 10.1093/oso/9780199670871.001.0001

Our experience with small angle solution scattering has been a journey to answer questions that other structural techniques defined. In that journey, we have had wonderful support from the specialist solution scattering community. We encourage others to explore the technique and experience the same comradeship and delight as biology is revealed. The field is large and has plenty of room for development.

Appendix

A.1 The Structure Factor Equation

Equation 2.3 in Chapter 2 provides a general way of calculating the complex diffraction pattern from a direct space distribution of electrons given by $\rho(\boldsymbol{r})$. We repeat this equation for convenience:

$$\boldsymbol{F}(\boldsymbol{q}) = \int \rho(\boldsymbol{r}) \exp(i\boldsymbol{q}\boldsymbol{r}) dV.$$

By taking into account that electrons in matter are organized into atoms, we can derive a particularly useful form of this expression called the structure factor equation. To begin, let us apply equation 2.3 to the case in which ρ_c represents the electron density distribution within a particular type of atom—say, carbon. For each, this distribution can be calculated from quantum mechanics and can be regarded as a known function. Thus, we can calculate f_c, the scattering by a single carbon atom, using

$$f_c(q) = \int_0^{\infty} \rho_c(\boldsymbol{r}) \exp(i\boldsymbol{q}\boldsymbol{r}) dV. \tag{A1.1}$$

Since all carbon atoms are the same, f_c is a universal function that can always be used to represent carbon scattering. It is given a special name, the atomic scattering factor. There is one atomic scattering factor for each atom type, and these are tabulated for convenient use in a wide variety of solution scattering and crystallographic programs. The equation above assumed that the origin of the atom—carbon, in this example—coincided with the origin of the coordinate system for the electron density. Let us now imagine a way of representing electron density in a molecule as the sum of the electron densities of its component atoms, centered at their own locations. In Chapter 2, we also derived the change in scattering from an object that undergoes a translation $\boldsymbol{r}_0$ in direct space. This change is a phase shift in the scattering wave of $-\boldsymbol{q}\cdot\boldsymbol{r}_0$. In the Euler notation, this means that the scattering factor from an atom located at $\boldsymbol{r}_0$ is

$$f(q) \exp(-i\boldsymbol{q}\boldsymbol{r}_0). \tag{A1.2}$$

Because scattering is linear process, the scattering from a molecule is simply the sum of the scattering from all of the component atoms summed as complex numbers. Thus,

$$\boldsymbol{F}(\boldsymbol{q}) = F(q) \exp(\alpha) = \sum_{i=1}^{N} f_j(q) \exp(-i\boldsymbol{q}\boldsymbol{r}_j). \tag{A1.3}$$

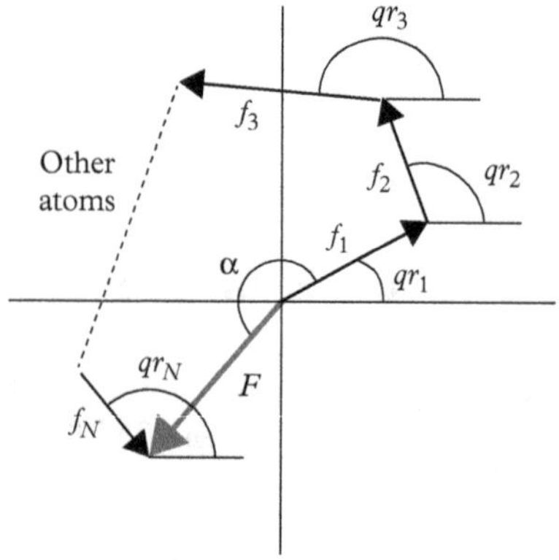

Figure A.1 Graphical illustration of equation A2.3.
DOI 10.1088/0953-4075/45/22/223001

Here, N is the number of atoms in the molecule, and $f_j(q)$ is the atomic scattering factor for the j-th atom. We use the scalar symbol q in lieu of the vector $\boldsymbol{q}$ because the spherical symmetry of atoms means that the direction of $\boldsymbol{q}$ does not matter.

Figure A.1 illustrates the meaning of this equation. Each term in the summation above can be represented as a vector in the complex plane, where the length of the vector is f_j, and the angle that the vector makes with the horizontal axis is qr_j. The vector sum of these individual terms yields the total scattering, $F(q)$.

The version of the structure factor equation that we have given above corresponds to continuous diffraction by a single particle. Essentially, the same equation is used to calculate scattering from crystals. In a book on solution scattering, an extended digression into crystallography is out of place, so we quote this equation without a detailed derivation. Readers are referred to texts on crystallography for more details.

In the crystal case, the summation above is carried out over all atoms in one unit cell of the crystal. In addition, the equation for crystals takes account of Bragg scattering, in which the periodic nature of the crystal causes full cancellation of scattering except for specific values of the vectors $\boldsymbol{q}$ that lie on the so-called reciprocal lattice. Diffraction from crystals thus produces diagrams where all the intensity is confined to spots. The crystal version is given by

$$\boldsymbol{F}(\boldsymbol{h}) = \sum_{i=1}^{N} f_j(\boldsymbol{h})\exp(-2\pi i \boldsymbol{h}\boldsymbol{x}_j). \tag{A1.4}$$

The differences in notation here arise from two sources. First, crystallographers normally use the scattering vector $\boldsymbol{s} = \boldsymbol{q}/2\pi$. The 2π factor propagates, producing the additional 2π in the exponential. Second, the coordinate system used in the above equation is also different: it is based on the lattice on which the crystal is built. As a simple illustration, we take a case where the lattice is cubic, so that the unit cell of the crystal is a cube defined by three orthogonal vectors $(\boldsymbol{a}, \boldsymbol{b}, \boldsymbol{c})$. The magnitude of these vectors corresponds to the edge length of the cube.

The position vector $\boldsymbol{x}_j = (x_j, y_j, z_j)$ is in *fractional* coordinates, not angstroms or nanometers. Thus, $\boldsymbol{r}$ and $\boldsymbol{x}$ are related via

$$\boldsymbol{r} = x\boldsymbol{a} + y\boldsymbol{b} + l\boldsymbol{c}. \tag{A1.5}$$

Finally, the vector $\boldsymbol{h}$, which corresponds to $\boldsymbol{q}$, is a triplet of integers (h, k, l) that serves to label each Bragg spot. And $\boldsymbol{h}$ is defined with respect to a diffraction space coordinate system called the reciprocal lattice, which is itself defined in terms of the unit cell vectors. In particular, the reciprocal lattice vectors $\boldsymbol{a}^\star$, $\boldsymbol{b}^\star$, $\boldsymbol{c}^\star$ are defined by

$$\boldsymbol{a}^\star = \frac{\boldsymbol{b} \times \boldsymbol{c}}{V}\ \boldsymbol{b}^\star = \frac{\boldsymbol{c} \times \boldsymbol{a}}{V}\ \boldsymbol{c}^\star = \frac{\boldsymbol{a} \times \boldsymbol{b}}{V}, \tag{A1.6}$$

where V is the volume of the unit cell. Note also that $\boldsymbol{a}\cdot\boldsymbol{a}^\star = \boldsymbol{b}\cdot\boldsymbol{b}^\star = \boldsymbol{c}\cdot\boldsymbol{c}^\star = 1$.

The Bragg condition arises when the scattering from each unit cell in the crystal is in phase with that from neighboring unit cells. This corresponds to a phase difference that is an integral number of cycles. Writing separately translations along the three crystal axes, we have

$$\boldsymbol{s}\cdot\boldsymbol{a} = h, \tag{A1.7}$$

$$\boldsymbol{s}\cdot\boldsymbol{b} = k, \tag{A1.8}$$

$$\boldsymbol{s}\cdot\boldsymbol{c} = l, \tag{A1.9}$$

where h, k, and l are the integers mentioned. We can use the reciprocal lattice vectors to write $\boldsymbol{s} = s_x\boldsymbol{a}^\star + s_y\boldsymbol{b}^\star + s_x\boldsymbol{c}^\star$. Inserting this above, we have

$$(s_x\boldsymbol{a}^\star + s_y\boldsymbol{b}^\star + s_z\boldsymbol{c}^\star)\cdot\boldsymbol{a} = s_x\boldsymbol{a}^\star\boldsymbol{a} = s_x = h, \tag{A1.10}$$

$$(s_x\boldsymbol{a}^\star + s_y\boldsymbol{b}^\star + s_z\boldsymbol{c}^\star)\cdot\boldsymbol{b} = s_y\boldsymbol{b}^\star\boldsymbol{b} = s_y = k, \tag{A1.11}$$

$$(s_x\boldsymbol{a}^\star + s_y\boldsymbol{b}^\star + s_z\boldsymbol{c}^\star)\cdot\boldsymbol{c} = s_z\boldsymbol{c}^\star\boldsymbol{c} = s_z = l, \tag{A1.12}$$

Thus, each of the spots or Bragg reflections appearing in an crystal X-ray diffraction diagram can be indexed by a triplet of integers, (h, k, l) that specifies its location in the reciprocal lattice.

A.2 Publication Guidelines

Publication guidelines exist for reporting small angle scattering data from biomolecules in solution (Jacques et al., 2012), and the rigor of those guidelines will no doubt increase. They serve as a guide both for authors and for those assessing the conclusions made from scattering experiments. The guidelines describe the equivalent of a crystallographic table, shown as Table A.1.

The table and guidelines establish the way in which a small angle scattering experiment should be presented to enable a reviewer and reader to assess the

Table A.1 *Recommended data-collection and scattering parameters to be reported in a solution scattering experiment (Jacques et al., 2012).*

Data-collection parameters	
Instrument	Specific instrument or beamline
Geometry	Slits, pinhole, monochromater, etc.
Wavelength (Å)	Wavelength (sometimes given in nanometers)
q range ($Å^{-1}$)	q range (sometimes given in nanometers)
Exposure time (s or min)	Sometimes a single exposure but often multiple exposures on a synchrotron source
Concentration range (mg/mL)	Upper and lower boundaries
Number of concentrations (ideally with concentrations listed)	Usually three or more for X-ray studies; for neutron studies, a single concentration is usually used with the sample also characterized with X-rays at several concentrations.
Structural parameters	
$I(0)$ from $P(r)$	For these parameters, consistency between them is important and with knowledge about the sample. For example, estimates of molecular mass can be obtained from $I(0)$.
R_g (Å) from $P(r)$	
$I(0)$ from Guinier	
R_g (Å) from Guinier	
D_{max} (Å)	Estimated as the value where $P(r)$ falls to zero.
Porod volume estimate ($Å^3$)	This can be used to estimate the molecular mass.
Dry volume calculated from sequence ($Å^3$)	
Molecular mass determination	
Partial specific volume	SEDNTERP (Laue et al., 1992) to estimate from protein sequence
Contrast	
Molecular mass from $I(0)$	Measured
Molecular mass from sequence	Calculated
Data processing	
Integration software	Usually specific to the instrument or beamline
Data analysis	These allow for reproduction of the result and may or may not apply to the particular study. They may identify particular problems (e.g., when the q range used is greater than that suitable for a particular program).
Ab initio analysis	
Validation and averaging	
Rigid body modeling	
Computation of model intensities	
Three-dimensional graphics representations	Software used.

validity of the conclusions taken from the data but also provide checks such that potential errors or sample problems may be identified. The first part of the table, data-collection parameters, is self-explanatory. This describes the manner in which the data was collected and is necessary to reproduce the results. The section on structural parameters describes measurements of invariants from the data. Some of these invariants can be calculated multiple ways (e.g., $I(0)$ and R_g), or compared to known values (e.g., Porod volume compared to calculated dry volume from the sequence). The section on software employed is more for completeness and reproducibility, but it can identify potential issues with known package limitations.

Jacques et al. (2012) recommend that parameters should be reported either normalized by macromolecule concentration or for each point in a concentration series with the sample-concentration values and with details as to how the scattering data was scaled (either to absolute values or relative to a known standard). Where multiple samples are being described, additional columns should be added to provide an easy comparison. The units indicated apply to both X-ray and neutron scattering. In the case of X-rays, scattering power and contrast values may also be reported as number of electrons (e) and e $Å^{-3}$, respectively.

Within the general text, it is recommended to supply evidence for the sample purity, concentration, and detail on how the sample blank is prepared. All scattering experiments, both X-ray and neutron, are sensitive to aggregation, and complementary techniques to monitor this are important. Data should be presented on this and can be important in cases of time-dependent dissociation or oligomerization.

Jacques et al. (2012) also provide guidelines on graphical information that should be provided for qualitative assessment. This includes a log plot of the scattering profile against q (with the predicted scattering if appropriate), and the corresponding $P(r)$ curve. Although not common, a log–log plot of the scattering profile is also useful, emphasizing as it does the information in the low-resolution part of the scattering and model fits to that experimental data. Additional representations that confirm the quality of the data include the Guinier and Kratky plots, which provide information on interparticle interactions and the folded state. This data is often evaluated subjectively, but robust statistical treatments are now available (Grant et al., 2015) and able to give a rapid, almost real-time assessment of data quality. Reliable conclusions require reliable data, and the information described allows the evaluation of the analysis and conclusions that result.

There have been steps to standardize small angle scattering data such that it is searchable and freely downloadable, much like X-ray data associated with crystallographic studies (Trewhella et al., 2013). At least two databases are available: the BIOISIS database and the Small-Angle Scattering Biological Data Bank (Hura et al., 2009; Valentini et al., 2015). These provide a resource not only to validate reported conclusions but also to provide data to refine and improve current methods. A community standard has been implemented with an ASCII format for the one-dimensional scattering data and implemented as sasCIF, an extension of the core crystallographic information file CIF (Kachala et al., 2016; Malfois & Svergun, 2000). The standard provides both scattering data and the meta-information necessary to understand the experiment and analyze it. Tools for manipulating the files are available. Most, if not all, current software makes use of these standards and, when developing new applications, it is recommended to incorporate them.

An updated set of guidelines was published by Trewhella et al. (2017). In addition to addressing sample quality, data acquisition, and reduction, the guidelines were extended to provide advice on specific types of SAXS (or SANS) studies. These included the inclusion of desmearing techniques for sources where the process is necessary, and for SEC–SAXS, a plot of $I(0)$ and R_g as a function of measurement time or frame, together with the corresponding UV trace. For measurements making use of contrast variation, a plot of normalized $\pm(I(0)/c)^{1/2}$ versus solvent scattering density should be included to show the total particle solvent match point along with transmissions at each contrast with controls for pure H_2O and D_2O. For assemblies of components, the M or V estimates from each contrast point should be included. Trewhella et al. (2017) also suggest a Stuhrmann plot (Stuhrmann et al., 1976) of R_g^2 versus $1/\Delta\rho$, which provides information on internal scattering densities within the scattering particle. Publication guidelines provide a good overview of the signatures associated with a well-executed and analyzed scattering experiment and are informative beyond their recommendations.

A.3 Troubleshooting

This extends and borrows from a table originally published by Skou et al. (2014).

Observation	Explanation	Solutions
Sample preparation		
Particulate matter seen in sample before experiment.	Denaturation, aggregation, or oligmerization.	Prepare sample fresh. Centrifuge sample. Use SEC methods after spin filtering.
Data collection		
Large difference in direct beam intensity between successive buffer or sample exposures.	Air bubbles in the sample or sample is moving out of the beam (if flowing or oscillating).	Use degassed solutions. Use a larger volume.
Buffer profiles differ before and after sample.	Protein residue may have built up on the sample chamber.	Clean sample chamber longer and/or with harsher cleaning solutions.
Buffer and protein solutions profiles do not converge at high q.	Potential buffer mismatch.	Use fresh buffer, sample, and repeat experiment. Check for possible contamination in capillary or in scattering pathway.
Buffer subtraction results in negative or under-subtracted signal at high q.	Buffer mismatch, poor intensity normalization, sample cell contamination, or instrument drift.	Repeat measurement with new buffer or perform a buffer exchange and repeat. Clean the sample cell. Check the beamline setup.

Sequential sample profiles do not superimpose.	Air bubbles or potential instrumentation issue (e.g., contamination from leak).	Check images for intensity difference (air bubble); use larger volume. Check the instrument and look for possible contamination in the scattering path.
Sample scattering profile appears to be increasing as a function of exposure.	Radiation damage.	Oscillate or flow sample. Use shorter exposure time. Consider free radical scavenger agents.
Scattering profile is extended and does not reach baseline at edge of detector.	Protein is too small for beamline configuration.	Reduce the sample-to-detector distance. Increase the energy.
Scattering profile is strong but only occurs near the beamstop.	Protein is too large for current beamline configuration.	Increase detector to sample distance. Decrease energy. Consider different instrument. SEC to remove higher-order oligomers.
Strong anisotropic scattering near beamstop.	Direct beam is not being fully stopped by beamstop.	Beamstop needs adjusting or beam needs slitting down. Slit scatter, guard slit adjustment required. Misaligned capillary/sample cell.
Data analysis		
The log(I) vs. log(q) plot of the background subtracted profile shows a steep increase at low q.	Aggregation, oligomerization, or radiation damage.	This is a warning sign that has to be investigated with further steps (Guinier analysis).
Guinier curve has an upturn at low q.	Aggregation, oligomerization, or radiation damage.	Perform an exposure-specific Guinier analysis. If there is no upturn in the first exposure, proceed with the next until an upturn is seen (radiation damage). Average the non-damaged data. If upturn in first exposure filter, centrifuge, or explore SEC methods. If the pH is close to the isolectric point (pI) reduce salt concentration or shift the pH away from the pI.
Guinier curve has a downturn at low q.	Interparticle repulsion.	Reduce protein concentration. Increase salt concentration.

(*continued*)

Observation	Explanation	Solutions
Not enough data points in the Guinier curve satisfying $qR_g < 1.3$.	Protein is too large for current beamline configuration.	Increase detector to sample distance. Decrease energy. Consider different instrument. SEC to remove higher-order oligomers.
Negative $P(r)$ at high r.	Interparticle repulsion.	Reduce protein concentration. Increase salt concentration.
Highly extended $P(r)$ at high r.	Aggregation, oligomerization, or potentially real structural feature.	Filter or centrifuge fresh sample. If the pH is close to the isolectric point (pI), reduce salt concentration or shift the pH away from the pI.
Slightly negative or positive $P(r)$ at high r.	Subtle interparticle effects or slight aggregation.	Use zero extrapolation to mitigate interparticle interactions.
Trends increase significantly or nonlinearly with increasing concentration.	Protein forming a mixture of oligomers.	Perform the experiment below or above the transition. Change solution conditions.
Model and data disagree.	Presence of dynamics or structural features previously constrained by crystal packing.	Investigate different oligomers, flexible domains, or the presence of multiple species.

Acronyms (Excluding Sample Names, Facilities, and Beamlines)

ALS	Advanced Light Source
APS	Advanced Photon Source
ATP	Adenosine triphosphate
BioCAT	Biophysics Collaborative Access Team
BPTI	Bovine pancreatic trypsin inhibitor
BSA	Bovine serum albumin
CCD	Charge-coupled device
CHESS	Cornell High Energy Synchrotron Source
CMC	Critical micelle concentration
DDM	*n*-Dodecyl-β-D-maltopyranside
DLS	Dynamic light scattering
dsRBD	Double-stranded RNA–binding domain
dsRNA	Double-stranded RNA
EOM	Ensemble optimization method
EQ-SANS	Extended q-range Small-Angle Neutron Scattering Diffractometer
ESRF	European Synchrotron Radiation Facility
FEN-1	Flap endonuclease 1
FRET	Fluorescence resonance energy transfer
FSC	Fourier shell correlation
FSS	Fluctuation solution scattering
HbI	Homodimeric hemoglobin
HFIR	High Flux Isotope Reactor
HPLC	High-pressure liquid chromatography
IDP	Intrinsically disordered protein
IEC	Ion-exchange chromatography
ILK	Integrin-linked kinase
ILL	Institut Laue–Langevin
JCSG	Joint Center for Structural Genomics
J-PARC	Japan Proton Accelerator Complex
LCLS	Linear Coherent Light Source
MALS	Multi-angle light scattering

MMPAD	Mixed mode pixel area detectors
NADPH	Nicotinamide adenine dinucleotide phosphate
NG6	Neutron guide 6
NG7	Neutron guide 7
NMR	Nuclear magnetic resonance
NRPS	Nonribosomal peptide synthetase
NSLS	National Synchrotron Light Source
OPAL	Open Pool Australian Lightwater
PAGE	Polyacrylamide gel electrophoresis
PCA	Principal component analysis
PCNA	Proliferating cell nuclear antigen
PCR	Polymerase chain reaction
PDB ID	Protein Data Bank ID number
PEG	Polyethylene glycol
PIP	PCNA-interacting motif
SANS	Small angle neutron scattering
SAXS	Small angle X-ray scattering
SDS	Sodium dodecyl sulfate
SEC	Size-exclusion chromotography
SIBYLS	Structurally Integrated Biology for Life Sciences
SINQ	Swiss Spallation Neutron Source
SLS	Static light scattering
SLS	Swiss Light Source
SNS	Spallation Neutron Source
SSRL	Stanford Synchrotron Radiation Lightsource
STAT	Signal transducer and activator of transcription
SVD	Singular value decomposition
TIP	*Thermococcales* inhibitor of PCNA
UOX	Urate oxidase
UV	Ultraviolet
WAXS	Wide angle X-ray scattering
XFEL	X-ray free electron laser

Glossary of Significant Terms

Term	Symbol	Definition
Chi-square goodness of fit	χ^2	A measure of the goodness of fit between two functions. Often used to compare the calculated scattering of a model to the experimental scattering profile.
Debye equation		Defines the spherically averaged intensity as a discrete sum of the elementary scattering elements and their scattering factors.
First data point	q_{min}	The first data point should be measured at $q_{min} \leq \pi/D_{max}$.
Forward scattering	$I(0)$	The scattering intensity at zero angle.
Form factor	$F^2(q)$ or $P^2(q)$	A one-dimensional function equal to the spherical average of the square of the molecular transform of a particle. This function contains the information about the size and shape of a particle.
Guinier plot		The Guinier plot is a plot of $\ln(I(q))$ against q^2. It can be used to calculate the radius of gyration independent of the absolute intensity and any model.
Guinier region		The low-q region of the scattering profile where the Guinier law is valid. Typically defined as $q < 1.3/R_g$ for globular molecules, decreasing to $q < 0.8/R_g$ for elongated molecules.
Information content	$N_{Shannon}$	The number of independent data points is given by the number of Shannon channels: $N_{Shannon} = (q_{max} - q_{min}) \times D_{max}/\pi$.
Invariant		The term given to values obtainable directly from SAXS profiles that are independent of modeling procedures (e.g., R_g and the Porod invariant).
Kratky plot		A plot of $q^2I(q)$ against q. Provides a qualitative assessment of the globularity and foldedness of a particle.
Maximum dimension	D_{max}	The maximum distance between scattering elements.
Molecular envelope		The surface representation of an object enclosing the entirety of its contents.
Pair distance distribution function	$P(r)$	A histogram of the interatomic pair distances present in a particle. Can be calculated as the Fourier transform of the scattering profile.
Particle surface area		The total surface area encompassed by a particle's molecular envelope.
Patterson function		The Fourier transform of the set of three-dimensional intensities. Corresponds to the autocorrelation function of the particle density in real space.

Phase and phase difference		The term 'phase' refers to the oscillation of a sinusoidal wave and is expressed in radians or degrees. For one whole cycle, the wave travels through 2πradians, or 360°. A particular phase can be used to calculate the amplitude at that point. The phase difference is the difference between the phases of two waves.
Porod invariant	Q	The integral of $I(q)q^2$ versus q. Provides a means of normalizing scattering profiles by their total scattering, to become independent of factors such as concentration of the particle.
Porod law		For globular molecules, the scattered intensity falls off proportionally to q^{-4}. This assumes uniform density and breaks down at high q values when atomic resolution information starts to contribute.
Porod plot		A plot of $I(q)$ versus q^4.
Porod–Debye plot		A plot of $I(q)q^4$ versus q^4.
Power law regime		The region where Porod's law breaks down and the scattered intensity falls of according to q^{-d}, where d is 4 for spherical particles, 2 for flat ellipsoids, 1 for needle-like objects, and ~5/3 for random coils.
Radius of gyration	R_g	The root-mean-square distance between all scatterers in a particle and the particle center of mass.
Spherical harmonics	$Y_{lm}(w)$	A series representation of a function in three-dimensions in spherical coordinates.
Structure factor	$F(\boldsymbol{q})$ or $S(q)$	The Fourier transform of particle density. A continuous, complex-valued function existing at every point in reciprocal space with a corresponding amplitude and phase ($F(\boldsymbol{q})$). Alternatively, in solution scattering, structure factor may refer to the function describing interparticle interactions in solution ($S(q)$).
Zero extrapolation		The process of extrapolating a set of scattering profiles collected at a series of concentrations to a theoretically infinite dilution to mitigate interparticle interactions.

Major Variables Defined

Symbol	Definition	Relation
2θ	Angle between incident and diffracted beam	
λ	Wavelength of the radiation	
$\boldsymbol{q}$	Position vector reciprocal space	$\lvert \boldsymbol{q} \rvert = 4\pi \sin(\theta)/\lambda)$
q	Magnitude of scattering vector	
$F(\boldsymbol{q})$	Structure factor amplitude at the scattering vector q	
$\rho(\boldsymbol{r})$	Scattering density	
$\boldsymbol{r}$	Vector distance between scattering centers	
$I(q)$	Scattering intensity	$I(q) = \lvert F(q) \rvert^2$
$\Delta\rho$	Scattering density difference between particle and solvent	
$P(r)$	The pair distribution function histograms the number of vectors of length r	$P(r) = 4\pi r^2 V\gamma(r)$
$D_{\max}$	Maximum linear dimension of a molecule	
N_{Shannon}	Number of Shannon channels	
$\gamma(r)$	The rotational average of $P(u, v, w)$; a three-dimensional function called the characteristic function	$V\gamma(r) = \int_0^\infty 4\pi q^2 I(q) \frac{\sin(rq)}{rq} dq$
N	Number of atoms (or scattering components, [e.g., beads])	
f	Scattering factor	
M	Molecular mass	
R_g	Radius of gyration	
Q	Porod invariant	
V	Particle volume	
S	Particle surface area (not to be confused with the position vector in reciprocal space)	
l_c	Correlation length	
V_c	Volume of correlation	
χ^2	Chi squared (model agreement with experimental data)	
$d_{\min}$	The minimum particle size that can be sampled	See section 6.6

d_{max}	The maximum particle size that can be sampled	See section 6.5
$\rho(x, y, z)$	The electron density at point (x, y, z)	
$\mathbf{F}(h, k, l)$	Complex structure factor at reciprocal space point (h, k, l)	
$F(h, k, l)$	Structure factor amplitude at reciprocal space point (h, k, l)	
$P(u,v,w)$	The three-dimensional Patterson function of a molecule or crystal	
$\gamma_0(r)$	The correlation function where the electron density difference is assumed to be constant, thus referring simply to the envelope of the particle.	

References

Abe, J., Hiyama, T. B., Mukaiyama, A., Son, S., Mori, T., Saito, S., Osako, M., Wolanin, J., Yamashita, E., Kondo, T. & Akiyama, S. (2015). Circadian rhythms. Atomic-scale origins of slowness in the cyanobacterial circadian clock. *Science* **349**, 312–316.

Abernathy, D. L., Grubel, G., Brauer, S., McNulty, I., Stephenson, G. A., Mochrie, S. G. J., Sandy, A. R., Mulders, N. & Sutton, M. (1998). Small-angle X-ray scattering using coherent undulator radiation at the ESRF. *J Synchrotron Radiat* **5**, 37–47.

Acerbo, A. S., Cook, M. J. & Gillilan, R. E. (2015). Upgrade of MacCHESS facility for X-ray scattering of biological macromolecules in solution. *J Synchrotron Radiat* **22**, 180–186.

Allaire, M. & Yang, L. (2011). Biomolecular solution X-ray scattering at the National Synchrotron Light Source. *J Synchrotron Radiat* **18**, 41–44.

Altieri, A. S., Ladner, J. E., Li, Z., Robinson, H., Sallman, Z. F., Marino, J. P. & Kelman, Z. (2016). A small protein inhibits proliferating cell nuclear antigen by breaking the DNA clamp. *Nucleic Acids Res* **44**, 10015.

Amenitsch, H., Bernstorff, S., Kriechbaum, M., Lombardo, D., Mio, H., Pabst, G., Rappolt, M. & Laggner, P. (1998). The small-angle X-ray scattering beamline at ELETTRA: A new powerful station for fast structural investigations on complex fluids with synchrotron radiation. *Nuovo Cimento D* **20**, 2181–2190.

Amenitsch, H., Bernstorff, S., Kriechbaum, M., Lombardo, D., Mio, H., Rappolt, M. & Laggner, P. (1997). Performance and first results of the ELETTRA high-flux beamline for small-angle X-ray scattering. *J Appl Crystallogr* **30**, 872–876.

Amenitsch, H., Bernstorff, S. & Laggner, P. (1995). High-flux beamline for small-angle X-ray-scattering at Elettra. *Rev Sci Instrum* **66**, 1624–1626.

Arnlund, D., Johansson, L. C., Wickstrand, C., Barty, A., Williams, G. J., Malmerberg, E., Davidsson, J., Milathianaki, D., DePonte, D. P., Shoeman, R. L., Wang, D., James, D., Katona, G., Westenhoff, S., White, T. A., Aquila, A., Bari, S., Berntsen, P., Bogan, M., van Driel, T. B., Doak, R. B., Kjaer, K. S., Frank, M., Fromme, R., Grotjohann, I., Henning, R., Hunter, M. S., Kirian, R. A., Kosheleva, I., Kupitz, C., Liang, M., Martin, A. V., Nielsen, M. M., Messerschmidt, M., Seibert, M. M., Sjohamn, J., Stellato, F., Weierstall, U., Zatsepin, N. A., Spence, J. C., Fromme, P., Schlichting, I., Boutet, S., Groenhof, G., Chapman, H. N. & Neutze, R. (2014). Visualizing a protein quake with time-resolved X-ray scattering at a free-electron laser. *Nat Methods* **11**, 923–926.

Bailey, D. C., Drake, E. J., Grant, T. D. & Gulick, A. M. (2016). Structural and functional characterization of aerobactin synthetase iuca from a hypervirulent pathotype of *Klebsiella pneumoniae*. *Biochemistry* **55**, 3559–3570.

Beno, M. A., Jennings, G., Engbretson, M., Knapp, G. S., Kurtz, C., Zabransky, B., Linton, J., Seifert, S., Wiley, C. & Montano, P. A. (2001). Basic Energy Sciences Synchrotron Radiation Center undulator sector at the advanced photon source. *Nucl Instrum Meth A* **467**, 690–693.

Bergmann, A., Orthaber, D., Scherf, G. & Glatter, O. (2000). Improvement of SAXS measurements on Kratky slit systems by Gobel mirrors and imaging-plate detectors. *J Appl Crystallogr* **33**, 869–875.

Bernado, P. (2010). Effect of interdomain dynamics on the structure determination of modular proteins by small-angle scattering. *Eur Biophys J* **39**, 769–780.

Bernado, P. & Blackledge, M. (2009). A self-consistent description of the conformational behavior of chemically denatured proteins from NMR and small angle scattering. *Biophys J* **97**, 2839–2845.

Bernado, P., Mylonas, E., Petoukhov, M. V., Blackledge, M. & Svergun, D. I. (2007). Structural characterization of flexible proteins using small-angle X-ray scattering. *J Am Chem Soc* **129**, 5656–5664.

Bernado, P., Perez, Y., Blobel, J., Fernandez-Recio, J., Svergun, D. I. & Pons, M. (2009). Structural characterization of unphosphorylated STAT5a oligomerization equilibrium in solution by small-angle X-ray scattering. *Protein Sci* **18**, 716–726.

Bernado, P. & Svergun, D. I. (2012). Analysis of intrinsically disordered proteins by small-angle X-ray scattering. *Methods Mol Biol* **896**, 107–122.

Berthaud, A., Manzi, J., Perez, J. & Mangenot, S. (2012). Modeling detergent organization around aquaporin-0 using small-angle X-ray scattering. *J Am Chem Soc* **134**, 10080–10088.

Bessman, N. J., Bagchi, A., Ferguson, K. M. & Lemmon, M. A. (2014). Complex relationship between ligand binding and dimerization in the epidermal growth factor receptor. *Cell Reports* **9**, 1306–1317.

Bijvoet, J. M. (1997). Phase determination in direct Fourier-synthesis of crystal structures (Reprinted from Proceedings of the Koninklijke Nederlandse Akademie van Wetenschappen, vol 52, pg 313–314, 1949). *P K Ned Akad Wetensc* **100**, 49–50.

Bilsel, O., Kayatekin, C., Wallace, L. A. & Matthews, C. R. (2005). A microchannel solution mixer for studying microsecond protein folding reactions. *Rev Sci Instrum* **76**, 014302.

Blanchet, C. E., Hermes, C., Svergun, D. I. & Fiedler, S. (2015). A small and robust active beamstop for scattering experiments on high-brilliance undulator beamlines. *J Synchrotron Radiat* **22**, 461–464.

Blanchet, C. E., Spilotros, A., Schwemmer, F., Graewert, M. A., Kikhney, A., Jeffries, C. M., Franke, D., Mark, D., Zengerle, R., Cipriani, F., Fiedler, S., Roessle, M. & Svergun, D. I. (2015). Versatile sample environments and automation for biological solution X-ray scattering experiments at the P12 beamline (PETRA III, DESY). *J Appl Crystallogr* **48**, 431–443.

Blanchet, C. E., Zozulya, A. V., Kikhney, A. G., Franke, D., Konarev, P. V., Shang, W., Klaering, R., Robrahn, B., Hermes, C., Cipriani, F., Svergun, D. I. & Roessle, M. (2012). Instrumental setup for high-throughput small- and wide-angle solution scattering at the X33 beamline of EMBL Hamburg. *J Appl Crystallogr* **45**, 489–495.

Blau, B., Clausen, K. N., Gvasaliya, S., Janoschek, M., Janssen, S., Keller, L., Roessli, B., Schefer, J., Tregenna-Piggott, P., Wagner, W. & Zaharko, O. (2009). The Swiss Spallation Neutron Source SINQ at Paul Scherrer Institut. *Neutron News* **20**, 5–8.

Boesecke, P. (2007). Reduction of two-dimensional small- and wideangle X-ray scattering data. *J Appl Crystallogr* **40**, S423–S427.

Bonnete, F., Ferte, N., Astier, J. P. & Veesler, S. (2004). Protein crystallization: Contribution of small angle X-ray scattering (SAXS). *J Phys IV* **118**, 3–13.

Bonnete, F. & Vivares, D. (2002). Interest of the normalized second virial coefficient and interaction potentials for crystallizing large macromolecules. *Acta Crystallogr D* **58**, 1571–1575.

Borsboom, M., Bras, W., Cerjak, I., Detollenaere, D., van Loon, D. G., Goedtkindt, P., Konijnenburg, M., Lassing, P., Levine, Y. K., Munneke, B., Oversluizen, M., van Tol, R.

& Vlieg, E. (1998). The Dutch-Belgian beamline at the ESRF. *J Synchrotron Radiat* 5, 518–520.
Boutet, S. & Williams, G. J. (2010). The Coherent X-ray Imaging (CXI) instrument at the Linac Coherent Light Source (LCLS). *New J Phys* 12, 0350024.
Broennimann, C., Eikenberry, E. F., Henrich, B., Horisberger, R., Huelsen, G., Pohl, E., Schmitt, B., Schulze-Briese, C., Suzuki, M., Tomizaki, T., Toyokawa, H. & Wagner, A. (2006). The PILATUS 1M detector. *J Synchrotron Radiat* 13, 120–130.
Brooks-Bartlett, J. C., Batters, R. A., Bury, C. S., Lowe, E. D., Ginn, H. M., Round, A. & Garman, E. F. (2017). Development of tools to automate quantitative analysis of radiation damage in SAXS experiments. *J Synchrotron Radiat* 24, 63–72.
Brosey, C. A., Ho, C., Long, W. Z., Singh, S., Burnett, K., Hura, G. L., Nix, J. C., Bowman, G. R., Ellenberger, T. & Tainer, J. A. (2016). Defining NADH-driven allostery regulating apoptosis-inducing factor. *Structure* 24, 2067–2079.
Bu, Z. M. & Engelman, D. M. (1999). A method for determining transmembrane helix association and orientation in detergent micelles using small angle X-ray scattering. *Biophys J* 77, 1064–1073.
Burger, V. M., Arenas, D. J. & Stultz, C. M. (2016). A Structure-free method for quantifying conformational flexibility in proteins. *Sci Rep* 6, 29040.
Caldwell, S. J. & Berghuis, A. M. (2012). Small-angle X-ray scattering analysis of the bifunctional antibiotic resistance enzyme aminoglycoside (6′) acetyltransferase-ie/aminoglycoside (2″) phosphotransferase-ia reveals a rigid solution structure. *Antimicrob Agents Chemother* 56, 1899–1906.
Cantor, C. R. & Schimmel, P. R. (1980). *Biophysical Chemistry Part. 2. Techniques for the Study of Biological Structure and Function.* San Francisco: Freeman.
Chacon, P., Moran, F., Diaz, J. F., Pantos, E. & Andreu, J. M. (1998). Low-resolution structures of proteins in solution retrieved from X-ray scattering with a genetic algorithm. *Biophys J* 74, 2760–2775.
Chayen, N. E., Helliwell, J. R. & Snell, E. H. (2010). *Macromolecular Crystallization and Crystal Perfection.* Oxford: Oxford University Press.
Chen, L., Wildegger, G., Kiefhaber, T., Hodgson, K. O. & Doniach, S. (1998). Kinetics of lysozyme refolding: Structural characterization of a non-specifically collapsed state using time-resolved X-ray scattering. *J Mol Biol* 276, 225–237.
Chen, Y., Tokuda, J. M., Topping, T., Meisburger, S. P., Pabit, S. A., Gloss, L. M. & Pollack, L. (2017). Asymmetric unwrapping of nucleosomal DNA propagates asymmetric opening and dissociation of the histone core. *Proc Natl Acad Sci U S A* 114, 334–339.
Chen, Y., Tokuda, J. M., Topping, T., Sutton, J. L., Meisburger, S. P., Pabit, S. A., Gloss, L. M. & Pollack, L. (2014). Revealing transient structures of nucleosomes as DNA unwinds. *Nucleic Acids Res* 42, 8767–8776.
Cherezov, V., Clogston, J., Misquitta, Y., Abdel-Gawad, W. & Caffrey, M. (2002). Membrane protein crystallization in meso: Lipid type-tailoring of the cubic phase. *Biophys J* 83, 3393–3407.
Cherezov, V., Fersi, H. & Caffrey, M. (2001). Crystallization screens: Compatibility with the lipidic cubic phase for in meso crystallization of membrane proteins. *Biophys J* 81, 225–242.
Cho, H. S., Schotte, F., Dashdorj, N., Kyndt, J., Henning, R. & Anfinrud, P. A. (2016). Picosecond photobiology: watching a signaling protein function in real time via time-resolved small- and wide-angle x-ray scattering. *J Am Chem Soc* 138, 8815–8823.

Chu, B., Fang, D. F. & Mao, Y. M. (2015). Instrumentation on multi-scaled scattering of bio-macromolecular solutions. *Int J Mol Sci* **16**, 10016–10037.

Classen, S., Hura, G. L., Holton, J. M., Rambo, R. P., Rodic, I., McGuire, P. J., Dyer, K., Hammel, M., Meigs, G., Frankel, K. A. & Tainer, J. A. (2013). Implementation and performance of SIBYLS: A dual endstation small-angle X-ray scattering and macromolecular crystallography beamline at the Advanced Light Source. *J Appl Crystallogr* **46**, 1–13.

Classen, S., Rodic, I., Holton, J., Hura, G. L., Hammel, M. & Tainer, J. A. (2010). Software for the high-throughput collection of SAXS data using an enhanced Blu-Ice/DCS control system. *J Synchrotron Radiat* **17**, 774–781.

Criswell, A. & Mooers, B. H. (2015). Structural studies of a double-stranded RNA from trypanosome RNA editing by small-angle X-ray scattering. *Methods Mol Biol* **1240**, 165–189.

Dauter, Z. & Wilson, K. S. (1994). Imaging plates in synchrotron and conventional X-ray crystallographic data collection. *Acta Phys Pol A* **86**, 477–486.

David, G. & Perez, J. (2009). Combined sampler robot and high-performance liquid chromatography: A fully automated system for biological small-angle X-ray scattering experiments at the Synchrotron SOLEIL SWING beamline. *J Appl Crystallogr* **42**, 892–900.

DeCaria, A. J. & Van Knowe, G. E. (2014). *A First Course in Atmospheric Numerical Modeling*. Madison: Sundog Publishing.

Dewhurst, C. D. (2008). D33: A third small-angle neutron scattering instrument at the Institut Laue Langevin. *Meas Sci Technol* **19**, 034007.

DiFabio, J., Chodankar, S., Pjerov, S., Jakoncic, J., Lucas, M., Krywka, C., Graziano, V. & Yang, L. (2016). The Life Science X-ray Scattering Beamline at NSLS-II. In Q. Shen and C. Nelson, eds., *Proceedings of the 12th International Conference on Synchrotron Radiation Instrumentation (Sri2015)*. Melville, NY: AIP, 030049.

Dinapoli, R., Bergamaschi, A., Henrich, B., Horisberger, R., Johnson, I., Mozzanica, A., Schmid, E., Schmitt, B., Schreiber, A., Shi, X. & Theidel, G. (2011). EIGER: Next generation single photon counting detector for X-ray applications. *Nucl Instrum Meth A* **650**, 79–83.

Donatelli, J. J., Zwart, P. H. & Sethian, J. A. (2015). Iterative phasing for fluctuation X-ray scattering. *Proc Natl Acad Sci U S A* **112**, 10286–10291.

Dong, B. Z., Sheng, W. J., Yang, H. L. & Zhang, Z. J. (1997). Status of the small-angle scattering station at the Beijing Synchrotron Radiation Facility. *J Appl Crystallogr* **30**, 877–879.

Dubuisson, J. M., Decamps, T. & Vachette, P. (1997). Improved signal-to-background ratio in small-angle x-ray scattering experiments with synchrotron radiation using an evacuated cell for solutions. *J Appl Crystallogr* **30**, 49–54.

Dumetz, A. C., Snellinger-O'Brien, A. M., Kaler, E. W. & Lenhoff, A. M. (2007). Patterns of protein–protein interactions in salt solutions and implications for protein crystallization. *Protein Sci* **16**, 1867–1877.

Dunker, A. K., Lawson, J. D., Brown, C. J., Williams, R. M., Romero, P., Oh, J. S., Oldfield, C. J., Campen, A. M., Ratliff, C. M., Hipps, K. W., Ausio, J., Nissen, M. S., Reeves, R., Kang, C., Kissinger, C. R., Bailey, R. W., Griswold, M. D., Chiu, W., Garner, E. C. & Obradovic, Z. (2001). Intrinsically disordered protein. *J Mol Graph Model* **19**, 26–59.

Duxbury, D., Heenan, R., McPhail, D., Raspino, D., Rhodes, N., Rogers, S., Schooneveld, E., Spill, E. & Terry, A. (2014). Performance characteristics of the new detector array for the SANS2d instrument on the ISIS spallation neutron source. *J Instrum* **9**, C12051.

Dyer, K. N., Hammel, M., Rambo, R. P., Tsutakawa, S. E., Rodic, I., Classen, S., Tainer, J. A. & Hura, G. L. (2014). *Structural Genomics: General Applications*, edited by Y. W. Chen. New York: Humana Press, 245–258.

Dyson, H. J. & Wright, P. E. (2005). Intrinsically unstructured proteins and their functions. *Nat Rev Mol Cell Biol* **6**, 197–208.

Dzananovic, E., Patel, T. R., Chojnowski, G., Boniecki, M. J., Deo, S., McEleney, K., Harding, S. E., Bujnicki, J. M. & McKenna, S. A. (2014). Solution conformation of adenovirus virus associated RNA-I and its interaction with PKR. *J Struct Biol* **185**, 48–57.

Edwards, J. S., Betts, L., Frazier, M. L., Pollet, R. M., Kwong, S. M., Walton, W. G., Ballentine, W. K.,3rd, Huang, J. J., Habibi, S., Del Campo, M., Meier, J. L., Dervan, P. B., Firth, N. & Redinbo, M. R. (2013). Molecular basis of antibiotic multiresistance transfer in *Staphylococcus aureus*. *Proc Natl Acad Sci U S A* **110**, 2804–2809.

Ellis, P. J., Cohen, A. E. & Soltis, S. M. (2003). Beamstop with integrated X-ray sensor. *J Synchrotron Radiat* **10**, 287–288.

Engelman, D. M. & Moore, P. B. (1972). New method for determination of biological quarternary structure by neutron-scattering. *Proc Natl Acad Sci U S A* **69**, 1997–1999.

Englich, U., Kriksunov, I. A., Cerione, R. A., Cook, M. J., Gillilan, R., Gruner, S. M., Huang, Q. Q., Kim, C. U., Miller, W., Nielsen, S., Schuller, D., Smith, S. & Szebenyi, D. M. E. (2011). Microcrystallography, high-pressure cryocooling and BioSAXS at MacCHESS. *J Synchrotron Radiat* **18**, 70–73.

Feigin, L. A. & Svergun, D. I. (1987). *Structure Analysis by Small-Angle X-Ray and Neutron Scattering*. New York: Plenum Press.

Fetler, L., Tauc, P., Herve, G., Moody, M. F. & Vachette, P. (1995). X-ray scattering titration of the quaternary structure transition of aspartate transcarbamylase with a bisubstrate analogue: Influence of nucleotide effectors. *J Mol Biol* **251**, 243–255.

Fischer, H., Neto, M. D., Napolitano, H. B., Polikarpov, I. & Craievich, A. F. (2010). Determination of the molecular weight of proteins in solution from a single small-angle X-ray scattering measurement on a relative scale. *J Appl Crystallogr* **43**, 101–109.

Flory, P. J. (1953). *Principles of Polymer Chemistry*. Ithaca: Cornell University.

Forster, F., Webb, B., Krukenberg, K. A., Tsuruta, H., Agard, D. A. & Sali, A. (2008). Integration of small-angle X-ray scattering data into structural modeling of proteins and their assemblies. *J Mol Biol* **382**, 1089–1106.

Franke, D., Kikhney, A. G. & Svergun, D. I. (2012). Automated acquisition and analysis of small angle X-ray scattering data. *Nucl Instrum Meth A* **689**, 52–59.

Franke, D., Petoukhov, M. V., Konarev, P. V., Panjkovich, A., Tuukkanen, A., Mertens, H. D. T., Kikhney, A. G., Hajizadeh, N. R., Franklin, J. M., Jeffries, C. M. & Svergun, D. I. (2017). ATSAS 2.8: A comprehensive data analysis suite for small-angle scattering from macromolecular solutions. *J Appl Crystallogr* **50**, 1212–1225.

Franke, D. & Svergun, D. I. (2009). DAMMIF, a program for rapid ab-initio shape determination in small-angle scattering. *J Appl Crystallogr* **42**, 342–346.

Fuchs, M. R., Sweet, R. M., Berman, L. E., Hendrickson, W. A., Chubar, O., Canestrari, N., Idir, M., Yang, L. & Schneider, D. K. (2014). NSLS-II biomedical beamlines for macromolecular crystallography, FMX and AMX, and for X-ray scattering, LIX: Current developments. *J Phys: Conf Ser* **493**, 012021.

Fujisawa, T., Inoue, K., Oka, T., Iwamoto, H., Uruga, T., Kumasaka, T., Inoko, Y., Yagi, N., Yamamoto, M. & Ueki, T. (2000). Small-angle X-ray scattering station at the SPring-8 RIKEN beamline. *J Appl Crystallogr* **33**, 797–800.

Fung, R., Hanna, A. M., Vendrell, O., Ramakrishna, S., Seideman, T., Santra, R., Ourmazd, A. & Ourmazd, A. (2016). Dynamics from noisy data with extreme timing uncertainty. *Nature* **532**, 471–475.

Furlong, E. J., Lo, A. W., Kurth, F., Premkumar, L., Totsika, M., Achard, M. E. S., Halili, M. A., Heras, B., Whitten, A. E., Choudhury, H. G., Schembri, M. A. & Martin, J. L. (2017). A shape-shifting redox foldase contributes to *Proteus mirabilis* copper resistance. *Nat Commun* **8**, 16065.

Garcia, P., Serrano, L., Durand, D., Rico, M. & Bruix, M. (2001). NMR and SAXS characterization of the denatured state of the chemotactic protein CheY: Implications for protein folding initiation. *Protein Sci* **10**, 1100–1112.

George, A. & Wilson, W. W. (1994). Predicting protein crystallization from a dilute-solution property. *Acta Crystallogr D* **50**, 361–365.

Gilbert, E. P., Schulz, J. C. & Noakes, T. J. (2006). "Quokka": The small-angle neutron scattering instrument at OPAL. *Physica B* **385–86**, 1180–1182.

Glatter, O. (1975). Indirect Fourier transformation:New method for data evaluation in small-angle scattering. *Acta Crystallogr A* **31**, S164–S164.

Glatter, O. (1977). Data evaluation in small-angle scattering: Calculation of radial electron-density distribution by means of indirect Fourier transformation. *Acta Phys Austriaca* **47**, 83–102.

Glatter, O. & Kratky, O. (1982). *Small Angle X-Ray Scattering*. London/New York: Academic Press.

Glinka, C. J., Barker, J. G., Hammouda, B., Krueger, S., Moyer, J. J. & Orts, W. J. (1998). The 30 m small-angle neutron scattering instruments at the National Institute of Standards and Technology. *J Appl Crystallogr* **31**, 430–445.

Goldenberg, D. P. & Argyle, B. (2014). Minimal effects of macromolecular crowding on an intrinsically disordered protein: A small-angle neutron scattering study. *Biophys J* **106**, 905–914.

Golub, G. H. & Van Loan, C. F. (1996). *Matrix Computations*, 3rd ed. Baltimore: Johns Hopkins University Press.

Graber, T., Anderson, S., Brewer, H., Chen, Y. S., Cho, H. S., Dashdorj, N., Henning, R. W., Kosheleva, I., Macha, G., Meron, M., Pahl, R., Ren, Z., Ruan, S., Schotte, F., Rajer, V. S., Viccaro, P. J., Westferro, F., Anfinrud, P. & Moffat, K. (2011). BioCARS: A synchrotron resource for time-resolved X-ray science. *J Synchrotron Radiat* **18**, 658–670.

Graceffa, R., Nobrega, R. P., Barrea, R. A., Kathuria, S. V., Chakravarthy, S., Bilsel, O. & Irving, T. C. (2013). Sub-millisecond time-resolved SAXS using a continuous-flow mixer and X-ray microbeam. *J Synchrotron Radiat* **20**, 820–825.

Graewert, M. A., Franke, D., Jeffries, C. M., Blanchet, C. E., Ruskule, D., Kuhle, K., Flieger, A., Schaefer, B., Tartsch, B., Meijers, R. & Svergun, D. I. (2015). Automated pipeline for purification, biophysical and X-ray analysis of biomacromolecular solutions. *Sci Rep* **5**, 10734.

Graewert, M. A. & Svergun, D. I. (2013). Impact and progress in small and wide angle X-ray scattering (SAXS and WAXS). *Curr Opin Struct Biol* **23**, 748–754.

Grant, T., Luft, J. R., Wolfley, J. R., Tsuruta, H., Martel, A., Montelione, G. T. & Snell, E. H. (2011). Small angle X-ray scattering as a complementary tool for high-throughput structural studies. *Biopolymers* **95**, 517–530.

Grant, T. D. (2018). *Ab initio* electron density determination directly from solution scattering data. *Nat Methods*, doi: 10.1038/nmeth.4581. [Epub ahead of print].

Grant, T. D., Luft, J. R., Carter, L. G., Matsui, T., Weiss, T. M., Martel, A. & Snell, E. H. (2015). The accurate assessment of small-angle X-ray scattering data. *Acta Crystallogr D* **71**, 45–56.

Grishaev, A. (2012). Sample preparation, data collection, and preliminary data analysis in biomolecular solution X-ray scattering. *Curr Protoc Protein Sci* **70**, 17.14.1–17.14.18.

Grishaev, A., Anthis, N. J. & Clore, G. M. (2012). Contrast-matched small-angle X-ray scattering from a heavy-atom-labeled protein in structure determination: Application to a lead-substituted calmodulin-peptide complex. *J Am Chem Soc* **134**, 14686–14689.

Guinier, A. (1939). La diffraction des rayons X aux tres petits angles; Application a l'etude de phenomenes ultramicroscopiques. *Ann. Phys. (Paris)* **12**, 161–237.

Guinier, A. & Fournet, G. (1955). *Small Angle Scattering of X-rays.* New York: Wiley Interscience.

Hamilton, W. A., Garrett, R. & Gilbert, E. P. (2009). Quokka: The small-angle neutron scattering instrument. *Neutron News* **20**, 24–28.

Hammersley, A.P. (2016). FIT2D: a multi-purpose data reduction, analysis and visualization program. *J Appl Crystallogr* **49**, 646–652.

Hammouda, B. (2016). *Probing Nanoscale Structures: The SANS Toolbox*, National Institute of Standards. Gaithersburg, MD: National Institute of Standards and Technology Center for Neutron Research.

Harpaz, Y., Gerstein, M. & Chothia, C. (1994). Volume changes on protein folding. *Structure* **2**, 641–649.

Heenan, R. K., Penfold, J. & King, S. M. (1997). SANS at pulsed neutron sources: Present and future prospects. *J Appl Crystallogr* **30**, 1140–1147.

Heenan, R. K., Rogers, S. E., Turner, D., Terry, A. E., Treadgold, J. & King, S. M. (2011). Small angle neutron scattering using Sans2d. *Neutron News* **22**, 19–21.

Heller, W. T., Lynn, G. W., Urban, V. S., Weiss, K. & Myles, D. A. A. (2008). The Bio-SANS small-angle neutron scattering instrument at Oak Ridge National Laboratory. *Neutron News* **19**, 22–23.

Heller, W. T., Urban, V. S., Lynn, G. W., Weiss, K. L., O'Neill, H. M., Pingali, S. V., Qian, S., Littrell, K. C., Melnichenko, Y. B., Buchanan, M. V., Selby, D. L., Wignall, G. D., Butler, P. D. & Myles, D. A. (2014). The Bio-SANS instrument at the High Flux Isotope Reactor of Oak Ridge National Laboratory. *J Appl Crystallogr* **47**, 1238–1246.

Henderson, S. J. (1996). Monte Carlo modeling of small-angle scattering data from non-interacting homogeneous and heterogeneous particles in solution. *Biophys J* **70**, 1618–1627.

Holdbrook, D. A., Burmann, B. M., Huber, R. G., Petoukhov, M. V., Svergun, D. I., Hiller, S. & Bond, P. J. (2017). A spring-loaded mechanism governs the clamp-like dynamics of the Skp chaperone. *Structure* **25**, 1079–1088.

Hong, X. G. & Hao, Q. (2009). Measurements of accurate X-ray scattering data of protein solutions using small stationary sample cells. *Rev Sci Instrum* **80**: 014303.

Hopkins, J. B., Katz, A. M., Meisburger, S. P., Warkentin, M. A., Thorne, R. E. & Pollack, L. (2015). A microfabricated fixed path length silicon sample holder improves background subtraction for cryoSAXS. *J Appl Crystallogr* **48**, 227–237.

Hua, W. Q., Wang, Y. Z., Zhou, P., Hu, T., Li, X. H., Bian, F. G. & Wang, J. (2017). Microfocus small-angle X-ray scattering at SSRF BL16B1. *Chinese Phys C* **41**, 048001.

Hura, G. L., Budworth, H., Dyer, K. N., Rambo, R. P., Hammel, M., McMurray, C. T. & Tainer, J. A. (2013). Comprehensive macromolecular conformations mapped by quantitative SAXS analyses. *Nat Methods* **10**, 453–454.

Hura, G. L., Menon, A. L., Hammel, M., Rambo, R. P., Poole, F. L.,2nd, Tsutakawa, S. E., Jenney, F. E., Jr., Classen, S., Frankel, K. A., Hopkins, R. C., Yang, S. J., Scott, J. W., Dillard, B. D., Adams, M. W. & Tainer, J. A. (2009). Robust, high-throughput solution structural analyses by small angle X-ray scattering (SAXS). *Nat Methods* **6**, 606–612.

Hutin, S., Brennich, M., Maillot, B. & Round, A. (2016). Online ion-exchange chromatography for small-angle X-ray scattering. *Acta Crystallogr D* **72**, 1090–1099.

Hynson, R. M. G., Duff, A. P., Kirby, N., Mudie, S. & Lee, L. K. (2015). Differential ultracentrifugation coupled to small-angle X-ray scattering on macromolecular complexes. *J Appl Crystallogr* **48**, 769–775.

Ibrahim, Z., Martel, A., Moulin, M., Kim, H. S., Hartlein, M., Franzetti, B. & Gabel, F. (2017). Time-resolved neutron scattering provides new insight into protein substrate processing by a AAA+ unfoldase. *Sci Rep* **7**, 40948.

Igarashi, N., Shimizu, N., Koyama, A., Mori, T., Ohta, H., Niwa, Y., Nitani, H., Abe, H., Nomura, M., Shioya, T., Tsuchiya, K. & Ito, K. (2013). New high-brilliance beamline BL-15A of the Photon Factory. *J Phys: Conf Ser* **425**, 072016.

Igarashi, N., Watanabe, Y., Shinohara, Y., Inoko, Y., Matsuba, G., Okuda, H., Mori, T. & Ito, K. (2011). Upgrade of the small angle X-ray scattering beamlines at the Photon Factory. *J Phys: Conf Ser* **272**, 012026.

Inoue, K., Doutch, J. & Terrill, N. (2013). The current status of small-angle X-ray scattering beam line at Diamond Light Source. *Bunseki Kagaku* **62**, 565–570.

Inoue, K., Oka, T., Suzuki, T., Yagi, N., Takeshita, K., Goto, S. & Ishikawa, T. (2001). Present status of high flux beamline (BL40XU) at SPring-8. *Nucl Instrum Meth A* **467**, 674–677.

Jacques, D. A., Guss, J. M., Svergun, D. I. & Trewhella, J. (2012). Publication guidelines for structural modelling of small-angle scattering data from biomolecules in solution. *Acta Crystallogr D* **68**, 620–626.

Jacques, D. A. & Trewhella, J. (2010). Small-angle scattering for structural biology: Expanding the frontier while avoiding the pitfalls. *Protein Sci* **19**, 642–657.

Jacrot, B. (1976). The study of biological structures by neutron scattering from solution. *Rep Prog Phys* **39**, 911–953.

Jacrot, B., Chauvin, C. & Witz, J. (1977). Comparative neutron small-angle scattering study of small spherical RNA viruses. *Nature* **266**, 417–421.

Jaksch, S., Martin-Rodriguez, D., Ostermann, A., Jestin, J., Duarte Pinto, S., Bouwman, W. G., Uher, J., Engels, R. & Frielinghaus, H. (2014). Concept for a time-of-flight small angle neutron scattering instrument at the European Spallation Source. *Nucl Instrum Meth A* **762**, 22–30.

Jeffries, C. M., Graewert, M. A., Blanchet, C. E., Langley, D. B., Whitten, A. E. & Svergun, D. I. (2016). Preparing monodisperse macromolecular samples for successful biological small-angle X-ray and neutron-scattering experiments. *Nat Protoc* **11**, 2122–2153.

Jeffries, C. M., Graewert, M. A., Svergun, D. I. & Blanchet, C. E. (2015). Limiting radiation damage for high-brilliance biological solution scattering: Practical experience at the EMBL P12 beamline PETRAIII. *J Synchrotron Radiat* **22**, 273–279.

Jeffries, C. M. & Svergun, D. I. (2015). High-throughput studies of protein shapes and interactions by synchrotron small-angle X-ray scattering. *Methods Mol Biol* **1261**, 277–301.

Jensen, M. R., Markwick, P. R. L., Meier, S., Griesinger, C., Zweckstetter, M., Grzesiek, S., Bernado, P. & Blackledge, M. (2009). Quantitative determination of the conformational properties of partially folded and intrinsically disordered proteins using NMR dipolar couplings. *Structure* **17**, 1169–1185.

Jimenez-Garcia, B., Pons, C., Svergun, D. I., Bernado, P. & Fernandez-Recio, J. (2015). pyDockSAXS: Protein–protein complex structure by SAXS and computational docking. *Nucleic Acids Res* **43**, W356–W361.

Jolliffe, I. T. (2002). *Principal Component Analysis*, 2nd ed. New York: Springer.

Jordan, A., Jacques, M., Merrick, C., Devos, J., Forsyth, V. T., Porcar, L. & Martel, A. (2016). SEC-SANS: Size exclusion chromatography combined in situ with small-angle neutron scattering. *J Appl Crystallogr* **49**, 2015–2020.

Joseph, J. S., Liu, W., Kunken, J., Weiss, T. M., Tsuruta, H. & Cherezov, V. (2011). Characterization of lipid matrices for membrane protein crystallization by high-throughput small angle X-ray scattering. *Methods* **55**, 342–349.

Kachala, M., Westbrook, J. & Svergun, D. (2016). Extension of the sasCIF format and its applications for data processing and deposition. *J Appl Crystallogr* **49**, 302–310.

Kam, Z. (1977). Determination of macromolecular structure in solution by spatial correlation of scattering fluctuations. *Macromolecules* **10**, 927–934.

Kam, Z. (1980). Reconstruction of structure from electron-micrographs of randomly oriented particles. *J Theor Biol* **82**, 15–39.

Kam, Z., Koch, M. H. J. & Bordas, J. (1981). Fluctuation X-ray-scattering from biological particles in frozen solution by using synchrotron radiation. *Proc Natl Acad Sci U S A* **78**, 3559–3562.

Kathuria, S. V., Guo, L., Graceffa, R., Barrea, R., Nobrega, R. P., Matthews, C. R., Irving, T. C. & Bilsel, O. (2011). Minireview: Structural insights into early folding events using continuous-flow time-resolved small-angle X-ray scattering. *Biopolymers* **95**, 550–558.

Kathuria, S. V., Kayatekin, C., Barrea, R., Kondrashkina, E., Graceffa, R., Guo, L., Nobrega, R. P., Chakravarthy, S., Matthews, C. R., Irving, T. C. & Bilsel, O. (2014). Microsecond barrier-limited chain collapse observed by time-resolved FRET and SAXS. *J Mol Biol* **426**, 1980–1994.

Kellermann, G., Vicentin, F., Tamura, E., Rocha, M., Tolentino, H., Barbosa, A., Craievich, A. & Torriani, I. (1997). The small-angle X-ray scattering beamline of the Brazilian Synchrotron Light Laboratory. *J Appl Crystallogr* **30**, 880–883.

Kikhney, A. G. & Svergun, D. I. (2015). A practical guide to small angle X-ray scattering (SAXS) of flexible and intrinsically disordered proteins. *FEBS Lett* **589**, 2570–2577.

Kim, J. G., Kim, T. W., Kim, J. & Ihee, H. (2015). Protein structural dynamics revealed by time-resolved X-ray solution scattering. *Accounts Chem Res* **48**, 2200–2208.

Kim, J. G., Muniyappan, S., Oang, K. Y., Kim, T. W., Yang, C., Kim, K. H., Kim, J. & Ihee, H. (2016). Cooperative protein structural dynamics of homodimeric hemoglobin linked to water cluster at subunit interface revealed by time-resolved X-ray solution scattering. *Struct Dynam-US* **3**, 023610.

Kim, T. W., Yang, C., Kim, Y., Kim, J. G., Kim, J., Jung, Y. O., Jun, S., Lee, S. J., Park, S., Kosheleva, I., Henning, R., van Thor, J. J. & Ihee, H. (2016). Combined probes of X-ray scattering and optical spectroscopy reveal how global conformational change is temporally and spatially linked to local structural perturbation in photoactive yellow protein. *Phys Chem Chem Phys* **18**, 8911–8919.

Kirby, N. M., Mudie, S. T., Hawley, A. M., Cookson, D. J., Mertens, H. D. T., Cowieson, N. & Samardzic-Boban, V. (2013). A low-background-intensity focusing small-angle X-ray scattering undulator beamline. *J Appl Crystallogr* **46**, 1670–1680.

Kirian, R. A. (2012). Structure determination through correlated fluctuations in X-ray scattering. *J Phys B* **45**: 223001.

Kirian, R. A., Schmidt, K. E., Wang, X., Doak, R. B. & Spence, J. C. (2011). Signal, noise, and resolution in correlated fluctuations from snapshot small-angle X-ray scattering. *Phys Rev E* **84**, 011921.

Knaapila, M., Svensson, C., Barauskas, J., Zackrisson, M., Nielsen, S. S., Toft, K. N., Vestergaard, B., Arleth, L., Olsson, U., Pedersen, J. S. & Cerenius, Y. (2009). A new small-angle X-ray scattering set-up on the crystallography beamline I711 at MAX-lab. *J Synchrotron Radiat* **16**, 498–504.

Knight, J. B., Vishwanath, A., Brody, J. P. & Austin, R. H. (1998). Hydrodynamic focusing on a silicon chip: Mixing nanoliters in microseconds. *Phys Rev Lett* **80**, 3863–3866.

Koerner, L. J., Gillilan, R. E., Green, K. S., Wang, S. & Gruner, S. M. (2011). Small-angle solution scattering using the mixed-mode pixel array detector. *J Synchrotron Radiat* **18**, 148–156.

Kohn, J. E., Millett, I. S., Jacob, J., Zagrovic, B., Dillon, T. M., Cingel, N., Dothager, R. S., Seifert, S., Thiyagarajan, P., Sosnick, T. R., Hasan, M. Z., Pande, V. S., Ruczinski, I., Doniach, S. & Plaxco, K. W. (2004). Random-coil behavior and the dimensions of chemically unfolded proteins. *Proc Natl Acad Sci U S A* **101**, 12491–12496.

Konarev, P. V., Petoukhov, M. V. & Svergun, D. I. (2001). MASSHA: A graphics system for rigid-body modelling of macromolecular complexes against solution scattering data. *J Appl Crystallogr* **34**, 527–532.

Konarev, P. V., Volkov, V. V., Sokolova, A. V., Koch, M. H. J. & Svergun, D. I. (2003). PRIMUS: A Windows PC-based system for small-angle scattering data analysis. *J Appl Crystallogr* **36**, 1277–1282.

Korndorfer, I. P., Dommel, M. K. & Skerra, A. (2004). Structure of the periplasmic chaperone Skp suggests functional similarity with cytosolic chaperones despite differing architecture. *Nat Struct Mol Biol* **11**, 1015–1020.

Kovalchuk, M. V., Blagov, A. E., Dyakova, Y. A., Gruzinov, A. Y., Marchenkova, M. A., Peters, G. S., Pisarevsky, Y. V., Timofeev, V. I. & Volkov, V. V. (2016). Investigation of the initial crystallization stage in lysozyme solutions by small-angle X-ray scattering. *Cryst Growth Des* **16**, 1792–1797.

Kozin, M.B, Svergun, D.I. (2001). Automated matching of high- and low-resolution structural models. *J Appl Crystallogr* **34**, 33–41.

Kratky, O. & Pilz, I. (1972). Recent advances and applications of diffuse X-ray small-angle scattering on biopolymers in dilute solutions. *Q Rev Biophys* **5**, 481–537.

Kratky, O. & Stabinger, H. (1984). X-ray small-angle camera with block-collimation system an instrument of colloid research. *Colloid Polym Sci* **262**, 345–360.

Krywka, C., Sternemann, C., Paulus, M., Javid, N., Winter, R., Al-Sawalmih, A., Yi, S. B., Raabe, D. & Tolan, M. (2007). The small-angle and wide-angle X-ray scattering set-up at beamline BL9 of DELTA. *J Synchrotron Radiat* **14**, 244–251.

Kutter, S., Weiss, M. S., Wille, G., Golbik, R., Spinka, M. & Konig, S. (2009). Covalently bound substrate at the regulatory site of yeast pyruvate decarboxylases triggers allosteric enzyme activation. *J Biol Chem* **284**, 12136–12144.

Kuwamoto, S., Akiyama, S. & Fujisawa, T. (2004). Radiation damage to a protein solution, detected by synchrotron X-ray small-angle scattering: dose-related considerations and suppression by cryoprotectants. *J Synchrotron Radiat* **11**, 462–468.

Kuzmanovic, D. A., Elashvili, I., Wick, C., O'Connell, C. & Krueger, S. (2003). Bacteriophage MS2: molecular weight and spatial distribution of the protein and RNA components by small-angle neutron scattering and virus counting. *Structure* **11**, 1339–1348.

Labrador, A., Cerenius, Y., Svensson, C., Theodor, K. & Plivelic, T. (2013). The yellow mini-hutch for SAXS experiments at MAX IV Laboratory. *J Phys Conf Ser* **425**: 072019.

Lafleur, J. P., Snakenborg, D., Nielsen, S. S., Moller, M., Toft, K. N., Menzel, A., Jacobsen, J. K., Vestergaard, B., Arleth, L. & Kutter, J. P. (2011). Automated microfluidic sample-preparation platform for high-throughput structural investigation of proteins by small-angle X-ray scattering. *J Appl Crystallogr* **44**, 1090–1099.

Lake, J. A. (1967). An iterative method of slit-correcting small angle X-ray data. *Acta Crystallogr* **23**, 191–194.

Lamb, J., Kwok, L., Qiu, X., Andresen, K., Park, H. Y. & Pollack, L. (2008). Reconstructing three-dimensional shape envelopes from time-resolved small-angle X-ray scattering data. *J Appl Crystallogr* **41**, 1046–1052.

Laue, T. M., Shah, B. D., Ridgeway, T. M. & Pelletier, S. L. (1992). Computer-aided interpretation of analytical sedimentation data for proteins. In S. E. Harding, A. J. Rowe & J. C. Horton, eds., *Analytical Ultracentrifugation in Biochemistry And Polymer Science.* Cambridge: The Royal Society of Chemistry. 90–125.

Li, Y. L., Beck, R., Huang, T., Choi, M. C. & Divinagracia, M. (2008). Scatterless hybrid metal-single-crystal slit for small-angle X-ray scattering and high-resolution X-ray diffraction. *J Appl Crystallogr* **41**, 1134–1139.

Li, Z. H., Wu, Z. H., Mo, G., Xing, X. Q. & Liu, P. (2014). A small-angle X-ray scattering station at Beijing Synchrotron Radiation Facility. *Instrum Sci Tech* **42**, 128–141.

Lindner, P. & Schweins, R. (2010). The D11 small-angle scattering instrument: A new benchmark for SANS. *Neutron News* **21**, 15–18.

Lipfert, J., Millett, I. S., Seifert, S. & Doniach, S. (2006). Sample holder for small-angle X-ray scattering static and flow cell measurements. *Rev Sci Instrum* 77: 046108.

Liu, D. Z., Hong, K. L., Gao, C. Y., Melnichenko, Y., Littrell, K., Smith, G. & Zhao, J. K. (2011). First data acquired on the extended Q-range small-angle neutron scattering (EQ-SANS) diffractometer at the Spallation Neutron Source. *J Appl Crystallogr* **44**, 1120–1122.

Liu, H., Morris, R. J., Hexemer, A., Grandison, S. & Zwart, P. H. (2012). Computation of small-angle scattering profiles with three-dimensional Zernike polynomials. *Acta Crystallogr A* **68**, 278–285.

Liu, H., Poon, B. K., Janssen, A. J. & Zwart, P. H. (2012). Computation of fluctuation scattering profiles via three-dimensional Zernike polynomials. *Acta Crystallogr A* **68**, 561–567.

Luft, J. R., Collins, R. J., Fehrman, N. A., Lauricella, A. M., Veatch, C. K. & DeTitta, G. T. (2003). A deliberate approach to screening for initial crystallization conditions of biological macromolecules. *J Struct Biol* **142**, 170–179.

Luft, J. R., Snell, E. H. & Detitta, G. T. (2011). Lessons from high-throughput protein crystallization screening: 10 years of practical experience. *Expert Opin Drug Del* **6**, 465–480.

Makowski, L., Bardhan, J., Gore, D., Rodi, D. J. & Fischetti, R. F. (2012). Multi-wavelength anomalous diffraction using medium-angle X-ray solution scattering (MADMAX). *Biophys J* **102**, 927–933.

Malaby, A. W., Chakravarthy, S., Irving, T. C., Kathuria, S. V., Bilsel, O. & Lambright, D. G. (2015). Methods for analysis of size-exclusion chromatography-small-angle X-ray scattering and reconstruction of protein scattering. *J Appl Crystallogr* **48**, 1102–1113.

Malfois, M. & Svergun, D. I. (2000). sasCIF: An extension of core Crystallographic Information File for SAS. *J Appl Crystallogr* **33**, 812–816.

Malmerberg, E., Kerfeld, C. A. & Zwart, P. H. (2015). Operational properties of fluctuation X-ray scattering data. *IUCrJ* **2**, 309–316.

Marchesini, S., He, H., Chapman, H. N., Hau-Riege, S. P., Noy, A., Howells, M. R., Weierstall, U. & Spence, J. C. H. (2003). X-ray image reconstruction from a diffraction pattern alone. *Phys Rev B* **68**, 140101.

Martel, A., Liu, P., Weiss, T. M., Niebuhr, M. & Tsuruta, H. (2012). An integrated high-throughput data acquisition system for biological solution X-ray scattering studies. *J Synchrotron Radiat* **19**, 431–434.

Martin, P. D., Malkowski, M. G., Box, J., Esmon, C. T. & Edwards, B. F. P. (1997). New insights into the regulation of the blood clotting cascade derived from the X-ray crystal structure of bovine meizothrombin des F1 in complex with PPACK. *Structure* **5**, 1681–1693.

Mathew, E., Mirza, A. & Menhart, N. (2004). Liquid-chromatography-coupled SAXS for accurate sizing of aggregating proteins. *J Synchrotron Radiat* **11**, 314–318.

Matsuoka, H. (1999). An introduction to the crystallographer's world. 1. An introduction to small-angle scattering. *Nihon Kessho Gakkaishi* **41**, 213–226.

Mattinen, M. L., Paakkonen, K., Ikonen, T., Craven, J., Drakenberg, T., Serimaa, R., Waltho, J. & Annila, A. (2002). Quaternary structure built from subunits combining NMR and small-angle X-ray scattering data. *Biophys J* **83**, 1177–1183.

McNeil, B. W. J. & Thompson, N. R. (2010). X-ray free-electron lasers. *Nat Photonics* **4**, 814–821.

McPhillips, T. M., McPhillips, S. E., Chiu, H. J., Cohen, A. E., Deacon, A. M., Ellis, P. J., Garman, E., Gonzalez, A., Sauter, N. K., Phizackerley, R. P., Soltis, S. M. & Kuhn, P. (2002). Blu-Ice and the Distributed Control System: Software for data acquisition and instrument control at macromolecular crystallography beamlines. *J Synchrotron Radiat* **9**, 401–406.

Meisburger, S. P., Thomas, W. C., Watkins, M. B. & Ando, N. (2017). X-ray scattering studies of protein structural dynamics. *Chem Rev* **117**, 7615–7672.

Meisburger, S. P., Warkentin, M., Chen, H., Hopkins, J. B., Gillilan, R. E., Pollack, L. & Thorne, R. E. (2013). Breaking the radiation damage limit with Cryo-SAXS. *Biophys J* **104**, 227–236.

Miao, J., Sayre, D. & Chapman, H. N. (1998). Phase retrieval from the magnitude of the Fourier transforms of nonperiodic objects. *J Opt Soc Am A* **15**, 1662–1669.

Millett, I. S., Doniach, S. & Plaxco, K. W. (2002). Toward a taxonomy of the denatured state: Small angle scattering studies of unfolded proteins. *Adv Protein Chem* **62**, 241–262.

Morse, P. M. & Feshbach, H. (1953). *Methods of Theoretical Physics*. New York: McGraw-Hill Book Company Inc.

Mylonas, E. & Svergun, D. I. (2007). Accuracy of molecular mass determination of proteins in solution by small-angle X-ray scattering. *J Appl Crystallogr* **40**, S245–S249.

Narayanan, T., Diat, O. & Bosecke, P. (2001). SAXS and USAXS on the high brilliance beamline at the ESRF. *Nucl Instrum Meth A* **467**, 1005–1009.

Né, F., Gazeau, D., Lambard, J., Lesieur, P., Zemb, T. & Gabriel, A. (1993). Characterization of an image-plate detector used for quantitative small-angle-scattering studies. *J Appl Crystallogr* **26**, 763–773.

Niedzialkowska, E., Gasiorowska, O., Handing, K. B., Majorek, K. A., Porebski, P. J., Shabalin, I. G., Zasadzinska, E., Cymborowski, M. & Minor, W. (2016). Protein purification and crystallization artifacts: The tale usually not told. *Protein Sci* **25**, 720–733.

Nielsen, S. S., Toft, K. N., Snakenborg, D., Jeppesen, M. G., Jacobsen, J. K., Vestergaard, B., Kutter, J. P. & Arleth, L. (2009). BioXTAS RAW, a software program for high-throughput automated small-angle X-ray scattering data reduction and preliminary analysis. *J Appl Crystallogr* **42**, 959–964.

Nyam-Osor, M., Soloviov, D. V., Kovalev, Y. S., Zhigunov, A., Rogachev, A. V., Ivankov, O. I., Erhan, R. V. & Kuklin, A. I. (2012). Silver behenate and silver stearate powders for calibration of SAS instruments. *J Phys: Conf Ser* **351**, 012024.

Oda, T., Aihara, T. & Wakabayashi, K. (2016). Early nucleation events in the polymerization of actin, probed by time-resolved small-angle X-ray scattering. *Sci Rep* **6**, 34539.

Orthaber, D., Bergmann, A. & Glatter, O. (2000). SAXS experiments on absolute scale with Kratky systems using water as a secondary standard. *J Appl Crystallogr* **33**, 218–225.

Owen, B. A., W, H. L. & McMurray, C. T. (2009). The nucleotide binding dynamics of human MSH2-MSH3 are lesion dependent. *Nat Struct Mol Biol* **16**, 550–557.

Pande, K., Hutchison, C. D. M., Groenhof, G., Aquila, A., Robinson, J. S., Tenboer, J., Basu, S., Boutet, S., DePonte, D. P., Liang, M. N., White, T. A., Zatsepin, N. A., Yefanov, O., Morozov, D., Oberthuer, D., Gati, C., Subramanian, G., James, D., Zhao, Y., Koralek, J., Brayshaw, J., Kupitz, C., Conrad, C., Roy-Chowdhury, S., Coe, J. D., Metz, M., Xavier, P. L., Grant, T. D., Koglin, J. E., Ketawala, G., Fromme, R., Srajer, V., Henning, R., Spence, J. C. H., Ourmazd, A., Schwander, P., Weierstall, U., Frank, M., Fromme, P., Barty, A., Chapman, H. N., Moffat, K., van Thor, J. J. & Schmidt, M. (2016). Femtosecond structural dynamics drives the trans/cis isomerization in photoactive yellow protein. *Science* **352**, 725–729.

Panjkovich, A. & Svergun, D. I. (2016). Deciphering conformational transitions of proteins by small angle X-ray scattering and normal mode analysis. *Phys Chem Chem Phys* **18**, 5707–5719.

Paris, O., Li, C. H., Siegel, S., Weseloh, G., Emmerling, F., Riesemeier, H., Erko, A. & Fratzl, P. (2007). A new experimental station for simultaneous X-ray microbeam scanning for small- and wide-angle scattering and fluorescence at BESSY II. *J Appl Crystallogr* **40**, S466–S470.

Park, H. Y., Qiu, X., Rhoades, E., Korlach, J., Kwok, L. W., Zipfel, W. R., Webb, W. W. & Pollack, L. (2006). Achieving uniform mixing in a microfluidic device: Hydrodynamic focusing prior to mixing. *Anal Chem* **78**, 4465–4473.

Pauw, B. R. (2014). Everything SAXS: Small-angle scattering pattern collection and correction (vol 25, 383201, 2013). *J Phys: Condens Matter* **26**: 239501.

Pedersen, J. S., Posselt, D. & Mortensen, K. (1990). Analytical treatment of the resolution function for small-angle scattering. *J Appl Crystallogr* **23**, 321–333.

Pelikan, M., Hura, G. & Hammel, M. (2009). Structure and flexibility within proteins as identified through small angle X-ray scattering. *Gen Physiol Biophys* **28**, 174–189.

Penczek, P. A. (2010). Fundamentals of three-dimensional reconstruction from projections. *Method Enzymol* **482**, 1–33.

Perez, J. & Koutsioubas, A. (2015). Memprot: A program to model the detergent corona around a membrane protein based on SEC-SAXS data. *Acta Crystallogr D* **71**, 86–93.

Perez, J., Vachette, P., Russo, D., Desmadril, M. & Durand, D. (2001). Heat-induced unfolding of neocarzinostatin, a small all-beta protein investigated by small-angle X-ray scattering. *J Mol Biol* **308**, 721–743.

Pernot, P., Round, A., Barrett, R., Antolinos, A. D. M., Gobbo, A., Gordon, E., Huet, J., Kieffer, J., Lentini, M., Mattenet, M., Morawe, C., Mueller-Dieckmann, C., Ohlsson, S., Schmid, W., Surr, J., Theveneau, P., Zerrad, L. & McSweeney, S. (2013). Upgraded ESRF BM29 beamline for SAXS on macromolecules in solution. *J Synchrotron Radiat* **20**, 660–664.

Pernot, P., Theveneau, P., Giraud, T., Fernandes, R. N., Nurizzo, D., Spruce, D., Surr, J., McSweeney, S., Round, A., Felisaz, F., Foedinger, L., Gobbo, A., Huet, J., Villard, C. & Cipriani, F. (2010). New beamline dedicated to solution scattering from biological macromolecules at the ESRF. *J Phys: Conf Ser* **247**, 012009.

Petoukhov, M. V., Franke, D., Shkumatov, A. V., Tria, G., Kikhney, A. G., Gajda, M., Gorba, C., Mertens, H. D., Konarev, P. V. & Svergun, D. I. (2012). New developments in the ATSAS program package for small-angle scattering data analysis. *J Appl Crystallogr* **45**, 342–350.

Petoukhov, M. V., Konarev, P. V., Kikhney, A. G. & Svergun, D. I. (2007). ATSAS 2.1: Towards automated and web-supported small-angle scattering data analysis. *J Appl Crystallogr* **40**, S223–S228.

Petoukhov, M. V. & Svergun, D. I. (2005). Global rigid body modeling of macromolecular complexes against small-angle scattering data. *Biophys J* **89**, 1237–1250.

Petoukhov, M. V., Weissenhorn, W. & Svergun, D. I. (2014). Endophilin-A1 BAR domain interaction with arachidonyl CoA. *Front Mol Biosci* **1**, 20.

Pham, N., Radajewski, D., Round, A., Brennich, M., Pernot, P., Biscans, B., Bonnete, F. & Teychene, S. (2017). Coupling high throughput microfluidics and small-angle x-ray scattering to study protein crystallization from solution. *Anal Chem* **89**, 2282–2287.

Polizzi, S. J., Walsh, R. M., Le Magueres, P., Criswell, A. R. & Wood, Z. A. (2013). Human UDP-alpha-D-xylose synthase forms a catalytically important tetramer that has not been observed in crystal structures. *Biochemistry* **52**, 3888–3898.

Pollack, L. (2011). SAXS studies of ion-nucleic acid interactions. *Annu Rev Biophys* **40**, 225–242.

Pollack, L. & Doniach, S. (2009). Time-resolved X-ray scattering and RNA folding. *Methods Enzymol* **469**, 253–268.

Pratt, A. J., Shin, D. S., Merz, G. E., Rambo, R. P., Lancaster, W. A., Dyer, K. N., Borbat, P. P., Poole, F. L.,2nd, Adams, M. W., Freed, J. H., Crane, B. R., Tainer, J. A. & Getzoff, E. D. (2014). Aggregation propensities of superoxide dismutase G93 hotspot mutants mirror ALS clinical phenotypes. *Proc Natl Acad Sci U S A* **111**, E4568–E4576.

Putnam, C. D., Hammel, M., Hura, G. L. & Tainer, J. A. (2007). X-ray solution scattering (SAXS) combined with crystallography and computation: defining accurate macromolecular structures, conformations and assemblies in solution. *Q Rev Biophys* **40**, 191–285.

Putnam, D. K., Lowe, E. W., Jr. & Meiler, J. (2013). Reconstruction of SAXS Profiles from Protein Structures. *Comput Struct Biotechnol J* **8**, e201308006.

Rajagopal, S., Schmidt, M., Anderson, S., Ihee, H. & Moffat, K. (2004). Analysis of experimental time-resolved crystallographic data by singular value decomposition. *Acta Crystallogr D* **60**, 860–871.

Rambo, R. P. & Tainer, J. A. (2010). Improving small-angle X-ray scattering data for structural analyses of the RNA world. *RNA* **16**, 638–646.

Rambo, R. P. & Tainer, J. A. (2013). Accurate assessment of mass, models and resolution by small-angle scattering. *Nature* **496**, 477–481.

Receveur-Brechot, V. & Durand, D. (2012). How random are intrinsically disordered proteins? A small angle scattering perspective. *Curr Protein Pept Sci* **13**, 55–75.

Regenfuss, P., Clegg, R. M., Fulwyler, M. J., Barrantes, F. J. & Jovin, T. M. (1985). Mixing liquids in microseconds. *Rev Sci Instrum* **56**, 283–290.

Rehm, C., Brule, A., Freund, A. K. & Kennedy, S. J. (2013). Kookaburra: The ultra-small-angle neutron scattering instrument at OPAL. *J Appl Crystallogr* **46**, 1699–1704.

Rehm, C. & de Campo, L. (2016). Kookaburra: The ultra-small-angle neutron scattering instrument at ANSTO. *Neutron News* **27**, 30–32.

Reimer, J. M., Aloise, M. N., Harrison, P. M. & Schmeing, T. M. (2016). Synthetic cycle of the initiation module of a formylating nonribosomal peptide synthetase. *Nature* **529**, 239–242.

Rodriguez-Ruiz, I., Radajewski, D., Charton, S., Phamvan, N., Brennich, M., Pernot, P., Bonnete, F. & Teychene, S. (2017). Innovative high-throughput SAXS methodologies based on photonic lab-on-a-chip sensors: Application to macromolecular studies. *Sensors* **17**, 1266.

Roessle, M. W., Klaering, R., Ristau, U., Robrahn, B., Jahn, D., Gehrmann, T., Konarev, P., Round, A., Fiedler, S., Hermes, C. & Svergun, D. (2007). Upgrade of the small-angle X-ray scattering beamline X33 at the European Molecular Biology Laboratory, Hamburg. *J Appl Crystallogr* **40**, S190–S194.

Rossmann, M. G. (1972). *The Molecular Replacment Method.* New York: Gordon & Breach.

Round, A., Brown, E., Marcellin, R., Kapp, U., Westfall, C. S., Jez, J. M. & Zubieta, C. (2013). Determination of the GH3.12 protein conformation through HPLC-integrated SAXS measurements combined with X-ray crystallography. *Acta Crystallogr D* **69**, 2072–2080.

Round, A., Felisaz, F., Fodinger, L., Gobbo, A., Huet, J., Villard, C., Blanchet, C. E., Pernot, P., McSweeney, S., Roessle, M., Svergun, D. I. & Cipriani, F. (2015). BioSAXS Sample Changer: A robotic sample changer for rapid and reliable high-throughput X-ray solution scattering experiments. *Acta Crystallogr D* **71**, 67–75.

Round, A. R., Franke, D., Moritz, S., Huchler, R., Fritsche, M., Malthan, D., Klaering, R., Svergun, D. I. & Roessle, M. (2008). Automated sample-changing robot for solution scattering experiments at the EMBL Hamburg SAXS station X33. *J Appl Crystallogr* **41**, 913–917.

Rozycki, B., Kim, Y. C. & Hummer, G. (2011). SAXS ensemble refinement of ESCRT-III CHMP3 conformational transitions. *Structure* **19**, 109–116.

Russel, D., Lasker, K., Webb, B., Velazquez-Muriel, J., Tjioe, E., Schneidman-Duhovny, D., Peterson, B. & Sali, A. (2012). Putting the pieces together: Integrative modeling platform software for structure determination of macromolecular assemblies. *PLoS Biol* **10**, e1001244.

Santos, S. P., Cuypers, M. G., Round, A., Finet, S., Narayanan, T., Mitchell, E. P. & Romao, C. V. (2017). SAXS structural studies of Dps from *Deinococcus radiodurans* highlights the conformation of the mobile N-terminal extensions. *J Mol Biol* **429**, 667–687.

Schindler, C. E. M., de Vries, S. J., Sasse, A. & Zacharias, M. (2016). SAXS data alone can generate high-quality models of protein-protein complexes. *Structure* **24**, 1387–1397.

Schmidt, M., Rajagopal, S., Ren, Z. & Moffat, K. (2003). Application of singular value decomposition to the analysis of time-resolved macromolecular X-ray data. *Biophys J* **84**, 2112–2129.

Schmidt, O., DeBartolo, J., Kurtz, C., Lee, B., Seifert, S., Winans, R. & Zuo, X. (2014). Variable q-range X-ray scattering chamber for chemical and materials science at the Advanced Photon Source. *J Phys: Conf Ser* **493**, 012010.

Schneidman-Duhovny, D., Hammel, M., Tainer, J. A. & Sali, A. (2016). FoXS, FoXSDock and MultiFoXS: Single-state and multi-state structural modeling of proteins and their complexes based on SAXS profiles. *Nucleic Acids Res* **44**, W424–W429.

Schneidman-Duhovny, D., Kim, S. J. & Sali, A. (2012). Integrative structural modeling with small angle X-ray scattering profiles. *BMC Struct Biol* **12**, 17.

Schneidman-Duhovny, D., Rossi, A., Avila-Sakar, A., Kim, S. J., Velazquez-Muriel, J., Strop, P., Liang, H., Krukenberg, K. A., Liao, M., Kim, H. M., Sobhanifar, S., Dotsch, V.,

Rajpal, A., Pons, J., Agard, D. A., Cheng, Y. & Sali, A. (2012). A method for integrative structure determination of protein-protein complexes. *Bioinformatics* **28**, 3282–3289.

Schroder, H. K., Willassen, N. P. & Smalas, A. O. (1998). Structure of a non-psychrophilic trypsin from a cold-adapted fish species. *Acta Crystallogr D* **54**, 780–798.

Schwemmer, F., Blanchet, C. E., Spilotros, A., Kosse, D., Zehnle, S., Mertens, H. D., Graewert, M. A., Rossle, M., Paust, N., Svergun, D. I., von Stetten, F., Zengerle, R. & Mark, D. (2016). LabDisk for SAXS: A centrifugal microfluidic sample preparation platform for small-angle X-ray scattering. *Lab Chip* **16**, 1161–1170.

Schwieters, C. D., Bermejo, G. A. & Clore, G. M. (2018). Xplor-NIH for molecular structure determination from NMR and other data sources. *Protein Sci* **27**, 26–40.

Scott, D. J., Patel, T. R. & Winzor, D. J. (2013). A potential for overestimating the absolute magnitudes of second virial coefficients by small-angle X-ray scattering. *Anal Biochem* **435**, 159–165.

Seifert, S., Winans, R. E., Tiede, D. M. & Thiyagarajan, P. (2000). Design and performance of a ASAXS instrument at the Advanced Photon Source. *J Appl Crystallogr* **33**, 782–784.

Shannon, C. E. (1948a). A mathematical theory of communication. *Bell Syst Tech J* **27**, 379–423.

Shannon, C. E. (1948b). A mathematical theory of communication. *Bell Syst Tech J* **27**, 623–656.

Shimizu, N., Mori, T., Igarashi, N., Ohta, H., Nagatani, Y., Kosuge, T. & Ito, K. (2013). Refurbishing of small-angle X-ray scattering beamline, BL-6A at the Photon Factory. *J Phys: Conf Ser* **425**, 202008.

Singleton, M. R., Isupov, M. N. & Littlechild, J. A. (1999). X-ray structure of pyrrolidone carboxyl peptidase from the hyperthermophilic archaeon *Thermococcus litoralis*. *Struct Fold Des* 7, 237–244.

Skou, M., Skou, S., Jensen, T. G., Vestergaard, B. & Gillilan, R. E. (2014). In situ microfluidic dialysis for biological small-angle X-ray scattering. *J Appl Crystallogr* **47**, 1355–1366.

Skou, S., Gillilan, R. E. & Ando, N. (2014). Synchrotron-based small-angle X-ray scattering of proteins in solution. *Nat Protoc* **9**, 1727–1739.

Smolsky, I. L., Liu, P., Niebuhr, M., Ito, K., Weiss, T. M. & Tsuruta, H. (2007). Biological small-angle X-ray scattering facility at the Stanford synchrotron radiation laboratory. *J Appl Crystallogr* **40**, S453–S458.

Sokolova, A., Christoforidis, J., Eltobaji, A., Barnes, J., Darmann, F., Whitten, A. E. & de Campo, L. (2016). BILBY: Time-of-flight small angle scattering instrument. *Neutron News* **27**, 9–13.

Sokolova, A. V., Volkov, V. V. & Svergun, D. I. (2003). Prototype of a database for rapid protein classification based on solution scattering data. *J Appl Crystallogr* **36**, 865–868.

Song, H., Fang, X. Y., Jin, L., Shaw, G. X., Wang, Y. X. & Ji, X. H. (2017). The functional cycle of Rnt1p: five consecutive steps of double-stranded rna processing by a eukaryotic RNase III. *Structure* **25**, 353–363.

Stiegler, A. L., Grant, T. D., Luft, J. R., Calderwood, D. A., Snell, E. H. & Boggon, T. J. (2013). Purification and SAXS analysis of the integrin linked kinase, PINCH, parvin (IPP) heterotrimeric complex. *PLOS ONE* **8**, e55591.

Stirling, P. C., Bakhoum, S. F., Feigl, A. B. & Leroux, M. R. (2006). Convergent evolution of clamp-like binding sites in diverse chaperones. *Nat Struct Mol Biol* **13**, 865–870.

Stockley, P. G., Stonehouse, N. J. & Valegard, K. (1994). Molecular mechanism of RNA phage morphogenesis. *Int J Biochem* **26**, 1249–1260.

Stuhrmann, H. B. (1970a). Interpretation of small-angle scattering functions of dilute solutions and gases. A representation of the structures related to a one-particle scattering function. *Acta Crystallogr A* **26**, 297–306.

Stuhrmann, H. B. (1970b). Ein neues Verfahren zur Bestimmung der Oberflächenform und der inneren Struktur von gelösten globulären Proteinen aus Röntgenkleinwinkelmessungen. *Z Phys Chem Neue Fol* **72**, 177–184.

Stuhrmann, H. B., Haas, J., Ibel, K., Wolf, B., Koch, M. H., Parfait, R. & Crichton, R. R. (1976). New low resolution model for 50S subunit of Escherichia coli ribosomes. *Proc Natl Acad Sci U S A* **73**, 2379–2383.

Sugiyama, M., Yagi, H., Ishii, K., Porcar, L., Martel, A., Oyama, K., Noda, M., Yunoki, Y., Murakami, R., Inoue, R., Sato, N., Oba, Y., Terauchi, K., Uchiyama, S. & Kato, K. (2016). Structural characterization of the circadian clock protein complex composed of KaiB and KaiC by inverse contrast-matching small-angle neutron scattering. *Sci Rep* **6**, 35567.

Svergun, D. I. (1992). Determination of the regularization parameter in indirect-transform methods using perceptual criteria. *J Appl Crystallogr* **25**, 495–503.

Svergun, D. I. (1999). Restoring low resolution structure of biological macromolecules from solution scattering using simulated annealing. *Biophys J* **76**, 2879–2886.

Svergun, D. I. & Koch, M. H. J. (2003). Small-angle scattering studies of biological macromolecules in solution. *Rep Prog Phys* **66**, 1735–1782.

Svergun, D. I., Koch, M. H. J., Timmins, P. & May, R. P. (2013). *Small Angle X-Ray and Neutron Scattering from Solutions of Biological Macromolecules*. Oxford: Oxford University Press.

Svergun, D. I., Petoukhov, M. V. & Koch, M. H. (2001). Determination of domain structure of proteins from X-ray solution scattering. *Biophys J* **80**, 2946–2953.

Svergun, D. I., Richard, S., Koch, M. H., Sayers, Z., Kuprin, S. & Zaccai, G. (1998). Protein hydration in solution: Experimental observation by X-ray and neutron scattering. *Proc Natl Acad Sci U S A* **95**, 2267–2272.

Svergun, D. I. & Stuhrmann, H. B. (1991). New developments in direct shape determination from small-angle scattering. 1. Theory and model calculations. *Acta Crystallogr A* **47**, 736–744.

Svergun, D. I., Volkov, V. V., Kozin, M. B. & Stuhrmann, H. B. (1996). New developments in direct shape determination from small-angle scattering. 2. Uniqueness. *Acta Crystallogr A* **52**, 419–426.

Takahashi, S., Yeh, S. R., Das, T. K., Chan, C. K., Gottfried, D. S. & Rousseau, D. L. (1997). Folding of cytochrome c initiated by submillisecond mixing. *Nat Struct Biol* **4**, 44–50.

Teixeira, S. C., Ankner, J., Bellissent-Funel, M. C., Bewley, R., Blakeley, M. P., Coates, L., Dahint, R., Dalgliesh, R., Dencher, N., Dhont, J., Fischer, P., Forsyth, V. T., Fragneto, G., Frick, B., Geue, T., Gilles, R., Gutberlet, T., Haertlein, M., Hauss, T., Haussler, W., Heller, W. T., Herwig, K., Holderer, O., Juranyi, F., Kampmann, R., Knott, R., Kohlbrecher, J., Kreuger, S., Langan, P., Lechner, R., Lynn, G., Majkrzak, C., May, R., Meilleur, F., Mo, Y., Mortensen, K., Myles, D. A., Natali, F., Neylon, C., Niimura, N., Ollivier, J., Ostermann, A., Peters, J., Pieper, J., Ruhm, A., Schwahn, D., Shibata, K., Soper, A. K., Straessle, T., Suzuki, U. I., Tanaka, I., Tehei, M., Timmins, P., Torikai, N.,

Unruh, T., Urban, V., Vavrin, R., Weiss, K. & Zaccai, G. (2008). New sources and instrumentation for neutrons in biology. *Chem Phys* **345**, 133–151.

Tian, F., Li, X. H., Wang, Y. Z., Yang, C. M., Zhou, P., Lin, J. Y., Zeng, J. R., Hong, C. X., Hua, W. Q., Li, X. Y., Miao, X. R., Bian, F. G. & Wang, J. (2015). Small angle X-ray scattering beamline at SSRF. *Nucl Sci Tech* **26**, 030101.

Toft, K. N., Vestergaard, B., Nielsen, S. S., Snakenborg, D., Jeppesen, M. G., Jacobsen, J. K., Arleth, L. & Kutter, J. P. (2008). High-throughput small angle X-ray scattering from proteins in solution using a microfluidic front-end. *Anal Chem* **80**, 3648–3654.

Trewhella, J. (2016). Small-angle scattering and 3D structure interpretation. *Curr Opin Struct Biol* **40**, 1–7.

Trewhella, J., Duff, A. P., Durand, D., Gabel, F., Guss, J. M., Hendrickson, W. A., Hura, G. L., Jacques, D. A., Kirby, N. M., Kwan, A. H., Pérez, J., Pollack, L., Ryan, T. M., Sali, A., Schneidman-Duhovny, D., Schwede, T., Svergun, D. I., Sugiyama, M., Tainer, J. A., Vachette, P., Westbrook, J. & Whitten, A. E. (2017). 2017 publication guidelines for structural modelling of small-angle scattering data from biomolecules in solution: An update. *Acta Crystallogr D* **73**, 710–728.

Trewhella, J., Gallagher, S. C., Krueger, J. K. & Zhao, J. (1998). Neutron and X-ray solution scattering provide insights into biomolecular structure and function. *Sci Prog* **81**, 101–122.

Trewhella, J., Hendrickson, W. A., Kleywegt, G. J., Sali, A., Sato, M., Schwede, T., Svergun, D. I., Tainer, J. A., Westbrook, J. & Berman, H. M. (2013). Report of the wwPDB Small-Angle Scattering Task Force: Data requirements for biomolecular modeling and the PDB. *Structure* **21**, 875–881.

Tria, G., Mertens, H. D., Kachala, M. & Svergun, D. I. (2015). Advanced ensemble modelling of flexible macromolecules using X-ray solution scattering. *IUCrJ* **2**, 207–217.

Tuukkanen, A. T., Kleywegt, G. J. & Svergun, D. I. (2016). Resolution of ab initio shapes determined from small-angle scattering. *IUCrJ* **3**, 440–447.

Tuukkanen, A. T. & Svergun, D. I. (2014). Weak protein-ligand interactions studied by small-angle X-ray scattering. *FEBS J* **281**, 1974–1987.

Valentini, E., Kikhney, A. G., Previtali, G., Jeffries, C. M. & Svergun, D. I. (2015). SASBDB, a repository for biological small-angle scattering data. *Nucleic Acids Res* **43**, D357–D363.

Villarreal, S. A., Pattanayek, R., Williams, D. R., Mori, T., Qin, X., Johnson, C. H., Egli, M. & Stewart, P. L. (2013). CryoEM and molecular dynamics of the circadian KaiB-KaiC complex indicates that KaiB monomers interact with KaiC and block ATP binding clefts. *J Mol Biol* **425**, 3311–3324.

Volkov, V. V. & Svergun, D. I. (2003). Uniqueness of ab initio shape determination in small-angle scattering. *J Appl Crystallogr* **36**, 860–864.

Walther, D., Cohen, F. E. & Doniach, S. (2000). Reconstruction of low-resolution three-dimensional density maps from one-dimensional small-angle X-ray solution scattering data for biomolecules. *J Appl Crystallogr* **33**, 350–363.

Walton, T. A. & Sousa, M. C. (2004). Crystal structure of Skp, a prefoldin-like chaperone that protects soluble and membrane proteins from aggregation. *Mol Cell* **15**, 367–374.

Wang, Z., Chen, X., Meng, L., Cui, K., Wu, L. & Li, L. (2014). A small-angle x-ray scattering system with a vertical layout. *Rev Sci Instrum* **85**, 125110.

Watanabe-Matsui, M., Matsumoto, T., Matsui, T., Ikeda-Saito, M., Muto, A., Murayama, K. & Igarashi, K. (2015). Heme binds to an intrinsically disordered region of Bach2 and alters its conformation. *Arch Biochem Biophys* **565**, 25–31.

Watkin, S. A. J., Ryan, T. M., Miller, A. G., Nock, V. M., Pearce, F. G. & Dobson, R. C. J. (2017). Microfluidics for small-angle X-ray scattering. In A. E. Ares, ed., *X-Ray Scattering*. Rijeka: Intech Europe, 89–104.

Wells, M., Tidow, H., Rutherford, T. J., Markwick, P., Jensen, M. R., Mylonas, E., Svergun, D. I., Blackledge, M. & Fersht, A. R. (2008). Structure of tumor suppressor p53 and its intrinsically disordered N-terminal transactivation domain. *Proc Natl Acad Sci U S A* **105**, 5762–5767.

Wernecke, J., Gollwitzer, C., Muller, P. & Krumrey, M. (2014). Characterization of an in-vacuum PILATUS 1M detector. *J Synchrotron Radiat* **21**, 529–536.

Wilkins, D. K., Grimshaw, S. B., Receveur, V., Dobson, C. M., Jones, J. A. & Smith, L. J. (1999). Hydrodynamic radii of native and denatured proteins measured by pulse field gradient NMR techniques. *Biochemistry* **38**, 16424–16431.

Wilkinson, C., Gabriel, A., Lehmann, M., Zemb, T. & Ne, F. (1993). Image plate neutron detector. *Proc SPIE* **1737**, 324–329.

Williamson, T. E., Craig, B. A., Kondrashkina, E., Bailey-Kellogg, C. & Friedman, A. M. (2008). Analysis of self-associating proteins by singular value decomposition of solution scattering data. *Biophys J* **94**, 4906–4923.

Wilson, W. W. (2003). Light scattering as a diagnostic for protein crystal growth: A practical approach. *J Struct Biol* **142**, 56–65.

Wilson, W. W. & DeLucas, L. J. (2016). Applications of the second virial coefficient: Protein crystallization and solubility. Corrigendum. *Acta Crystallogr F* **72**, 255–256.

Xing, X. Q., Liu, Q., Wang, W., Zhang, K. H., Li, T., Cai, Q., Mo, G., Cheng, W. D., Wang, D. H., Gong, Y., Chen, Z. J., Qiu, D. W. & Wu, Z. H. (2013). Shape evolution with temperature of a thermotolerant protein (PeaT1) in solution detected by small angle X-ray scattering. *Proteins* **81**, 53–62.

Yamamoto, M., Kumasaka, T., Fujisawa, T. & Ueki, T. (1998). Trichromatic concept at SPring-8 RIKEN beamline I. *J Synchrotron Radiat* **5**, 222–225.

Yang, C., van der Woerd, M. J., Muthurajan, U. M., Hansen, J. C. & Luger, K. (2011). Biophysical analysis and small-angle X-ray scattering-derived structures of MeCP2-nucleosome complexes. *Nucleic Acids Res* **39**, 4122–4135.

Yang, L. (2005). The X21 SAXS instrument at NSLS for studying macromolecular systems. *Macromol Res* **13**, 538–541.

Yang, L. (2013). Using an in-vacuum CCD detector for simultaneous small- and wide-angle scattering at beamline X9. *J Synchrotron Radiat* **20**, 211–218.

Yang, S., Blachowicz, L., Makowski, L. & Roux, B. (2010). Multidomain assembled states of Hck tyrosine kinase in solution. *Proc Natl Acad Sci U S A* **107**, 15757–15762.

Yigzaw, Y., Hinckley, P., Hewig, A. & Vedantham, G. (2009). Ion exchange chromatography of proteins and clearance of aggregates. *Curr Pharm Biotechnol* **10**, 421–426.

Zaccai, N. R., Sandlin, C. W., Hoopes, J. T., Curtis, J. E., Fleming, P. J., Fleming, K. G. & Krueger, S. (2016). Deuterium labeling together with contrast variation small-angle neutron scattering suggests how Skp captures and releases unfolded outer membrane proteins. *Methods Enzymol* **566**, 159–210.

Zanini, F., Lausi, A. & Savoia, A. (1999). The beamlines of ELETTRA and their application to structural biology. *Genetica* **106**, 171–180.

Zhang, Z., Carrillo, J. M. Y., Ahn, S. K., Wu, B., Hong, K. L., Smith, G. S. & Do, C. (2014). Atomistic structure of bottlebrush polymers: Simulations and neutron scattering studies. *Macromolecules* **47**, 5808–5814.

Zhao, J., Hoye, E., Boylan, S., Walsh, D. A. & Trewhella, J. (1998). Quaternary structures of a catalytic subunit-regulatory subunit dimeric complex and the holoenzyme of the cAMP-dependent protein kinase by neutron contrast variation. *J Biol Chem* **273**, 30448–30459.

Zhao, J. K., Gao, C. Y. & Liu, D. (2010). The extended Q-range small-angle neutron scattering diffractometer at the SNS. *J Appl Crystallogr* **43**, 1068–1077.

Zheng, W. & Tekpinar, M. (2011). Accurate flexible fitting of high-resolution protein structures to small-angle X-ray scattering data using a coarse-grained model with implicit hydration shell. *Biophys J* **101**, 2981–2991.

Index